The Viable System Model

The Viable System Model

Interpretations and Applications of Stafford Beer's VSM

Edited by

Raúl Espejo
Aston Business School
University of Aston

and

Roger Harnden
Doctoral Programme, Aston Business School
University of Aston

JOHN WILEY & SONS
Chichester ● New York ● Brisbane ● Toronto ● Singapore

Wiley Editorial Offices

John Wiley & Sons Ltd, Baffins Lane, Chichester,
West Sussex PO19 1UD, England

John Wiley & Sons, Inc., 605 Third Avenue,
New York, NY 10158–0012, USA

Jacaranda Wiley Ltd, G.P.O. Box 859, Brisbane,
Queensland 4001, Australia

John Wiley & Sons (Canada) Ltd, 22 Worcester Road,
Rexdale, Ontario M9W 1L1, Canada

John Wiley & Sons (SEA) Pte Ltd, 37 Jalan Pemimpin #05–04,
Block B, Union Industrial Building, Singapore 2057

Library of Congress Cataloging-in-Publication Data:

The viable system model: interpretations and applications of Stafford Beer's VSM
edited by Raul Espejo and Roger Harnden.
    p.   cm.
  Bibliography: p.
  Includes index.
  ISBN 0 471 92288 9
  1. Organizational effectiveness. 2. Communication in
organizations. 3. Cybernetics. 4. System analysis. I. Espejo,
Raul. II. Harnden, Roger.
  HD58.9.V53 1989
  658.4—dc20   89-8916
                                           CIP

British Library Cataloguing in Publication Data:

The viable system model: interpretations and applications of Stafford Beer's VSM
1. Systems. Mathematical models
I. Espejo, Raul II. Harnden, Roger
003'.0724

ISBN 0 471 92288 9

Typeset by Acorn Bookwork, Salisbury, Wiltshire
Printed and bound in Great Britain by
Biddles Ltd, Guildford and King's Lynn

Contents

v

Part Three: Methodology and Epistemology

Part Four: Critical Views

Contributors

R. Anderton   *Department of Systems, University of Lancaster, Bailrigg, Lancaster, UK*

S. Beer   *34 Palmeston Square, Toronto, Ontario, M6G 2S7 Canada*

M.U. Ben-Eli   *The Cybertec Consulting Group, Inc., New York, New York, USA*

G. Britton   *Mechanical Engineering Department, University of Canterbury, Christchurch 1, New Zealand*

R. Espejo   *Aston Business School, University of Aston, Birmingham, B4 7ET, UK*

R. Foss   *Yew Tree House, School House Lane, Aylsham, Norfolk, NR11 6EX, UK*

R. Harnden   *117 High Street, Blaenau Ffestiniog, Gwynedd, LL41 3AX, North Wales*

B. Holmberg   *Director Corporate Development, ASSI, 105 22 Stockholm, Sweden*

M. Jackson   *Department of Management Systems and Sciences, University of Hull, Cottingham Road, Hull, HU6 7RX, UK*

A. Leonard   *33-R Richroad, Greenbelt, Maryland 20770, USA*

H. McCallion   *Mechanical Engineering Department, University of Canterbury, Christchurch 1, New Zealand*

F. Waelchli
(deceased)   *Department of Defense, Defense Systems, Management College, Fort Belvoir, Virginia 22060 5426, USA*

Preface

Brain of the Firm and *The Heart of Enterprise* were published during the 1970s. In them, Stafford Beer unfolded his model of the organizational structure of Viable Systems. Both books were received with interest but had limited impact. In general, they were considered insightful but not practical. Over the years Beer received many requests to make this work more accessible to practitioners, and this led him to write the volume *Diagnosing the System for Organizations*, which was published in 1985. This book was intended as a practical guide to enable any manager to organize and manage his/her tasks effectively.

At Aston University around the same time, the Operational Research and Systems Group made Beer's work one of the cornerstones both for research and teaching practice. The work commenced under the leadership of the late Professor Steven Cook and continued after his death, with the invaluable support of John Watt. Results of the related research appeared in a variety of publications, but in particular the Aston Management Centre Working Paper Series became a key outlet. In parallel, the teaching was producing several important master and doctoral theses.

However, it was only after the publication of *Diagnosing the System for Organizations* that at Aston we sensed an increasing interest in management cybernetics. This was to lead in January 1986 to the setting up of a two-day workshop at the Manchester Business School, with the participation of more than 20 scientists and practitioners, including Stafford Beer himself. Participants were asked to prepare a position paper and the two days witnessed a most remarkable succession of presentations. Of all the papers presented at that meeting, the one by Richard Foss was the most original and well researched. Indeed, it was 'The organization of a Fortress Factory' (see this volume) that provided us with the impetus to embark on the task of putting together a book about the Viable System Model.

The original aim was for a truly transdisciplinary book with philosophical, biological, artistic, social and managerial contributions. Unfortunately we were unable to achieve this over-zealous ambition and the book evolved towards discourse concerning the management of social organizations. The exception to this evolving identity was Foss's paper, which was retained both as a tribute to its source inspiration, and as a reminder that the management of complexity is not restricted to the human domain alone. Nature's wisdom can be a source of inspiration for our own activities.

More than three years will have passed by the time of the publication of this volume. In the meantime, interest in Beer's work has continued to grow, and a wide range of people from all over the world continue to write to enquire about references and to ask advice as to how to most profitably use the Viable System Model. Many such requests have not been answered in the deserved depth. Day-to-day pressures have inhibited adequate responses, but we hope that this volume might be seen as a global answer to all who have approached Aston, and by their approaches have encouraged us to keep up the momentum.

The contributions to the book come from different parts of the world and there has been virtually no attempt to level out the styles and idiosyncrasies of the various authors. Some may find this diversity of forms of presentation untidy. However, we ourselves feel that the result actually contributes to the value of the volume as a whole, and enhances particular strengths of individual chapters. For example, Stafford Beer likes to hand-draw his own exhibits, and we decided to retain this style. Another instance is the contribution by Bengt Holmberg. In order to highlight the source of his chapter—a paper given by Holmberg, accompanied by a brochure distributed to the workforce of the ASSI Group—we decided to retain the actual brochure as an Appendix instead of attempting to incorporate it into the text.

This volume is not intended to be a tribute to Beer's work. Rather, we see it as a companion volume to *Brain, Heart* and *Diagnosing*. In spite of the great practical value of the last of these books, the editors of the present work feel that many people still have difficulties in understanding and making use of the Viable System Model. We hope and indeed expect that this volume will contribute to make it more accessible. In particular we would like to see this book help to build a bridge between people like ourselves and those working in a variety of roles in organizational theory and management practice. We think that the Viable System Model is a rich conceptual model that has much to offer to the understanding of organizations and to the practice of management.

The holistic nature of cybernetics offers both a powerful way of thinking about complex situations and an unparalleled means of handling complexity. In the contemporary world people are perceiving more and more interdependencies in their activities, and powerful ways to handle this increased complexity are in short supply. This volume addresses such a need, and as such is relevant to a wide range of people. We would like to offer it to scientists as much as to practitioners. The former group includes not only management scientists and organization experts but also information scientists and computer experts. The latter group embraces all kinds of managers at all levels in the organization, both in public and private enterprises. All of them have to cope with complexity and this book should help them to learn how to cope with it more effectively. But foremost, the chapters in this book should provide invaluable reading for students on a wide range of university courses, from traditional MBAs to the Operational Research and System Analysis courses.

Sadly, during the production of this book, one of the contributors, Fred Waelchli, died unexpectedly; his loss will continue to bear on wider circles.

Finally, we would both like to express our thanks to Dr Allenna Leonard for her close reading of the manuscript, and her valuable comments.

RAUL ESPEJO   *Birmingham*
ROGER HARNDEN   *May 1989*

The Viable System Model: Interpretations and Applications of Stafford Beer's VSM
Edited by R. Espejo and R. Harnden
© 1989 John Wiley & Sons Ltd

Introduction

The purpose of this book is to offer insights into Stafford Beer's Viable System Model (Beer, 1972, 1979, 1981, 1985) both by discussion and by demonstration. However, it is more than a compilation of case studies. In approaching the task, the editors distinguished an opportunity to undertake a rather more ambitious and interesting project. After all, however interesting a case study might be in itself, it always finds itself within some context: not merely the real-world context encompassing the phenomena serving as the source for the study, but the theoretical context which sharpens the particular tools used to undertake the analytical work.

This approach has allowed us to put together a book capable of being read along several distinct dimensions, and thus of interest to a variety of audiences. The reader might choose to treat the volume as a unified whole, a play between the logic of the Viable System Model and various different interpretations and uses of the model. Alternatively, readers primarily concerned with practical issues of implementation will find it easy to pick out and to concentrate on those chapters which have bearing on their own special interests and concerns, and to ignore those which appear either too esoteric or theoretical for their taste. We hope that this rarely happens, as for ourselves—one primarily concerned with the problems of methodology, the other more concerned with theoretical issues—the experience of collating the whole volume and sorting out our own thoughts, changed and deepened our insights into cybernetics in general, and Beer's work in particular. Our ambition is that this experience might be shared by others.

In spite of the detail and richness of some of the applications, it has to be stressed that the methodological implications of much of Beer's original work have yet to be worked out in detail. However, this surely is no shortcoming or fault. Rather the opposite. Methodological development can never occur in isolation, as if in some vacuum, but, as practical problems are overcome, the lived-experience feeds back into the embryo theoretical foundations. Indeed, one of the interesting aspects of putting the book together was to discover that there is really no such thing as some 'fixed', 'unyielding' seminal body of work, forming rock-solid foundations and just requiring the erection of some superstructure. The theoretical 'foundations' turned out to be first terms in an unfolding conversation very much alive and evolving in the present, though having its historical origins in books like *Cybernetics and Management* (Beer, 1959 and *Decision and Control* (Beer, 1966). Such a conversation entails exchanges and introduces themes literally not available to the original discourse, itself trapped in history.

This particular volume gives evidence not just of Beer's own ideas, but the way that seeds contained in his original work, though falling on a variety of different terrains—some barren, some fertile—have given birth to a range of interpretations and uses. Although readers will discern a coherent theme throughout the pages, they will not discover a single, unified view of the world. On the contrary, they are likely to discover a spectrum of views, ranging from positivistic to hermeneutic, not necessarily fully consistent with each other, although 'fitting' one particular context. Indeed, while we take responsibility for the contents of this volume, in the sense that it contains a set of coherent and well-articulated chapters, the papers do not necessarily express our own views. A shared *model* does not entail a monolithic vision, as will become clear as the different chapters are perused. The views expressed unfold a *conversation*, not a monologue.

For readers who might previously have found some of the ideas expressed by Beer himself interesting and stimulating, yet tantalizingly out of reach, or who might have banged their heads against a brick wall in efforts to put theory into practice only to give up in frustration and disgust, the papers in this volume offer a new and rewarding promise—a wide variety of viewpoints about putting theory into practice. Let's face it, this is very much an invitation to join in a conversation about matters which have not only intellectual and practical importance but also political and moral implications. We return to these issues in the final chapter.

Concepts

We open the book with a paper by Stafford Beer. 'The Viable System Model: its provenance, development, methodology and pathology' was first published in 1984 in the *Journal of the Operational Research Society*, and was intended by Beer as a synthesis of his unfolding of the VSM. In our view, it offers a suitable platform to launch the conversation about itself.

The paper reflects on the history, nature and status of the VSM—making reference to most of the publications in the field at its time of writing. The methodology used to derive the model is itself a model of clarity. Beer describes how he transformed his early insights about mathematical invariances in the management of complexity into the form of the VSM. This model, Beer argues, is *not* derived by analogy from the human central nervous system, but represents the isomorphisms which underlie any viable system, natural or artificial, biological or social.

This is a strong claim, and while the applications offered in this volume support it, they also help to qualify the assumptions behind it. The issue concerns whether the principles entailed by the VSM are necessary and sufficient for viable social organizations to exist. Is it not the case that we can point to organizations that, while appearing to be viable, manifestly do not comply with the principles of viability?

Beer also reflects upon some of the perceived limitations of the VSM: how is the theory affected by the fact that the elementary parts of a social organization are purposeful people, with free will, rather than cells or organs with no free will? Is it not the case that in individuals as biological organisms there is no inheritance of acquired characteristics, while this is an important characteristic of human activity systems?

These important questions are ones that we ourselves return to in the closing chapter of the book.

Finally Beer addresses the problem of using the VSM as a diagnostic tool in order to identify organizational pathology. While many of the pathologies that the VSM permits one to bring to light might coincide with those turned up by alternative methodologies, there are others which are likely to remain invisible when other methodologies are used. Beer argues that to recognize these normally invisible problems is not just a matter of peering 'out there' with better spectacles, but requires a paradigmatic change. One of the key arguments of this book is that such a change is already overdue.

Ron Anderton's chapter, 'The need for formal development of the VSM', will strike a chord for many readers, because people do have difficulties in actually using the VSM. As the title suggests, his argument is that the VSM is at a stage where it needs a more formal development, and this lack is one of the main reasons that Beer's work has not entered the mainstream of contemporary thought. After presenting a concise summary of Beer's ideas, Anderton sets out to pinpoint the difficulties commonly encountered in using the ideas: how to measure complexity? how to decompose the organization's total task? what is the VSM a model of?

Anderton is careful to locate the VSM in neither the positivistic (control centered) or the hermeneutic (interpretative) paradigms. In fact he seems to be saying that Beer's ideas have evolved in parallel to these developments and perhaps offer an alternative avenue. However, he argues that such uncertainties will only be resolved when a more formal model is developed, one more accessible to analysis and criticism.

The chapter by Fred Waelchli, 'The VSM and Ashby's Law as illuminants of historical management thought', looks at the historical emergence of different management theories. Mapping these various approaches on to Ashby's Law of Requisite Variety, Waelchli asks whether there may not in fact be a set of universal laws of organization. Cybernetics appears to offer a new way to think about such an issue. In the context of the unfolding conversation of this book, this chapter points towards a positivistic perspective. Where a more interpretative approach would tend to locate the invariances suggested by Ashby in the *conventions* adopted by a particular community of practitioners, Waelchli suggests that such invariances underlie the existence of social organizations as objective phenomena.

Raul Espejo's paper, 'The VSM revisited', unfolds the VSM from first principles. For this purpose he discusses the Law of Requisite Variety, and stresses the need to think in terms of *residual variety* rather than in terms of variety itself. The author defines the structure of an organization primarily in terms of two sets of mechanisms: the first monitors control, while the second concerns adaptation of organizational tasks. Espejo develops a discourse and terminology somewhat different from Beer's, in order to highlight what he considers to be the crucial functional invariances. The discussion is focused upon the relationships between these functions more than upon the functions themselves. The outcome is a set of criteria for organizational effectiveness. This paper offers not only a logical description of the VSM, but also some methodological hints as to how to put such a logic into practice.

Applications of the VSM

Raul Espejo's second contribution, 'P.M. Manufacturers: the VSM as a diagnostic tool', is an application of the VSM as it was unfolded in the previous chapter. This application was previously published in the context of information systems design. In the context of this book its purpose is to demonstrate the use of the VSM as a diagnostic tool. The outcome of the case study is a detailed list of the diagnostic points. This application illustrates how to define recursion levels and how to assess the adequacy of control and adaptive mechanisms in a clear and accessible manner. Encapsulating a range of the principles entailed in the VSM and showing how to make them operational, it is widely used in teaching around the world.

In 'The organization of a Fortress Factory', Richard Foss rigorously applies the VSM to the description of a non-human system. While human activity systems require more sophisticated strategies to handle complexity than do 'natural' systems, Foss argues that observation of the evolution of biological systems offers insight into viability, by providing examples of coherent coupling between autonomous parts as they realize one particular organization. By mapping the VSM on to such a system, Foss demonstrates how it provides a handle for his own analytical tools as a scientist, regardless of the fact that the system-in-focus does not involve human notions such as goal, purpose, value or belief. We stress that, while this application provides insight into viability and demonstrates how the Law of Requisite Variety can be instanced in nature, it is not intended as an example of how to make sense of human activity systems.

'Application of the VSM to the trade training network in New Zealand' reflects on the outcome of a professional report by Graeme Britton and Harry McCallion. The case assumes a degree of understanding of the institutional and political context in which the study was made, the original for the chapter being a working paper rather than an academic exercise. It demonstrates the VSM as a diagnostic tool in a multi-institutional setup: the training network for the electrical/electronics industry in New Zealand. The first part of the chapter usefully relates the model to well-established organizational theory. In the view of the authors, the model permits one to take into account highly turbulent environments, at the same time as providing powerful diagnostic and predictive insight. In the second part of the paper, the authors focus on the details of their system of concern. They successfully provide a rich description of the interactions of this system with its environment. This application is complex, and to be fully appreciated deserves a serious consideration of the details of the training system they are concerned with.

Allenna Leonard offers a novel application of the VSM. 'Application of the VSM to commercial broadcasting in the United States' uses the model as a tool to discuss an area of policy. For this purpose, she hypothesizes the embedding of a broadcasting station in several larger systems. The author makes apparent the implications of these multiple embeddings. She works out the existing structural arrangements on the assumption that the elucidation of several viable systems is desirable, and also works out some of the consequences of these structures not complying with the laws of viability. The chapter thus provides an instance of the VSM as a tool for policy analysis.

Stafford Beer's second contribution is 'The evolution of a management cybernetics process'. This paper is Beer's account of his work with a mutual insurance company over a period of nine years. An uncompleted report of this work was published in *The Heart of Enterprise* (Beer, 1979). The presence of the contribution in this book has two justifications: firstly, it completes an interesting story; secondly, it now appears in the context of several other applications and methodological contributions. In the final analysis, we believe that this chapter is one of the richest contributions in the book. It can be read from multiple points of view, and in particular it offers insights about the practical use of the model and the process of intervention. It highlights the difficulties of producing necessary change when the channel capacity of the mechanisms to produce these changes is inadequate.

Bengt Holmberg refers to a company in the heavy chemical industry (the ASSI Group in Sweden), with 8000 people distributed in 60 workplaces and a turnover of about a billion dollars. 'Developing organizational competence in a business' is an outstanding testimony of the VSM as a organizational development tool. People in ASSI moved from a functional to a decentralized structure, and in the course of the transformation discovered the need to be more conscious of organizational processes. the 'organization' in itself became an issue.

Today, the VSM is taken as a standard reference to guide cooperation and coordination in the ASSI Group. Beer's model is a regular part of their management training, at all levels from foremen to chief executive. This being the case, a major concern is to teach people to delegate and distribute information. The Appendix to this paper is of particular interest, containing a most useful description of the VSM which reflects the company's interpretation of the model. Discussion of the model is focused on the meaning of the functions rather than on the meaning of interactions between them, and in this respect can be usefully compared with the ideas in Chapter 4. This is a powerful practical implementation of the VSM.

'Strategic planning and management reorganization at an academic medical center' offers a good example of using the VSM as a reference to study a problem. Michael Ben-Eli demonstrates the use of the VSM as a 'guide' in the diagnosis and design of a problem situation. The author uses a phenomenological approach in the writing up; this is an instance of someone involved in a shift in management paradigm away from the positivistic and towards a more hermeneutic perspective. Ben-Eli makes a powerful use of the model even if his purpose was only to use it as an additional tool to produce insights for him as the analyst and not for the clients themselves.

However, the account unfolded by Ben-Eli provides a salutary lesson. It makes clear that it is not enough to demonstrate the logic of an analytical approach in order to get the insights it makes visible accepted. The messy 'real world' and our conceptual aspirations about it are always distinct. As practitioners and teachers, we must never lose sight of this fact.

Methodology and epistemology

Stafford Beer's last contribution is the paper 'National government: disseminated regulation in real time, or "How to run a country" '. This chapter reflects on the information system requirements in contemporary organizations. The opportunity for reflection was given to Beer by his recent work in Uruguay as part of a United Nations funded operation. The VSM offers a common rubric to think about and design information systems for all kinds of organizations. It provides a template to map all kinds of structures, and offers the chance to detect inadequate structures.

The chapter offers both a methodology and the tools for systems design. At the methodological level it makes apparent that information systems are totally intertwined with organization structure; the development of information systems not supported by the design of effective organization structures is totally inadequate. As for the information itself, it needs to be aggregated at the right level of recursion and needs to be in real time. The technology available today, in the form of hardware and software, permits the management of organizations in real time. While Beer suggests concrete products to support his vision, the main strength of this chapter is the vision itself. It is this vision that has been strongly criticized by a variety of people (see the examples given in the chapter by Jackson).

Raul Espejo's third contribution is methodological. 'A cybernetic method to study organizations' evidences a shift to the hermeneutic view of the world, and thence suggests a more flexible interpretation and use of the model. In this paper the author makes apparent that the study of organizations depends on the purposes that relevant viewpoints ascribe to it. Depending on these purposes, different models will define criteria of effectiveness for the same situation. Also it is possible to construct these models using different modes; this chapter focuses upon the distinction between diagnostic and design modes of study.

'Outside and then: an interpretative approach to the VSM' discusses the significance of System Four, the 'Intelligence function'. Roger Harnden makes clear that System Four is a modelling facility rather than a 'model of' anything. It is as such that it increases the viability of an enterprise. The argument starts from the Conant/Ashby Theorem which states that any good regulator of a system must be a good model of that system, and considers the implications of this theorem in the light of contemporary developments and discourse. The chapter explicitly gives an account of the VSM that is compatible with a hermeneutic perspective, shifting away from the notion that the model might be a literal representation of some objective, scientistic reality. Harnden feels that this is the way to make the model both more accessible and effective in its implementation.

Critical views

Mike Jackson's 'Evaluating the managerial significance of the VSM' aims to establish a reflective conversation between supporters and critics of the VSM, by presenting both cases. The model is assessed both with reference to the *premises* that must underpin effective managerial practice together with the *purposes* that managerial action serves, and the manner in which these purposes are determined.

This is a serious, densely argued paper which goes some way to exploring the problematical operational, logical and political issues in terms of which management cybernetics is both applauded and attacked. There *are* important questions concerning whether Beer's own model is seen to be intrinsically undemocratic, and to reduce the rich variety of human social interaction, or whether the promise of distributed power and information is actually met. This chapter is a valuable and impartial summary of such controversial debates, and refers widely to relevant literature.

In the final chapter, 'The VSM: an ongoing conversation', the volume editors offer their insight into what is entailed for an effective implementation of the model. This involves 'bedding' methodology within the context of a particular epistemology or way of speaking. The chapter thus offers a way forward.

In summary

This volume presents a range of challenging insights as well as giving witness to practical methodological progress. The underlying theme, or paradigm, concerns a holistic view of social organization, transcending formal functions and structures, and instead focusing upon perceived relationships and networks of social interaction. As is made apparent from a reading of the chapters, such a holistic approach in no manner reduces the capacity to act decisively and apply rigorous analytical procedures.

References

Beer, S. (1959). *Cybernetics and Management*. Oxford: English Universities Press.
Beer, S. (1966). *Decision and Control*. London: John Wiley.
Beer, S. (1972). *Brain of the Firm*. London: Allen Lane.
Beer, S. (1979). *The Heart of Enterprise*. Chichester: John Wiley.
Beer, S. (1981). *Brain of the Firm*, 2nd Edition. Chichester: John Wiley.
Beer, S. (1985). *Diagnosing the System for Organizations*. Chichester: John Wiley.

Part One
Concepts

The Viable System Model: Interpretations and Applications of Stafford Beer's VSM
Edited by R. Espejo and R. Harnden
Published 1989 by John Wiley & Sons Ltd

1

The Viable System Model: its provenance, development, methodology and pathology*

Stafford Beer†

President of the World Organization for Systems and Cybernetics

It took the author 30 years to develop the Viable System Model, which sets out to explain *how systems are viable*—that is, capable of independent existence. He wanted to elucidate the laws of viability in order to facilitate the management task, and did so in a stream of papers and three (of his ten) books. Much misunderstanding about the VSM and its use seems to exist; especially its methodological foundations have been largely forgotten, while its major results have hardly been noted. This paper reflects on the history, nature and present status of the VSM, without seeking once again to expound the model in detail or to demonstrate its validity. It does, however, provide a synopsis, present the methodology and confront some highly contentious issues about both the managerial and scientific paradigms.

Provenance

At the end of my military service, I spent a year from the autumn of 1947 to that of 1948 as an army psychologist running an experimental unit of 180 young soldiers (a moving population, 20 of them changing every

*Reprinted by permission from the *Journal of the Operational Research Society*, vol. 35, pp. 7–26. Copyright © 1984: Operational Research Society Ltd.
†Visiting Professor of Cybernetics, Manchester University Business School; and Professor of Social System Sciences, University of Pennsylvania, the Wharton School.

fortnight). All these men were illiterate, and all had been graded by a psychiatrist as psychopathological personalities. They could not write a letter home, nor read a newspaper, and such sums as 4 + 3 = ? often had them fooled. But they could debate with great energy and verbal facility if not felicity; they could play darts—'21 that's 15 and a double 3 to go'; and they could state the winnings on a horse race involving place betting and accumulators with alacrity and accuracy, and apparently without working it out. They had their own conception of discipline, involving terrorism and violence in the barrack room, which met every desideratum of a military unit in its ends, though not in its means.

I had a background in philosophy first and psychology second; the latter school had emphasized the role of the brain in mentation and of quantitative approaches in methodology. The analytical models that I now developed, the hypotheses set up and tested, were thus essentially neurophysiological in structure and statistical in operation. The behavioural models derived mainly from experience: I had a background in the Gurkha Rifles too. What made these people, unusual as they were, tick—and be motivated and be adaptive and be happy too (for most of them were)? And how did the description of individuals carry over into the description of the whole unit, for it seemed so to do: every one of many visitors to this strange place found it quite extraordinary as an organic whole. It simply was not just a unit housing a population of unusual soldiers. The first regimental sergeant major asked for a posting.

This was the empirical start of the subsequent hypothesis that there might be *invariances* in the behaviour of individuals, whether they be 'normal' or not, and that these invariances might inform also the peer group of individuals, and even the total societary unit to which they belong. In the early 'fifties this theme constantly emerged in my operational research work in the steel industry: I used then to refer to the structure of 'organic systems'. So the viable systems model (VSM) dates back 30 years. I pursued it through neurocybernetics and social science, through the invention and study of cybernetic machines, through the mathematics of sets and stochastic processes, and at all times through the OR fieldwork in industry and government. The quest became to know *how systems are viable*; that is, how they are 'capable of independent existence'—as the dictionary has it. By the time my first book on management cybernetics was published, I had also mapped a set-theoretic model of the brain on to a company producing steel rods, and published the basis of the whole approach (Beer, 1959, 1960).

The set-theoretic model proved difficult for people to understand, and eventually a streamlined version of the model appeared called *Brain of the Firm*, using neurophysiological terminology instead of mathematics (Beer, 1972). Some commentators were offended by this and called the model analogical—despite my denials and explanations (see later) that this was so. Hence, in a still later book a new version of the VSM was developed from first principles, called *The Heart of Enterprise*, in the belief that the necessary and sufficient conditions of viability had by now been established (Beer, 1979).

The invariances that I had finally unearthed were stated; and the central principle of recursion (that every viable system contains and is contained in a viable system) stood duty as the explanation of all the observational evidence that had begun to accumulate from the military experience onward. Moreover, I developed a topological version of the original set-theoretic algebra that it seemed no-one would study properly. The drawings were now rigorous mathematics in themselves in that they offered explicit homomorphic mappings of any one VSM recursion on to the next—as may be seen in the simplified version at Figure 4. (In 1972 the drawings had given an indication of the recursion theorem and relied on the independently published mathematics.)

Throughout its development, and to this day, the VSM has been in a process of continuous testing and verification. Meanwhile, however, the whole approach had its most significant and large-scale application during 1971–73 in Allende's Chile. As an outcome of this experience, five new chapters were added to *Brain*, and the overhauled and extended text was republished (Beer, 1981). Thus (the new) *Brain* and *Heart* stand, as complementary volumes, for the theory of the viable system and its 'laws' in management cybernetics, and a trilogy has been completed with *Diagnosing the System* (Beer, 1985). Commentators often imply that I am obsessed with this model. Well, the quest to establish *how systems are viable* and its 30-year pursuit have certainly been demanding. Even so, the three books mentioned are only three out of ten. The philosophy of science that I was simultaneously developing is expounded in *Decision and Control*, and it is from this that I draw the following methodology and apply it to the VSM (Beer, 1966).

The methodology of topological maps

When we notice similarities between two different systems, for instance between the regulatory system of an individual and a group, or between a brain and a firm, the comparison often begins in a literary manner. There is the simile: 'management communications are like the nervous system, in that . . .'. There is the more direct metaphor: 'the real muscle of the plant is the cogging mill'. Such comparisons may help to convey insights, although everyone knows better than to take them too seriously. But as perception of the two systems deepens, and perhaps observations are taken, we may come to hold conceptual models of both systems that become exciting and helpful. This stage is easily recognized, because we find that some circumstance that we understand in one system throws light on a parallel circumstance in another. It is now worth 'drawing analogies'; on the other hand, everyone knows that 'analogies may be carried too far'.

The process continues, and begins to have the marks of a scientific method, when we try to develop rigorous formulations of the two conceptual models. (Figure 1 refers.) These will each be a *homomorphic mapping*, insofar as many elements in the system that is conceptually modelled will map on to one element in a rigorous model. All falling apples, and not only the particular falling apple observed by Newton, obey the law of gravitation: we *select* those mappings that exhibit mathematical *invariance*. And if we travel to Pisa, we find Galileo (who died in the year

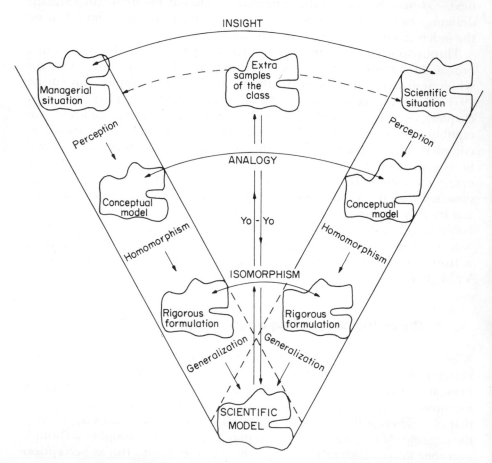

Figure 1. Beer's account of scientific modelling

that Newton was born) supposedly dropping not–apples from the leaning tower, but determining a *constant* none the less.

Now what happens if we map the two rigorous formulations of orchard systems and Pisa systems on to each other? If we find invariances between the two systems, then these are isomorphic mappings, one-to-one in the elements selected as typifying systemic behaviour in some selected but important way. The generalized system that comes out of this process, which applies to all systems of a particular class, is a scientific model—in the case just considered, a model of gravitation. The generalization of some behaviour invariably and invariantly exhibited by the system as interpreted through this systemic model we usually call a law. Nonetheless, we have made a selection; we have reduced systemic variety through our homomorphisms. But that is the very business of scientific discovery. In fact, every system can be mapped on to any other system *under some transformation*; thus Ashby was wont to say that the Rock of Gibraltar makes a good model of the brain, if your interest is exclusively in spatio-temporal extensity.

Considering these matters coolly, and handling them in a world which upholds a particular paradigm that does not compare rocks to brains, is not in easy matter. The precise difficulty that most people have arises when a breach of taxonomy is offered as between social systems, individual people and artifacts. The amalgam is seen as essentially different from the unity, and the animate as essentially different from the inanimate. But these were among the major paradigmatic distinctions that were explicitly questioned by the founders of cybernetics in the 'forties. Certainly my own methodology, especially as it relates to the class of viable systems, makes its mappings quite happily across these boundaries. Witness the very title of the most formal statement of the method: *The World, the Flesh and the Metal* (Beer, 1965). An extract from this paper, giving a group-theoretic analysis of the modelling methodology, will be found in Appendix 1.

Having said all this, there is no way of 'proving' a model: the by now classical criterion of 'falsifiability' remains instead. As experience of the VSM grew, as its format was made tidier, and as others became involved, more and more viable systems were mapped on to the model: the invariances held. The methodology at this point may be described as the *yo-yo technique*. That is to say: we have constructed a VSM by mapping (let's say) a brain on to a firm and now wish to test a second, third, and so on viable system against the scientific model. We run down the chain of similes, analogies and homomorphs with one of these fresh systems until the isomorph is reached, testing the insights and invariances as appropriate on the way; then we return up the chain with another fresh system; then down again, and so on—hence the yo-yo metaphor (rather than model, note). Other scientists around the world have confirmed the VSM in various modes and situations, most but not all of them managerial. A note about these activities appears at Appendix 3.

On mapping and measuring complexity

Although we may derive a model in the manner shown, and although we may develop confidence in it through many applications over a long period, practical activity requires more than this. The management of any viable system poses the problem of managing complexity itself, since it is complexity (however generated) that threatens to overwhelm the system's regulators. This is very obvious in biological systems, wherein there are no self-proclaimed 'managers'; but in social systems too complexity tends to overwhelm those managers whose activities are not seriously directed towards viability but to short-term goals such as profit. A precise measure of systemic complexity had been proposed as *variety*, meaning the number of distinguishable elements in a system, or by extension the number of distinguishable systemic states (Ashby, 1965). The problem of controlling this variety is daunting indeed, if all distinguishable states are equally likely. But they are not.

We are used to suppose the variety in social systems is kept under control by a legislative mode of regulation that restrains variety proliferation. But, as Ashby learned from biological systems, something more subtle underlies any such technique. The notion of a 'coenetic variable' explains the delimitation of the variety of environmental circumstances and of apparently regulatory responses at the same time (Sommerhoff, 1950). Sommerhoff wrote (see Figure 2): Coenetic (pronounced 'sennetic', from the Greek meaning 'common') variables simultaneously delimit variety as shown, so that trajectories of the system *converge* on to a subsequent occurrence. Sommerhoff called this 'directive correlation'. The schematic diagram exemplifies what I later called 'intrinsic control': in the very process of disturbing environmental circumstances, the coenetic variable evokes a response that converges on an adaptive outcome.

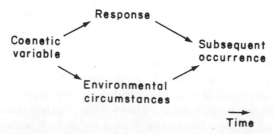

Figure 2. Sommerhoff's account of 'directive correlation

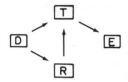

Figure 3. Ashby's account of 'requisite variety'

Ashby for his part had developed a schematic treatment based on Shannon's notation (Shannon and Weaver, 1949; see Figure 3). D stands for disturbance, and is equated by Ashby with the coenetic variable. E is still the outcome set, which is exhausted by good and not-good subsets (in relation to viability). T is a table of the transformations which D will undergo to generate E, and is equated by Ashby with the environmental circumstances of Summerhoff. But now Ashby is taking note that R may, after all, directly influence T in its task of modifying E.

He argues thus. If R's state is always to have the same effect on T, whatever state D may adopt, then the variety of E will be the same as the variety of D. But if R may adopt two states, then the variety at E can be halved. And so on. 'If the variety in the outcomes is to be reduced to some assigned number, or assigned fraction of D's variety, R's variety *must* be increased to at least the appropriate minimum. *Only variety in R's moves can force down the variety in the outcomes.*' This is the famous Law of Requisite Variety.

Now it is clear that if D is a coenetic variable, so that R and T are directively correlated, then the variety of the outcomes E will be constrained. Since in both biological and social systems there may be coenetic variables that are unrecognized as such, this would account for a more regulated system than the unrecognizing observer would have any right to expect. Even so, and as Ashby says:

> 'variety comes to the organism in two forms. There is that which threatens the survival of the gene pattern—the direct transmission by T from D to E. This part must be blocked at all costs. And there is that which, while it may threaten the gene-pattern, can be transformed (or re-coded) through the regulator R and used to block the effect of the remainder (in T).'

The model of any viable system, VSM, was devised from the beginning (the early 'fifties) in terms of sets of interlocking Ashbean homeostats. An industrial operation, for example, would be depicted as homeostatically balanced with its own management on one side, and with its market on the other. But both these loops would be subject to the Law of Requisite Variety. Since the variety generated by the market would obviously be greater than the industrial operation could contain, then 'this part must be blocked at all costs', as Ashby has said. This became in my first book (Beer, 1959):

> 'Often one hears the optimistic demand: "give me a *simple* control system; one that cannot go wrong". The trouble with such "simple" controls is that they have insufficient variety to cope with variety in the environment. Thus, so far from not going wrong, they cannot go right. Only variety in the control system can deal successfully with variety in the system controlled.'

  This understanding came from down-to-earth experience as the production controller of a steelworks. By the same token, just as proliferating incoming variety must be blocked at all costs, so must outgoing managerial variety be enhanced—by transformation or recoding through the regulator R, as Ashby said. Looking at the variety-disbalanced homeostats of the VSM, I wrote:

> 'Each part-system provides unlimited variety . . . It is the function of intelligence to tap that variety, to organize it, to *select*. . . . What is needed, is the amplification of the primary selection.'

  It has always seemed to me that Ashby's Law stands to management science as Newton's Laws stand to physics; it is central to a coherent account of complexity control. 'Only variety can destroy variety.' People have found it tautologous; but all mathematics is either tautologous or wrong. People have found it truistic; in that case, why do managers constantly act as if it were false? Monetary controls do not have requisite variety to regulate the economy. The Finance Act does not have requisite variety to regulate tax evasion. Police procedures do not have requisite variety to suppress crime. And so on. All these regulators could be redesigned according to cybernetic principles, as I have argued *passim* (Beer, 1975, especially).

  For present purposes, however, I seek only to show how Ashby's Law was derived, and how it at once suggested to me that if variety were not requisite in a regulatory homeostat, then either the greater variety must be attenuated, or the lesser variety must be amplified, or both. This conclusion does not appear to be novel, as has been suggested, but to be sanctioned by Ashby's own words quoted above. Certainly my own applications and extensions of homeostatic's theory in management went beyond Ashby in treating the box called T, supposedly a 'table', as a *black box*—that is to say that the box contains a table that is not available to inspection (something that I had learned in military OR, for foes do not care to make their transformation rules manifest). But Ashby was the doyen of black boxes too.

  What was perhaps novel, for the record, was the recognition that in the VSM homeostats requisite variety applies in three distinct ways: to the blocks of variety homeostatically related, to the channels carrying information between them, and to the transducers relaying information across boundaries. Statements about these came to constitute my first three Principles of Organization (Beer, 1979; see Appendix 2). Ashby saw his Law as bearing particularly on the second question, that of channel capacity, probably because he had derived it from Shannon's communication model—which deals with the transmission of information. Indeed he comments that Shannon's Tenth Theorem is a special case of the law of Requisite Variety. Next, and unsurprisingly, he had no difficulty in accepting the identification of transduction as a particular aspect of transmission, and one especially important in management work. But Ashby was not satisfied that requisite variety could be contemplated in terms of relative blocks of variety generators, as my First Principle proposed. Again, it is probable that only information transmission gave operational meaning to requisite variety in his eyes; but in arguing (as he sometimes did) that therefore he had done no more than generalize the Tenth Theorem, I think that he seriously under-rated his own discovery.

Since Ashby was a psychiatrist, I put the counter-case thus. We have a set of mental illnesses, evidently of very high variety—since maybe no two people ever had exactly the same syndrome. There arises quite naturally, and this is an example of requisite variety exerting itself in informational terms, a vast number of 'names' for these illnesses; that is, if we allow that descriptive qualifiers for such generic terms as 'schizophrenia' abound. Unfortunately, however, there is no more than a handful of treatments available: psychoanalysis, convulsive therapy, tranquilization, deep narcosis, surgical intervention . . . it is difficult to continue. It follows that all the amplification of channel and transduction variety in the naming is not to the purpose when it comes to managing the illness. Since the syndromes must be mapped homomorphically on to a low-variety therapeutic map, Ashby's Law asserts itself *regardless* of the operational format that is followed.

The point is important in any management process. For just as large numbers of strategies for regulating a firm or an economy can be invented to provide requisite variety, only to be proven useless because they cannot be conveyed through low-variety channels and transducers (and Ashby liked to point this out), so high-variety channels cannot enhance low-variety inputs—unless they contain the intrinsic generative power to be amplified because of the way they are organized *inside* the block. A map-reference has this quality, for instance, and so does a personal file; the policy to 'cut all stocks or costs by 10 per cent' does not.

Limitations

Analogies have limitations; but in a real sense a scientific model as defined should have few—because the transformations it covers are listed and are exactly specified. The problem with analogies is to delineate the contexts in which they are supposed to hold, and then to run the risk that elements will unexpectedly turn up in one system that have no analogues in the other. These dangers are not encountered with scientific models that are properly mapped.

To take an obvious example: Newton's theory of gravitation works very well inside the solar system, give or take the perihelium of Mercury. In a spatiotemporal system that is much larger, Newton must be adjusted by relativity theory. We have, in short, to nominate the context, to fix the boundaries. Now a viable system survives under considerable perturbation because it can take avoiding action, because it can acclimatize, because it accommodates, because it is adaptive, and so on. But put a human in a box, suck out all the air, and s/he dies. We know this, and do not make a lot of fuss about it, because it is an agreed aspect of the definition of viability that there should be a rather closely controlled environment. If we send an astronaut into space, therefore, we equip him with a space suit. We shall certainly not say that our whole conception of viability is faulty because s/he must wear one. On the contrary, one of the most useful products of the manned space programme was its exact specification of a life support system; this indeed fixes the physiological boundaries of viability, though (interestingly) not the psychological boundaries.

Secondly, as to elements which may be recognized in one system and not in another, let us remember that the methodology deals with formal homomorphic mappings and nominates invariances. Anything not so mapped, and anything not determined as a constant, will not be a topic of concern. If it becomes such a topic after the modelling has been done, then its mappings will have to be tested.

Two limitations of the VSM are matters of importance, but they propose no serious misgivings when examined in context and under invariance. The first is often brought up, sometimes in hysterical fashion, by those who notice that people may be the basic elements of a so-called viable system under the VSM rubric. *People* (they say) have free will. Yes, maybe; but people also have constraints laid upon their variety by upbringing, or by the roles that they agree to play in a social unit like a firm. It is true that, for example, the liver cannot resign and be replaced by one less gnarled, but what about it? What matters is the functioning of an element, under whatever constraints that the job entails: not the identity of the element itself. And this is just as well for freedom-lovers—let them by all means get out, if the system is oppressive towards them, and they can. It will make no difference to the viable system, *unless* the element has special properties that cannot be replaced. Well, this is simply a matter of nominating what elements in the mapping are to count as invariances. I have known businesses fail because one man was lost, and he accounted for 85 per cent of sales. There is nothing surprising in that. So if the heart of an employee stops beating, that finishes *him* as a viable system. At the next level of recursion, whether that is considered to be his firm or his family or his church or anything else, his loss as an element of this next viable system may or may not be important to its viability. He may simply be replaced; or perhaps *that* system will die too. Obviously, all this will be of high significance to those concerned; but it has no methodological significance to the scientific model within which invariant mappings have been specified in advance.

The second limitation is of more interest, although it can be handled by similar arguments, because it seems to me to be a limitation of society itself rather than a limitation of the model. In either case, it has never been raised with me by anyone at all—at least, not in the terms that are used here. A major battle in biology concerning the possible inheritance of acquired characteristics in the individual, as conceived by Lamarck, seems to have been settled in recent years by microbiologists. There is no such inheritance, for genetic information is always carried by nucleic acid to inform the protein molecule—and never the reverse. In society, however, that is in the social group, there clearly *is* an inheritance of acquired characteristics. Therefore a major difference emerges as between the VSM of the individual and the VSM of society to constitute, at least on first sight, a limitation of the model.

However, as we saw earlier in discussing Ashby and requisite variety, there must always be a barrier (at T) to block the effects of proliferating variety (at D); otherwise results (at E) will reflect the full input variety—and are likely to be quixotic. It seems that in the case of the individual, the gene pool is protected by the encoding of the transformation table (at T). In the case of society, stability in subsequent generations must be ensured by the collaboration of the response with the transformation table (Ashby's R-and-T interaction). Experience shows that this always happens. There is always an element of tradition in the directive correlation of society—that is to say that the transformation table is acting as a block; and there is always an element of novelty coming through from recent outcomes (at E) by regulatory feedback (through R)—that is to say that the response function is acting as an amplifier. So the model can cope with these divergences. The question is whether society itself gets the (R,T) admixture right. Even if it does, it appears to be short of damping mechanisms to prevent uncontrollable oscillations—but that is another story, covered later in System Two of the VSM itself.

The viable system model (VSM)

According to the cybernetic model of any viable system, there are five necessary and sufficient subsystems interactively involved in any organism or organization that is capable of maintaining its identity independently of other such organisms within a shared environment. This 'set of rules' will therefore apply to an organism such as a human being, or to an organization consisting of human beings such as the State. The comparison is made not by way of analogy, but, as has already been explained, because the rules were developed to account for viability in any survival-worthy system at all.

In very brief, the first subsystem of any viable system consists of those elements that *produce* it (they are the system's autopoietic generators, to use Maturana's terminology). These elements are themselves viable systems. In the limit, the citizens constitute the System One of the State. I say 'in the limit', because the citizens first produce communities and firms, cities and industries, and other viable agglomerations, which are themselves all elements to be included in the State. So a full account of the matter (see *The Heart of Enterprise*) will show how systems of increasing complexity are nested within each other like so many Russian dolls or Chinese boxes to produce the whole. Mention was made at the outset (under 'Provenance') of the discovery of the theorem of recursion, and this is where it belongs. 'In a recursive organizational structure, any viable system contains, and is contained in, a viable system.' Out of a five-fold systemic set constituting a viable system, says the model, System One is always a viable system itself. The topology is clearly visible in Figure 4, where (in the first place) one complete viable system fills the page. Inspection will show the five interacting subsystems labelled ONE, TWO, and so on, in capital letters. Among these may be discerned two Systems ONE (there could be more), each of which *contains* a complete viable system displayed at a 45 degree angle.

The whole-page viable system is shown as interacting (see above) in a precisely defined way with its environment through both its Systems ONE, and through its System FOUR, and not otherwise. Equally, the embedded viable systems are shown as interacting in exactly the same way with local environments that are peculiar to each of them—although they are (inevitably) subsets of the whole-page environment. It is vital to understand that the topology of recursion demands an exact replica in each case. In the drawing, the only discrepancy is that the connection between System 4 in the second System ONE and its sub-environment has not been completed, as its twin in the first System is correctly completed, for obvious graphical reasons.

Brief annotations are made in the diagram to indicate the roles of the five subsystems. To enlarge on these within the compass of this paper is not possible without trivializing the elaborate functions of every box and every line, and the reader wishing to investigate the theory itself must be referred to the companion volumes *Brain* and *Heart* previously mentioned. Some

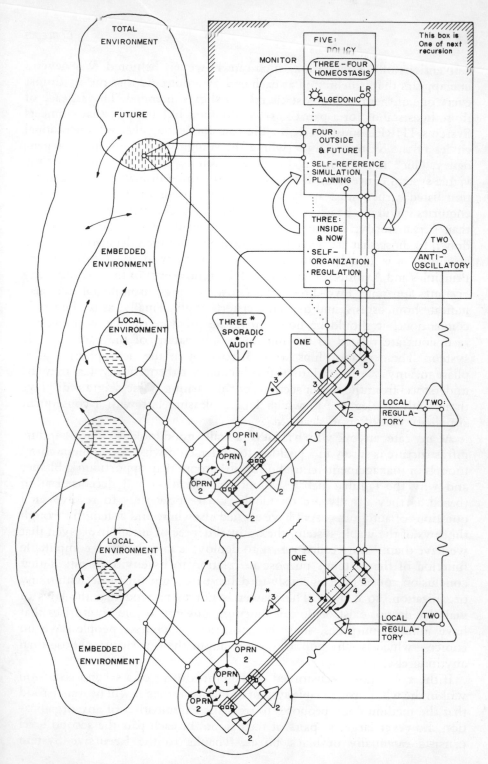

Figure 4. The Viable System Model (VSM) showing recursive embedments

conventions of the diagram as such can, however, be noted. Whenever a line appears that is delimited at each end by a dot (and that means almost every one of them) a homeostatic relationship is intended. That is, each of those lines stands for a pair of arrows looking like the pair that connects Systems THREE and FOUR, or the pair connecting the two operational circles of the Systems ONE (where the squiggly line indicates a dependency which may be strong or weak depending on the purposes of the viable system concerned). It follows that the three Principles of Organization listed earlier must apply to each of hundreds of them. In practical enquiries it will not be necessary to investigate every such homeostat more than perfunctorily; with experience, the consultant's attention becomes drawn to those that are defective or unstable, and then detailed analysis is essential. It is largely from the insights thereby derived through the Principles and Axioms (of Appendix 2) that the power of the methodology becomes apparent. Two-directional arrows in the environmental areas also indicate homeostasis, and the same criteria apply. Finally, as to the iconic conventions, it should be noted how the subsystems of the embedded recursions are related to the matching subsystems of the parent viable system. These relationships were discovered in the neurophysiological phase of my work, and first formulated set-theoretically (as may be understood from the earlier sections of this writing). The diagram displays them with an elegance in which I take pleasure but with a consequent simplicity that often misleads.

At any rate, anyone who has taken a really good look at Figure 4 and its infrastructure is surely in a position to understand why purely hierarchical models of management are useful for little more than apportioning blame, and why the familiar debates about centralization and decentralization (based as they usually are on ideology) are powerless to resolve vital questions of autonomy. As I have argued elsewhere and at length through the laws of the viable system, the optimal degree of autonomy (given that we have the measure called *variety* to deploy) is in principle a computable function of the system's purpose in relation to its environment. If that conclusion appears outrageously to delimit freedom for those within any organization, so it does. The answer is not to pretend that the laws of viability do not exist or do not function, but not to join enterprises that have inimical purposes. In some countries, of course, people have no choice—which is why politics is more about the capacity to choose than anything else.

If the eye is now accustomed to the embedment of a second recursion within the whole-page viable system of Figure 4, then it will be understood that the methodology proposes to treat the examination of any organization, however large, in pairs of recursions. In each pair the second level consists essentially of black boxes. Thanks to the Recursive System

Theorem, however, each of these black boxes can next be elevated to 'whole-page' treatment—whereupon a new recursion of viable system embedments will be disclosed. The methodology resembles the movement of a magnifying glass and an illuminating spotlight down the chain of embedments so the accustomed eye of which I was speaking may now review Figure 4 with its pair of recursions so far described, and discover the outsize square box at the top right-hand side which is the management element of System ONE of the *next higher* recursion; it may also discover the rudiments of the level of recursion *next below* the embedment orginally discussed. Thus Figure 4 can be regarded as indicating four levels of recursion out of an arbitrary series (which descends to cells and molecules and ascends to the planet and its universe), of which the middle two recursions receive complete iconic representation.

This is not a claim that an account of a viable system's recursive embedment is ever unique, despite its progression to infinity in both directions, because each viable system figures in an infinite number of chains. Rather is it a manifestation of Hegel's Axiom of Internal Relations: the relations by which terms (or in this case, recursions) are related are an integral part of the terms (or recursions) they relate. Incidentally, if we put the Self as a viable system in the centre of the sphere generated by the infinite set of its recursive chains, then we have a model of selfhood that both expands to embrace the universe and also shrinks to a vanishing grain of sand—a model familiar in oriental philosophy.

This thought leads us conveniently to the recognition that the boundaries of any viable system are arbitrary, as is the number 'five' of its subsystems. The 'fiveness' was due to my efforts to establish the necessary and sufficient conditions of viability, and five was their number; it might have been otherwise, if I had used a different rubric. What could not have been otherwise is the fact of the *logical closure* of the viable system by 'System Five', whatever its number: only this determines an *identity*. Nominating the components of System Five in any application is a profoundly difficult job because the closure identifies self-awareness in the viable system. 'What business are we in?' asks the Manager. But who are 'we'? Shareholders, employees, managers, directors, customers, taxmen, environmentalists . . . all these have different answers to offer. 'What business is the self in?'—see above.

I have repeatedly told the story (for instance, see *Brain*) of how President Salvador Allende in the Chile of 1972 told me that System Five, which I had been thinking of as himself, was in fact the people. Then perhaps the president embodies the people; or perhaps the presidency is overtaken by a gang of thugs, as was to happen in 1973. (For some recent discussions of this example, see Beer, 1983.) At any rate, it is clear that the determination of closure, and thus the recognition of identity and self-awareness, in any viable system is an outstanding example of the observer's imputations of *purpose* to that system that are probably idiosyncratic. There are ideological traps: for example, the biggest confusion in which I was ever profession- ally involved concerned the purpose of a health system, to which there are as many answers as interests involved. There are teleological fallacies: think once again of selfhood . . .

These difficulties are not indications that the VSM 'doesn't work': the model does not create the problems that it makes explicit. Rather does it enable managers and their consultants alike to elaborate policies and to develop organizational structures in the clear understanding of the recur- sions in which they are supposed to operate, and to design regulatory systems within those recursions that do not pretend (as do so many of those we employ) to disobey the fundamental canons of cybernetics.

The pathology of the viable system

Many people dislike to see the word 'pathology' written in such a context as this, because the theory of the viable system may be dealing with societary units, or even with such entirely inanimate systems as computer- based communication networks. Some of these people would be placated if the word in the title were set in inverted commas. The fact is, however, that either we have a theory of viability, meaning 'capable of independent existence', or we have not. The possibility of such a theory is anti- paradigmatic within the subculture, true; but that paradigm is overdue for change: see Capra (1982). The risk of making mistakes under any method- ology of analogy is great, true; but we have been at pains to show that an heuristic such as the yo-yo technique is in search of a mathematical invariance that transcends analogy. A viable system made of metal could be melted down, true, and one made of people could be disbanded, true; but the feotus of eight months is the classic example of a viable system, and many conditions of existence are attached to its capability for independence too. In short, the opponents of 'biological analogies' are often the first to misapply them when they try to make their own case, thanks to an uncritical belief in the properties of protein-based machines which in fact work only within rather narrow physiological limits.

According to these cybernetic enquiries, practised, as has been said, in many countries over many years, viable systems of all kinds are subject to breakdown. Such breakdowns may be diagnosed, simply in the fact that some inadequacy in the system can be traced to malfunction in one of the five subsystems, where in turn one of the cybernetic features that compose the rules (cf. Appendix 2) will be found not to be functioning. To continue unabashed with medical-sounding talk that is in fact wholly appropriate to the cybernetics of viability: the etiology of the disorder may be traced, a prognosis may be prepared, and antidotes (even surgery) may be prescribed.

Subjectively speaking, confidence in the VSM as applied to societary systems derives not so much from the fact that the pathology of the viable system can be investigated with ease, as from the *speed* with which the diagnosis can be made. The knowledgeable user may expect to 'home in' on (say) half-a-dozen causes of concern within a day or so of exposure to the real-life system, and it is a frequent experience to find such danger points when they have been deliberately concealed out of embarrassment or self-serving: they tend to signal themselves. Interestingly enough, such incidents tend to enhance the confidence not only of the VSM-er, but of the client management itself.

A question often asked is this: if we are dealing with an organization that exists, that is actually there to be investigated, then surely it is by definition a viable system—and nothing remains to be said? This is where the pathological vocabulary becomes so useful. The fact that the societary system is there does not guarantee that it will always be there: its days may well be numbered, and many have been the 'buggy-whip' companies to prove it. The fact that it is there does not prove that it is effectively there, witness universities, nor efficiently there, witness hospitals. Monoliths and monopolistic systems in particular (such as these two) often operate at the margins of viability, creaking and choking like the valetudinarian organizations that they are. Moreover, many such are operating at such an enormous cost that they are becoming less and less viable in front of everyone's eyes.

One of the main reasons for this, particularly in the social services, is that people looking for cheaper ways of doing things attempt to repeal the Law of Requisite Variety itself. Policing, for example, whether by the police themselves in terms of crime, or by environmental agencies in terms of pollution, or by health scanners of pre-symptoms, often fails to recognize that only variety can absorb variety. A great many examples are reproduced elsewhere (Beer, 1975).

Next, there are four diagnostic points made in a learned journal (*Brain and Strategy*, 1983). All four have been expounded in my own writings, but not I think with such pith; therefore I take leave to reproduce them here as direct quotations.

1. *Is management presiding over a 'viable system'?*

If any of Beer's five necessary functions are removed from, say, a subsidiary, then its abilities to operate successfully may well be killed.

This could perhaps involve taking away a subsidiary's freedom to invest its financial surpluses or removing its sales function, for example.

2. Does subsystem Five truly represent the entire system within the context of larger, more comprehensive and more powerful systems?

If this function, or subsystem, is unable to find a way to represent the essential qualities of the whole system to the larger meta-system, then the system's survival is in question.

3. Do managers often fail to understand the need for subsystems Two and Four?

Business people have little difficulty recognizing the need for subsystems One, Three and Five. If Two is missing, activity in One can turn deadly and self-defeating as units fight for resources and against entropy; if Four is missing, Three and Five can collapse into each other, leaving the critical Five subsystem a mere functionary.

4. Do the Three, Four and Five subsystems need to form a Three–Four–Five subsystem to encourage 'synergy' and interactivity?

Without a constant interaction and exchange of information between these three functions, Three is vulnerable to 'narrow tunnel' syndrome and Four is exposed to the perils of 'flights of imagination'.

Not only are these points extremely cogent and penetrating, they well illustrate how the structure and the language of the model make possible the expression of elaborate and/or subtle comments in very few words. Let me add a few remarks on each of the indicated pathologies, drawn from experience.

(i) Subsidiaries that are 'taken over' are always painstakingly assured that their individuality will be preserved, their autonomy respected, and so on. After all, the argument (very plausibly) goes, your individuality, your reputation, your goodwill, your people are all assets for which we have paid hard cash—naturally we shall nurture them. This is poppycock—although it is often believed by the takeover bidder himself. A study of the embedment of the new System One in terms of the Law of Cohesion (see Appendix 2) will reveal how the inter-connectivity between the subsystems of the two recursions inevitably takes up variety from the new subsidiary. In the VSM, 'autonomy' is a precisely defined term, and it does *not* mean zero interference. Incidently, if the taking-over company makes the mistake of leaving intact all the new subsidiary's variety (or of handing over too much variety to an old subsidiary), this company is very likely to be the subject of a reverse takeover bid.

(ii) This is an issue of identity. The work here reported has repeatedly encountered situations in which all manner of adjustments have been necessary to make the viable system secure in a changing environment. That is, adaptation is evoked (in those situations) as a key characteristic of viability, and much change ensues. Will the system still be able to recognize itself? More particularly, will others be able to recognize it? Philosophers used to ask whether 'this apple' were still 'this apple' after a large bite had been taken out of it. . . . *The Heart of Enterprise* includes a highly sophisticated Test of Identity with this point in mind.

(iii) The collapse of Five into Three (in the effective absence of Four) is made particularly likely insofar as Five people have usually been promoted from Three. They are uncomfortable as demi-gods with no clear duties beyond being wise and pleasant. Thus, when something goes wrong in System Three (or even One), they are likely to dive down into the problem that they understand so well—never to emerge again. They may be seen around, but only as their previous Three incarnation—erratic and abrasive as ever. But the collapsed metasystem is a special pathology. It is a decerebrate cat, pinned out, intravenously fed. It responds *reactively*, from the autonomic command centres at Three, and is incapable of planning and foresight (Four) and will and judgment (Five). But it will react to prods by a reflex kicking-back. With no apologies to those complaining about biological metaphors, who knows an organization that is a decerebrate cat?

(iv) The attention drawn to this problem is well merited. It is the intellectual springboard for recognizing the value of an operations room, or (a better term) management centre. In such an 'environment of decision', as I have called it, the Three–Four–Five metasystem has a chance to find its own cohesion, and to operate in a nutrient medium.

Obviously it would be possible to comment on every feature of the viable system from the standpoint of its pathology. But that would be boring; and perhaps the above discussion of some already profound points sufficiently gives the flavour of the pathologist's commentary.

But it may be worth ending with a suggestion which this discussion seems naturally to propose—in medical practice, there is such a thing as post mortem examination. Much knowledge of viable systems has been gained by the study of those that are viable no more. I have done some work of this kind, but only as the result of being fortuitously present at the deathbed. The suggestion would be that a small team of organizational pathologists should be formed, ready to rush to the scene of any incipient organizational demise. Of course, these people would not be loitering about, waiting for something to happen. They would be organized more like a lifeboat crew.

The first imperative would be to resuscitate the moribund victim. Failing that, however, a post mortem would be performed before rigor mortis had set in, and before those nearest to the deceased had closed in like the vultures they often emulate. I have certainly noticed many times how history is rewritten in these circumstances with breathtaking speed. It happens with people too.

APPENDIX 1: The theory of the model in operational research[1]

If we call the set M of elements a the totality of world events which we propose to examine, then the systemic configuration of events which we know about is a sub-set A of set M. If we call the set N of elements b the totality of systemic science, then the configuration of system which we ourselves understand is a sub-set B of set N. The process of creating a systemic model may then be described as a mapping f of A into B. By this I mean that for every element $a \in A \subset M$ there exists a corresponding element $b \in B \subset N$, and thus $b = f(a)$. The image of the sub-set A, namely, $f(A) \subset N$, is the model. If we are able to exhaust the elements of A and to nominate their images in B, we have every hope of creating an isomorphic model. This means that there exists a complete inverse image of B under mapping f in M, so that $f(A) \subset N = f^{-1}(B) \subset M$. This is the state of affairs, expressed group-theoretically, which the operational research man is trying to reach.

Now an isomorphism is important because it preserves the structure of the original group in the mapping. Typically, if it is possible to perform additions inside set M, those additions will remain valid when the same operations are performed on the images of their elements in set N. It is this persistence of relationship when the mapping is done which makes a model operate as a model. So, if a_1 and a_2 when added together equal a_n in set M, it can be shown that $f(a_1)$ plus $f(a_2)$ must equal $f(a_n)$ in set N. Now comes the interesting comment. The conditions can be set up in which the same answer $f(a_n)$ in set N is obtained from the mapping f whether the trans-

[1]Extract from *The World, the Flesh and the Metal* (the 1964 Stephenson Lecture). Reprinted by permission from *Nature*, vol. 205, no. 4968, pp. 223–231. Copyright © 1965: Macmillan Journals Ltd.

formation is effected before or after the mapping occurs. That is to say, we may either add the original elements in M and transform the answer under f, or we may transform the original elements first and then add them. The result will be the same. Formally: $f(a_1 + a_2) = f(a_1) + f(a_2)$. When one group is mapped into another group and this condition is generally fulfilled, the mapping is called homomorphic.

These elementary definitions are included so that the argument can be made quite clear. Because it is possible to coalesce elements of M before transforming them, without losing the capability of a mapping to preserve structural relationships as discussed, it is clear that a homomorphism may have fewer elements than its inverse image. In the case of the model, then, the mapping of A into B turns out to be a mapping *on to* a sub-group of B. Isomorphism turns out to be a special case of homomorphism, in that $f(A) \subset B$ turns out to mean $f(A) = B$: the one–one correspondence of elements with which we begin is maintained. But for any other sub-group of B other than B itself, homomorphism involves a many–one correspondence, and the inverse mapping $f^{-1}(B)$ will not exhaust the elements of A.

It is suggested, then, that the models of big systems that we entertain are homomorphisms of those systemic characteristics of the big system that we can identify. The homomorphic group $f(A) \subset B \subset N$ is the particular model we use. It is in practice extremely difficult to include in this model all the features recognized in A, and typically we do make the many–one reductions mentioned. Thus, for example, we undertake production costings as if the behaviour of all three shifts in a works were indistinguishable, and as if two similar products were identical, and as if materials were consistently uniform—although we actually know that none of these simplifications is true. Then the effectiveness of the model as predictive depends on the choice of an effective transformation by which to map. If we add up the outputs of three shifts and then transform the answer by some mapping into the model, it is no use supposing that any calculation, comparison or prediction undertaken in the model can be worked backwards through an inverse mapping which will distinguish between the shifts. On the other hand, it is necessary to handle only a third of the elements we know about inside the model. A definite choice has been made to jettison modelling-power in favour of economy in the recording and handling of data. This is acceptable, so long as the choice is deliberate rather than accidental, and so long as it is remembered as a limitation in the model.

Secondly, however, there is a further loss of modelling power in the facts that A is a sub-set of M and B is a sub-set of N. Now an interdisciplinary team of scientists can minimize the losses of modelling power due to $B < N$. Because such a team can examine all the major sub-sets of N before deciding to use one specific group B; it may even experiment with other groups too. But the losses due to $A < M$ are more serious, and may be disastrous to the exericise. For if what we recognize in a big system is not what is really important about its systemic character, the ability to predict A may not help much in M. In other words, A is itself a homomorphic mapping M, and one which by definition we cannot properly specify. Remember that $M - A$ was acknowledged to be systemically unrecognized from the start. We may know that our knowledge of a big system does not exhaust it, without having the faintest idea of the character of the knowledge that is missing.

It is hoped that this attempt somewhat rigorously to formulate what goes on in model-building will prove helpful in pin-pointing what we can and cannot do. The ordinary operational research exercise works, and we can see why. It is possible to advance what we understand about a stock-holding system, for example, to the point where A approaches M asymptotically. It is possible to examine most B of N, which is to say most scientific approaches to the scientific totality of understanding about such systems. If we know what the stockholding system has to do, if (as the operational research man would say) we can define its criteria of success or objective function, then we can define a homomorphic mapping f of $A \simeq M$ on to $B \simeq N$ which preserves the stochastic relationships in which we are interested. More especially, we can do this in a way that the inverse image of B under mapping f yields a set $f^{-1}(B)$ of elements in the real system M which are useful.

The difficulties about doing successful operational research in various circumstances can now be made quite specific. First, the modelling will not on the average work well if $N - B$ is large: this happens if the operational research team is not corporately versatile. Secondly, the modelling will not work at all unless f is well defined: this entails good empirical research into what the system really has to do. Thirdly, the predictions of the model will be of no actual use if a modelled outcome $\phi(b_1 \text{----} b_n)$ turns out to have a pragmatically undiscriminating inverse $f^{-1}(\phi)$ in A. This also entails good empirical research into the forms of many–one reduction. Fourthly, the modelled predictions though useful will not exert what could be called control unless the $M \rightarrow A$ homomorphism captures the systemic character of the big system *in extenso*. This again appears to be a matter for good empirical research, although there is more to say.

Contrary to increasingly current belief, then, operational research is empirical science above all. The mathematical models dreamed up in back rooms are useless unless they can meet the four kinds of difficulty enumerated, and this cannot be done remotely from the world.

APPENDIX 2: Glossary of rules for the viable system[1]

Aphorisms

The first regulatory aphorism
  It is not necessary to enter the black box
  to understand the nature
  of the function it performs.   (p. 40)

The second regulatory aphorism
  It is not necessary to enter the black box
  to calculate the variety
  that it potentially may generate.   (p. 47)

Principles

The first principle of organization
  Managerial, operational and environmental varieties,
  diffusing through an institutional system, tend to equate;
  they should be designed to do so with minimum damage to
  people and to cost.   (p. 97)

The second principle of organization
  The four directional channels carrying information between
  the management unit, the operation, and the environment
  must each have a higher capacity to transmit a given amount
  of information relevant to variety selection in a given time
  than the originating subsystem has to generate it in that
  time.   (p. 99)

The third principle of organization
  Wherever the information carried on a channel capable of
  distinguishing a given variety crosses a boundary, it
  undergoes transduction; the variety of the transducer must
  be at least equivalent to the variety of the channel.   (p. 101)

The fourth principle of organization
  The operation of the first three principles must be cyclically
  maintained through time without hiatus or lags.   (p. 258)

[1]Extract from *The Heart of Enterprise* (Beer, 1979) to which book the page numbers refer.

Theorem

Recursive system theorem
In a recursive organizational structure, any viable system
contains, and is contained in, a viable system.   (p. 118)

Axioms

The first axiom of management
The sum of horizontal variety disposed by n operational
elements
equals
the sum of vertical variety disposed on the six vertical
components of corporate cohesion.   (p. 217)

The second axiom of management
The variety disposed by System Three resulting from the
operation of the First Axiom
equals
the variety disposed by System Four.   (p. 298)

The third axiom of management
The variety disposed by System Five
equals
the residual variety generated by the operation of the Second
Axiom.   (p. 298)

Law

The law of cohesion for multiple recursions of the viable system
The System One variety accessible to System Three of
Recursion x
equals
the variety disposed by the sum of the metasystems of
Recursion y for every recursive pair.   (p. 355)

APPENDIX 3: Some applications of the Viable System Model

Applications of the VSM by its author during the evolution and verification of the model have been so many and so widespread as to defy a proper

listing. For the record, however, the range of amenable organizations ought to be indicated, leaving case histories to the published papers and books. Small industrial businesses in both production and retailing, such as an engineering concern and a bakery, come to mind; large industrial organizations such as the steel industry, textile manufacturers, shipbuilders, the makers of consumer durables, paper manufacturers are also represented. Then there are the businesses that deal in information: publishing in general, insurance, banking. Transportation has figured: railways, ports and harbours, shipping lines. Education, and health (in several countries), the operation of cities, belong to studies of services. Finally comes government at all levels—from the city, to the province, to the state and the nation-state itself—and the international agencies: the VSM has been applied to several.

In this opening paragraph we have been talking of one man's work. Obviously, then, these were not all major undertakings, nor is 'success' claimed for massive change. On the other hand, none of these applications was an academic exercise. In every case we are talking about remunerated consultancy, and that is not a light matter. The activities did not necessarily last for very long either, since speedy diagnosis is a major contribution of the whole approach. On the other hand, some of them have lasted for years. Undoubtedly the major use of this work to date was in Chile from 1971–73: five chapters ending the second edition of *Brain* describe it in full (Beer, 1981). As this is written, however, a new undertaking on a similar scale is beginning in another country. On the question of what constitutes 'success' in consulting; reference may be made to page 211 of this book.

Of other people's work in the field of managerial cybernetics that has made application of the VSM, first mention must go to Raul Espejo. He has given his own account of the 1971–73 Chilean application that we undertook together (Espejo, 1980*a*). Since then, his teaching and research at Aston University in England has been centred on the VSM, and outcomes have been published in several articles and papers (especially Espejo, 1978, 1980*b*). His diagnoses have been profound, and he is adding to the corpus of theory.

The number of senior degrees, including doctorates, that have employed the VSM under Espejo's direction is already in double figures. Professor David Mitchell's teaching has generated a similar number of postgraduate theses using the VSM at Concordia University in Quebec, as has that of Professor Manuel Mariña at the Central University of Venezuela. Several more have emerged from Brunel University, under the direction of Professor Frank George. In the United States, Professors Richard Ericson and Stuart Umpleby (at George Washington University), Professor Barry Clemson (at the Universities of Maryland and of Maine), and Professor William Reckmeyer (at San José State University) have all made extensive use of this teaching, and others from Australia to India have reported similarly.

At Manchester University in the Business School, Geoffrey Lockett (directing the doctoral programme) has sponsored whole-week 'experiences' of the VSM; and Professor Roger Collcutt has invented a unique pedagogic framework whereby MBS students undertake projects to apply the VSM to functional management, subsequently to merge the insights gained into a general management picture. Another novel development has been made by Ronald H. Anderton in the Systems Department of Lancaster University: practical applications of the VSM in the form of project work have for some years been an important part of his *under*graduate teaching.

A veritable kaleidoscope of applications of the VSM has been presented by Dr Paul Rubinyi in Canada. From penological systems to health services in the public sector, from oil companies to wheat cooperatives in the private sector, and from provincial planning to air transportation in federal government: every kind of organization has been mapped, in virtually continuous work over the last 13 years.

Other separate applications in Canada include the work of Walter Baker, Raoul Elias and David Griggs on the Fisheries and Marine Service, which took unique advantage of managerial involvement, and that of Raoul Elias for Gaz Metropolitain. David Beatty has used the model for educational planning in Ontario, and I believe that it has been in independent action on the West Coast as well (Baker, Elias and Griggs, 1978).

In Latin America, Professor Jorge Chapiro is a leading exponent of the VSM who consults over the whole spectrum of industrial and governmental management in several countries.

In Australia, applications in an insurance company have been made by J. Donald de Raadt; in Switzerland Dr Peter Gomez has used the VSM in a publishing company, making an interesting experiment in melding this methodology with the 'root definitions' of Professor Peter Checkland (Gomez, 1982). In wider fields still we find a useful VSM application in Finland by Dr S. Korolainen to ekistics (Korolainen, 1980); and David Noor has published 'A viable system model of scientific rationality' as a working paper from the University of Western Ontario.

On the strictly biological side, but not from the original neurophysiological perspective, Dr Richard Foss in England has made many mappings: for example, on the Eukaryote cell, the annual plant and the honeybee colony. He has found the VSM to hold in such diverse systems; and he is extending the work to the slime mould *Dictyoltelium*, to lichens and to vertebrates, considering both the evolution and ontogeny of each system.

It does appear that the VSM has sufficient generality to justify its origin as an attempt to discover *how systems are viable*; and that it also generates considerable power to describe and predict, diagnose and prescribe. No systematic archive of applications has been kept: perhaps it would be helpful to start one. These notes are compiled from such recollections and records as happen to be to hand.

References

Ashby, W.R. (1965). *Introduction to Cybernetics*. London: Chapman & Hall.

Baker, W., Elias, R. and Griggs, D. (1978). 'Managerial involvement in the design of adaptive systems', in *Management Handbook for Public Administration* (ed. Sutherland, J.W.). New York: Van Nostrand Reinhold.

Beer, S. (1959). *Cybernetics and Management*. London: English Universities Press.

Beer, S. (1960). 'Towards the cybernetic factory', in *Principles of Self Organization* (symposium). Oxford: Pergamon Press.

Beer, S. (1965). 'The world, the flesh and the metal', *Nature*, **205**, 223–31.

Beer, S. (1966). *Decision and Control*. Chichester: John Wiley.

Beer, S. (1972). *Brain of the Firm*. Harmondsworth: Allan Lane.

Beer, S. (1975). *Platform for Change*. Chichester: John Wiley.

Beer, S. (1979). *The Heart of Enterprise*. Chichester: John Wiley.

Beer, S. (1981). *Brain of the Firm*, 2nd edn. Chichester: John Wiley.

Beer, S. (1983). 'A reply to Ulrich's "Critique of pure cybernetic reason: the Chilean experiment with cybernetics" ', *Journal of Applied Systems Analysis*, **10**, 115–19.

Beer, S. (1985). *Diagnosing the System for Organizations*. Chichester: John Wiley.

Brain and Strategy (1983). vol. 4, issue 9.

Capra, F. (1982). *The Turning Point*. New York: Bantam Books.

Espejo, R. (1978). Multi-organizational strategies; an analytical framework and case, in *Applied General Systems Research: Recent Developments and Trends* (ed. Klir, G.). New York: Plenum Press.

Espejo, R. (1980a). 'Cybernetic praxis in government: the management of industry in Chile 1970–1973', *Journal of Cybernetics*, **11**, 325–38.

Espejo, R. (1980b). 'Information and management: the cybernetics of a small company', in *The Information Systems Environment* (eds. Lucas, H., Land, F., Lincoln, T. and Supper, K.). Amsterdam: North Holland.

Gomez, P. (1982). 'Systems methodology in action', *Journal of Applied Systems Analysis*, **9**, 67–85.

Korolainen, S. (1980). *On the Conceptual and Logical Foundations of the General Theory of Human Organizations*. Helsinki School of Economics.

Shannon, C. and Weaver, W. (1949). *The Mathematical Theory of Communication*. University of Illinois Press.

Sommerhoff, G. (1950). *Analytical Biology*. London: Oxford University Press.

The Viable System Model: Interpretations and Applications of Stafford Beer's VSM
Edited by R. Espejo and R. Harnden
© 1989 John Wiley & Sons Ltd

2

The need for formal development of the VSM

Ron Anderton

*Department of Systems and Information Management,
University of Lancaster*

In spite of its importance and range of applicability Beer's theory of viable systems has entered the mainstream of academic work only to a limited extent. The argument of this chapter is that this passage would be facilitated by more formalization of the theory and that this should be attempted.

Wiggles and boxes

Alan Watts, the Californian guru-figure of the 'sixties, was, in spite of some personal eccentricities, an unusually effective expositor of ideas derived from his studies of Eastern religions, particularly the Chinese philosophy of Tao. Many of these are interesting and relevant in the context of 'systems thinking' and particularly the paradoxes to which this often gives rise. In one of Watts' lectures he describes the experience of looking out of an aircraft window on a West to East Coast flight. At first the journey was over the Sierra, over the rivers and lakes of the Rocky Mountains. He saw the lines followed by the rivers and occasional roads as forming an intricate and complex arboreal pattern (what the Taoists call 'li'); the contours allowed no other possibility. The natural forms were *wiggly*. As the plane flew on and hills fell away into the Great Plains country, into Kansas and Oklahoma, the effects of human activity became apparent and the landscape became dominated, not by wiggles, but by rectangles. The grids of the town streets, the geometrically shaped fields and the straight intercity highways showed up as a simple diagram-like imposition of box patterns. No wiggles but clear, disciplined, rather dull oblongs. (We call, Watts pointed out, a certain kind of person a square.)

It occurs to me that a similar contrast illuminates differences in approach to systems work. It can be seen in the style of picture systems people like to draw. The systems engineer, and the computer systems designer, like to draw complicated but neat arrangements of blocks connected with straight lines and arrows. They use stencils and rulers as best they can but are pleased when professional draughtsmen or computer graphics produce more perfect versions for their books and reports. Their world has become a diagram. Others—of those I have worked with Stafford Beer is certainly one, Peter Checkland another—go at it differently. They insist that their freehand drafts are reproduced unchanged with the slightly wiggly lines retained. They prefer cloud shapes to boxes. Intuitively they feel that something of the richness of their propositions, of the organic origins of what they seek to represent, is lost by a conversion into perfect grids, squares and other geometrical artefacts. Parallel feelings are associated with styles of language: austere, controlled, dry, logical on the one hand; metaphorical and richly various on the other. The use of the words 'hard' and 'soft' to describe systems approaches normally refers to a somewhat different distinction, but the choice of words reflects not unrelated attitudes.

An argument

In what follows I shall consider Stafford Beer's work on the Viable System Model (VSM) taken, in its most important aspect, as a theory of organization. I want to argue that this gloriously rich and wiggly piece of work (high compliments) is at a stage where it needs some more formal (box-like) development and this will advance its understanding and use.

Specifically the argument will go like this: Stafford Beer's work on the principles of 'viable systems' and their application to problems of organizational design constitutes material of great interest and importance, both practically and theoretically. However, in spite of extensive and brilliantly written publications, a large, diverse and enthusiastic following, and a significant number of attempts, often very successful, at application, it remains true that many people, including highly intelligent ones, attracted by the ideas, find them most difficult to grasp; they understand them superficially and find them cogent but when it comes to the point of detailed practical use they seem to slip away. The work has failed thus far to enter the mainstream of intellectual development, whether in the so-called sciences of administration or management or in those theoretical parts of sociology or biology to which it has relevance. This is not to say that it has not received attention, only that it is usually regarded as peripheral, as idiosyncratic, not as part of a central corpus of material.

Solution of the problem which the above suggests would be be made easier if a more explicit version of the underlying theory were to be propounded, one designed to facilitate careful critical analysis of the propositions and arguments it contains. This would provide a complement to other material already available which aims at developing intuitive understanding or is otherwise polemical and persuasive in its intention.

Now of course each of these points may be challenged. Some may dispute my assessment of the power and profundity of the ideas themselves. Others may say that the ideas *are* understood or that if they are not found much in academic curricula then so much the worse for academics; many influential ideas are so neglected—take those of Freud for instance. And many would doubt that the third point proposing more formal developments would result in anything more valuable than additions to the pile of unread PhD theses.

So what follows is a variety of comments in support of these propositions. The theme is a complex one and I shall approach it from several angles. No attempt is made actually to *supply* the theoretical development which, on this argument, is both missing and needed. I should also make it clear that I am not writing either as a wiggler or, I trust, a square. It is the balance and exchange between the two that matters, and I am arguing for, at this particular time, more concentration on the formal development.

The ideas

I must start then by establishing what the ideas are that need this treatment and why I believe them to be important. Their distinction arises from a rare combination of intellectual depth and emergence from intimate involvement with the world of practical affairs. It is this that allows them to offer significant insight in situations of major concern.

One such concern which frequently arises is that of structuring the activities of large, complex organizations, whether they be public institutions—the NHS for instance—or private enterprises, such as an ICI or a British Aerospace. Decisions on these significant matters are quite often made on the advice of consultancy firms, normally US-based, whose favoured principles go through cycles of fashion changing according to the success of their business rivals. But more often the choice is made by individuals or small groups and is based either on intuition and hunch, or from memories of a personally experienced previous organizational change which proved to be successful, or expediently in relation to the power concentrations which happen to be in place at the time.

It is possible to exaggerate the importance of decisions about organizational structure and no doubt occasionally too much time and energy is put into them at the expense of other pressing matters. But their importance is very high and it is extraordinary how flimsy the basis on which they are made often is. Theories of Organization of course exist, but none perhaps combines, in the way Beer's does, operational usability with a clear theoretical base.

It may anchor this discussion if I refer to two situations which I have encountered recently, in each case as a result of supervising postgraduate students engaged in extended project work with 'real world' organizations using a variety of techniques and other methodologies as part of a course of management education. In each one I devoted a few days to considering, in terms of the VSM, the complicated situation with which I had become familiar. I will describe these circumstances very briefly and then return to their VSM interpretation later on.

The first project was concerned with a group of operating companies each with a few hundred employees which made particular products and distributed them through a number of trade and retail outlets in a sector of the UK consumer market. The group was itself owned by a large European-based multinational company. It was lightly managed by the UK head office and an executive board of directors located there, but with minimal staff support. The executive group's concern was that the operating companies were excessively 'product-orientated', geographically and to an extent 'culturally' separated, and exhibited duplication in their operations. More significantly the set of products held a minority share of a market which was not growing, or only slowly, so that the return on investment required by the group's owners was not expected to be maintained. A new strategic direction was called for and this the existing structures and mechanisms was failing to generate. New proposals being made included the adoption of Strategic Business Units to overlay the Operating companies together with centralized functional direction to provide more rationalized control of the various operations. There were fears of loss of morale and effectiveness owing to the uncertainty accompanying the anticipated change and with the breakdown of the strong 'cultural' groupings associated with the longstanding product companies. There was plainly a threat to viability.

The second project was also with a privately owned group of commercial and manufacturing companies, mainly small (100 employees or less) in the North of England. Although the companies could be roughly categorized into a number of business areas, the linkages between them seemed small. They were controlled by a small holding company in which the principal owners played a leading part. The companies were arranged in 'divisions' with a divisional MD accountable for each. An elaborately detailed form of budgetary control had been set up following a long external consultancy study, and large volumes of data, mainly on financial performance and variance from budget, flowed weekly up to the company MDs, the division MDs and on to the group MD. The student project was initiated by one of the divisional MDs who, in spite of this wealth of numbers, had difficulty handling them and wondered whether they were of an appropriate kind. As in the first project a need for strategic development of the group was sensed, but time and relevant information seemed to be lacking.

Here then are two examples, ones which could be multiplied endlessly, of dilemmas faced by those with responsibilities for organizational control and adaptation, but who, as we have said, receive little soundly based advice on their resolution. But the success of vital enterprises of many kinds depends on the way in which this is done. If Beer's ideas can be of help in this structural task then their importance is not in doubt.

What are these ideas? The dustcover of Beer's latest work (*Diagnosing the System for Organizations*) mentions that he has published over 200 items, including nine books. They cannot be summarized here. Fortunately a relatively small number of themes recur and form a core. Let us review them.

Right at the centre is a consideration of the conditions under which the creation and maintenance of recognizable *order* is possible. Anything interacting in the world is subject to disturbances tending to upset its recognizable features. They can only be preserved if a compensating response can be generated for each disturbance nullifying the change that would otherwise result. Hence the Law of Requisite Variety proposed by Ross Ashby in his theory of regulation: 'only variety can absorb variety'. The variety of the controller must match the variety of the controlled.

In Beer's interpretation and development of Ashby's work applied to the field of complex human activity, the central question becomes: How is it possible for multiple control elements, human or mechanical, each one possessing only limited powers of perception, computation and action, to achieve the enormous tasks of regulation needed to achieve complex purposes, or even any kind of identifiable continuity—that is to say, stability—in turbulent, noisy, and sometimes aggressively competitive environments? The general answer is by making them subject to appropriately organized systems of constraints. How to determine what is appropriate is, on this view, the topic of the theory of organization. How can low-variety elements be connected together to form a high-variety coordinate whole? Only if this is done can purposes be efficiently achieved in high-variety environments.

Beer sees this task of organization as one of achieving a massive selection in which states of affairs inconsistent with its continuation are identified and action to change those states are initiated. Twin problems of establishing external adaptation and internal coherence must be simultaneously and continuously solved. To organize is thus to arrange for complex regulation. If an organization is to be *viable* then this system of constraints must continue to provide effective regulation even though its environment may change.

Beer's main result is that a viable system structure is one that can be broken down into a number of component structures which also have the form of a viable system, together with a system for controlling relations between these component systems. We thus have a total structure which is defined recursively, to borrow a term from mathematics and computing, and can therefore be conceived as forming a hierarchy. A strong version of this theory would assert (roughly) that for any real structure to be viable in an environment it must be mappable on to this model structure.

We can interpret this in the familiar terms of a business in this way. Every viable business must be made up of a collection of viable businesses together with a system which manages the relations between these lower-level businesses so that this collection is viable. Thus a large part of the variety absorption required for ensuring viability is pushed down to the next level of recursion and the remaining task is manageable. The whole procedure is repeated at that level.

A consequence of this structuring is that at each level there are two types of activity. The first, done by what we call the collection of viable subsystems, is that of actually doing the operational work of producing whatever it is which constitutes the identity of the whole. The second can be described as meta-systemic to the collection. It does not take part directly in the production activities of the lower-level collection. It is to be understood as a controller of both the internal relations between the viable subsystems and the relation of the whole to their environment.

The model can be interpreted as two types of discourse carried on it correspondingly distinct languages. The first type asks separately, for each of the lower-level 'viable subsystems': Given the current state of the environment, what operational action needs to be taken by the subsystem using present constraints, structures and rules? The second type asks four meta-questions about the whole collection of subsystems: Are they co-ordinated to avoid internal interaction (oscillation)? Is each constrained in such a way that 'synergy' and effective resource usage is generated? Is the particular decomposition one which will continue to be viable in anticipated future environments? Is the whole satisfying constraints which may be being passed down from higher recursion levels?

The whole structure is seen then as an organized set of interlocking controllers. Viability requires each controller's criterion to be satisfied. It follows, taking this control perspective, that each controller, if it is to achieve balance in time with the rate of corresponding change in its environment, must have a model of that environment to guide its search.

Beer defines his scheme (using graphical representation extensively) with five necessary component systems (corresponding to the five questions above) at each level of recursion. He exploits for purposes of exposition the student's knowledge of real 'systems' which exhibit viability; from neurophysiology and also from the actual workings of successful enterprises such as multidivisional corporations.

To give a rigorous abstract account of this scheme is a demanding exercise. And yet interpretations in terms of a particular organizational situation are surprisingly accessible.

To illustrate this we may return to the two examples mentioned earlier. It was possible in these terms to see each as a structural problem, partly of dubious 'variety engineering' to use Beer's phrase and partly of inadequate metasystemic development.

In the first situation there were inadequate arrangements at the metasystemic level to design and implement the radical strategic changes necessary for survival (although at least the need for change had been detected). The proposals that had been made divided the whole group up, not into smaller free-standing businesses with new sustainable 'missions', but into a mixture of functional controls and product-responsible units arranged in a kind of matrix structure. This scheme if attempted would result in an immense problem of managing the interactions between the various activities. The effect on morale and motivation would also have been unfortunate with no new foci of loyalty to replace the old product-oriented groupings. Incidentally it was useful to see how company culture or 'the way we do things around here' was important as a variety reducer: it made people's reactions more predictable.

In the second situation, although a hierarchical scheme of group, division, company, operating department had been named, its implementation did not realize the intention. Each level was preoccupied with the scrutiny of the operations below it, particularly so as to detect deviation from the budgetary plan. The proper tasks at each level—formulating adaptive strategies and identifying synergistic opportunities for the level immediately below—were being neglected because attention was concentrated on operational control, often directed two or three levels below. The control information flowing up was noisy and unfiltered, so that it could not perform its proper task of pointing to where high-level attention was needed (e.g. through detailed investigation). This could only be done in one or two areas at a particular time: the information system needed to indicate where these should be—it didn't. Beer's models make the distinctions clear. At one level there are a number of separate operational systems, each interacting with their particular local environments. That is their concern. At the higher level (the metalevel) the concern is sharply different: it is with the *relations* between these operational systems; and it is whether the operational systems *considered together* will meet and continue to meet some higher purpose. Only by concentrating on tasks appropriate to each level can overload be avoided. The models shows what these tasks are and how they must be linked together by the flow of information.

Thus in each case, a little understanding of the principles of viability enabled quick comment to be made on the inadequacy of plausible organizational schemes. But can this kind of insight be developed into detailed, engineered, organizational designs? And what is the status of the organizational prescriptions which are made? Beer says to the first 'Yes it can', and to the second 'The prescription is derived from exceptionless cybernetic laws'.

If he is correct in these responses then plainly this is a matter of high significance, both practically and theoretically. It becomes correspondingly important to formulate the theory in terms which allow it to be analysed, criticized and fully understood.

Difficulties

What I want to consider in this section are the difficulties, theoretical and practical, which sympathetic and intelligent students of these ideas stumble against. Of course in many, perhaps most, cases this will be the result of a simple failure to understand what has already been expounded in published work, or of a lack of the application experience from which personal learning can develop. In other cases there may be genuine ambiguities or lacunae in the available accounts. The more formal treatments which this chapter is advocating may help distinguish the two.

Some of the difficulties are apparently practical ones. How do you measure variety? What is a significant difference? How can you anticipate what will be significant? How is the decision made on an appropriate decomposition of a total task into lower-level systems? Is that decision made externally by a system designer, or is it being made continuously *inside* the system? If a decision to change is made how is it implemented; how does the metasystem induce discrete structural change?

These questions then become theoretical ones. Some of them concern the way in which the organism must continuously produce itself, be 'autopoietic' in Maturana's terms. Others result from the fact that real organizations are implemented largely by human agents, constrained, for most of the time, by rules and procedures, but capable of attaching to events a variety of meanings and significances not shared by other agents in the organization, and supremely capable of thinking at multiple, changing logical levels.

Perhaps both of these sorts of theoretical questions reduce to one: what precisely is the viable system a model of? Not, it would seem, a set of unproblematic entities like 'Sales Office' or 'British Steel' connected by message flows: not, indeed, any kind of direct image or picture of the organization of the kind naïve readers of the accounts may take it to be.

Beer's ideas have developed over a number of years during which there have been a number of shifts in the characteristic ways in which problems in management system theory have been solved. In the 'sixties there was great interest (of course there still is!) in machine and organismic metaphors and in particular in schemes of automatic control. The contribution of the cyberneticists, led by Beer, was to make it clear that the control of human organizations involved constraining variety of totally different orders of magnitude to those dealt with even in the most complex pieces of engineering. The central issue was how could this immense amount of variety be coped with. Later, in the 'seventies, other perspectives were often taken (by Checkland, for example). Behaviour in organizations was seen as following from human intentions and only to be understood in terms of the perceptions and meanings that corresponded to them. An organization was not seen as a unified organism integrated as an identifiable whole. A situation could be, and was, interpretable in many different ways. Some of these could be in part orchestrated: where this had not been done 'issues' arose, the resolution of which constituted much of the significant activity in the situation. The interest of the investigators or intervenors was in the phenomenology of the situation, less in the cybernetics of control. This change in emphasis was regarded as a paradigm shift.

I would not wish to place the VSM within either the control paradigm, although that is where it had its origin, nor within what is sometimes called the hermeneutic or interpretive tradition. Perhaps the main theme of VSM work as it has developed is distinct from either and is concerned with the conditions under which autonomy can be exhibited within certain environmental constraints. The issue might be put as follows. Human behaviour transcends rules. Humans make rules; sometimes they break rules. Mental life can only be described, in other than trivial cases, in terms of intentional characteristics—beliefs, preferences, purposes. But they can only achieve these purposes through structures—sets of constraints—which exist in the world. So we distinguish between purposeful systems, what we see when we take what Dennett calls 'the intentional stance'; and real-world structures which have properties that may be studied and described in quite other ways. We have the purpose of commuting between London and Brighton: we use the structure of railway lines and scheduled trains. As a bridge must be designed to have the property of supporting a crossing load, so Beer shows us how an organizational structure—the necessary requirement for achieving purposes—can be given its essential property: viability.

My purpose here is not to rehearse the arguments in this continuing intellectual debate, but to point out how confusion between these different positions may lead to difficulties in understanding of what Beer's theory of viable systems is, in fact, intended to assert. And again to suggest that a more formal treatment might make matters easier to be clear about.

Towards a formal model

What then might a more formal VSM be like, one sufficiently free of wiggles to be suitably precise for analysis and criticism? Is it indeed possible usefully to represent as one object, bound to be relatively simple, another which *by definition* is extremely complex? Of course many (all) models simplify; a tennis ball can model the whole planet. But in our case is it not the consequences of complexity itself which is to be studied and would that not be necessarily lost, leaving metaphor as the only descriptive recourse? Beer's own device of recursion suggests one important way out. The model can be defined by defining one step in a process which can then develop models of any required complexity. This can be done either for one level in the hierarchy or for one event in a process developing through time.

The word 'model' is used widely in systems talk and not always carefully. Here I use it to refer to an object which can stand alone and possess its own properties. The object may be a physical one or it may be defined as a purely mathematical structure. It will have parts (elements) and relations between these parts will be defined. This abstract object may then be given an interpretation in that its elements and relations can be matched to corresponding parts and connections in another object. A satisfactory matching enables the first object to be a model of the second and to share some of its properties. An object may be used as theoretical model of something observable or constructable in the real world and new theorems stated or derived from the model which have interpretations in the real world.

In all this we sharply distinguish descriptions of the model from descriptions of corresponding real-world phenomena. It is always temptingly convenient to blur this distinction by using the names of real-world objects to refer to the parts of the model which correspond to those real-world objects.

For analytical purposes it may be most useful to establish a correspondence between a set-theoretic object (the model) and linguistic sentences which can use the names of entities in the world (these can be given the precise forms of predicate logic if necessary).

This is technical work for which I claim no expertise. But I do not believe that arcane analysis using difficult symbolic notions is likely to be called for. The names of real-world elements can be used (for mnemonic purposes) and put in quotation marks for example. What is important is that the model is distinct from the modelled. ('The map is not the territory'.) The model can and should be completely defined: the reality it may correspond to cannot. The ideal is a simple, austere account of an uninterpreted model only then followed by rich and imaginative interpretation.

What kind of abstract object is the VSM or what could it be? It can be constructed formally in set-theoretic terms (which may for most purposes be displayed as diagrams) which can generate properties of the kind Ashby and Beer demonstrate in homeostats. Particular relations are computed and a process of change continues until these satisfy some criteria. The interlocking homeostats are the building blocks of the model and are arranged in a particular recursively describable pattern so that the familiar five-system picture emerges. The model can be realized on a computer if that is helpful. But it is still entirely abstract with no interpretation.

Essentially the problem addressed with the model is this. Given a number of 'objects' which can be interpreted as existing in an 'environment', using 'resources' provided by this environment, satisfying externally defined constraints; how can these objects be assembled to form a new object capable of satisfying other constraints and existing in a wider environment? If an answer to this is found, and Beer's theory claims to provide one, then it may be applied recursively to solve problems of any required complexity.

The approach, of course, is plainly implicit and often explicit in Beer's own writings. But usually accounts of the model and its interpretation are deliberately confused. They are developed together for purposes of exposition. Brilliantly, Beer darts from model to meaning and back again, persuasively taking the reader through his complex development, drawing on the reader's experiences, generalizing from them, building up understanding of the intricate design.

Suppose then in a complementary version this development was unraveled. Here would be the starkly defined, initially meaningless model with its own characteristic properties. Next, quite separately we interpret it, matching each part and connection to entities in the world (or more exactly to sentences describing them). This would be a systematic procedure with no place for metaphor.

If such an abstract model were available, and in these days a computable form is probably the most convenient, a rigorous theory of viability with this as its theoretical model could be analysed and a series of application studies started. It could become, in other words, the subject of an academic research programme. This could be developed through 'action research' projects to develop methodology for diagnosis and design. Like most theories it would not very often be used directly in its formal dress.

Practitioners would continue to learn informally the craft of application. A well developed theory would, however, aid practice by improving the language used to describe organizational phenomena. For example it would provide reminders of the need to distinguish different language types corresponding to hierarchical levels, and in other ways make easier the resolution of perceived ambiguities and confusions which trouble beginning, and sometimes advanced, students of the subject. More positively a clear uninterpreted statement of the theory might encourage new applications in non-managerial areas. Psychology (or neuropsychology) and the History of Ideas are two which come to mind.

The proposal I make here is not intended to tame an unruly body of imaginative insights so they become manageable, caged in an academic zoo. What I want to see is more attention given to new kinds of exposition and to their critical analysis. The potential they have for increasing our understanding of what (following Gregory Bateson) may be called the ecology of ideas is very large. But a great deal of work must be done by many people if this is to be made real. The corpus of ideas must itself become an autonomous Viable System. I have argued that a version which separates model from interpretation will encourage independent developments in several disciplines. The rigorous model permits one kind of selection in the intellectual evolutionary process (another is concerned with weeding out models which are found not to be practically useful).

Just as a science develops through an interplay between theory and practice, so theory itself advances by a succession of intuitive and formal representations. The infinitely variable, fascinating and beautiful, wiggles of evolving organizational form will never be straightened out. Carefully drawn simple maps may help us understand its flow and occasionally guide its direction. Never did a generation need more than ours to appreciate complexity, its aesthetic, its ecology, the delicate balance between stability and ability to change, its addictive, trapped pathological forms. To achieve this, rigour *and* imagination are needed. Beer has shown us a way to this: it needs to be followed.

References

Ashby, W.R. (1964). *Introduction to Cybernetics*. London: Methuen.
Bateson, G. (1973). *Steps to an Ecology of Mind*. London: Paladin.
Checkland, P.B. (1981). *Systems Thinking, Systems Practice*. Chichester: John Wiley.
Dennett, D. (1987). *The Intentional Stance*. Cambridge, Mass.: MIT Press.
Maturana, H. and Varela, F. (1980). *Autopoiesis and Cognition: Realization of the Living*. Dordrecht: Reidel.

The Viable System Model: Interpretations and Applications of Stafford Beer's VSM
Edited by R. Espejo and R. Harnden
© 1989 John Wiley & Sons Ltd

3

The VSM and Ashby's Law as illuminants of historical management thought

Fred Waelchli[†]

Defense Systems Management College, Fort Belvoir, Virginia*

In this chapter it is argued that Ashby's Law of Requisite Variety, corner-stone of Beer's Viable System Model, is also a root law of organizations. Manifestations of the Law are everywhere visible in historical and contemporary management theory and practice, from the early thinking of Fayol and Taylor, through the 'Human Behavior' movement, to today's 'excellent' organizations of Drucker, Peters, Ouchi, Kanter, Cifford, Cavanagh, and others.

Prologue

On a starred night Prince Lucifer uprose.
....
He reached a middle height, and at the stars,
Which are the brain of heaven, he looked and sank.
Around the ancient track marched, rank on rank,
The army of unalterable law.

George Meredith, *Lucifer in Starlight*

Do we live in a universe governed by immutable law? The question is as old as mankind, and on its answer hang all of our science and philosophy.

A marked difference between the 'hard' or natural sciences and the 'softer' social sciences is that natural scientists tend to answer this question

[†]Deceased.
*Also University of Maryland, University College Graduate School.

51

in the affirmative, while the social scientists are not so sure. Physicist Leon Lederman marks one end of a philosophical spectrum:

> '. . . a single and economical law of nature, valid throughout the universe for all time. The quest for such a unified scientific law has been undertaken and advanced by all nations and all creeds. Indeed, the idea of the unity of science has been a major force in developing the unity of humanity . . .'. (Lederman, 1984, p. 40)

The middle ground, not surprisingly, is well populated, including John Warfield's 'Typology of laws' (1986) for the sciences, and Jay Galbraith (1977) in organization theory. Certain organizational contingency theorists occupy the other extreme:

> 'One of the consequences of [the contingency view] is a rejection of simplistic statements concerning universal principles of organization design and management practice.' (Kast and Rosenzweig, 1979, p. 115)

> 'The contingency approach denies the universal assumption and pragmatically relates the environment to appropriate management concepts and techniques.' (Luthans, 1976, p. 54)

The young discipline of management has never pretended to the rigor of physics, but some of our prominent intellectual progenitors did apparently share Lederman's faith. As a sample, Frederick W. Taylor (1911), Henri Fayol (1916), Mooney and Reiley (1931), and Lyndall Urwick (1937, 1947) all searched for, found, and prescribed universal principles or 'one best ways' to manage work and the organization. It has been relatively recently, perhaps beginning with Lawrence and Lorsch (1967) (or maybe with Elton Mayo, 1933), that organization theorists have backed away from the concept of universal law, and now think more in terms of contingencies and laws of the situation.

In still more recent times, the nascent discipline of managerial cybernetics (largely the product of Stafford Beer and W. Ross Ashby), and the Viable System Model (VSM) of Beer, seem now to proffer an avenue of thought that may lead organization theory back toward Lederman; back toward the possibility of universal laws of organizations. Managerial cybernetics and the VSM further articulate a framework that seems to illuminate and order the diverse, brief history of management thought.

This paper explores that framework. The exploratory device is a question: *Is it possible that the myriad of actual, potential and contingent approaches to organization theory are really varied manifestations of the same law or set of laws?* I speculate that the answer is 'yes', and that Ashby's Law of Requisite Variety, the foundation stone of managerial cybernetics, and core of the VSM, is one such law. My avenue into the discussion is a provocative observation by Stafford Beer.

Introduction

'It has always seemed to me that Ashby's law stands to management science as Newton's Laws stand to physics; it is central to a coherent account of complexity control.' (Beer, 1984, p. 11)

Beer here directly suggests that a thesis he hints at continuously throughout his many works on management is true; that Ashby's Law of Requisite Variety (Ashby, 1956, 1960) is, in fact, an Iron Law of management.

To scholars weaned on Koontz's (1961) 'Management theory jungle', and its many floral cousins, this will seem rather a sweeping claim. Succeeding paragraphs will argue, however, that Beer is conservative; that beyond the mechanism of complexity control, Ashby's law also plays a central role in the coherent ordering of historical management philosophy and practice.

If these contentions are valid, we should find manifestations of Ashby's Law as recurrent themes in management principles, theories and actions. The purpose of this paper is to explore briefly historical and contemporary management thought for evidence relevant to that proposition.

In the first section of the paper I briefly recapitulate and discuss Ashby's Law; first as a law of systems and cybernetics, and then as a centerpiece of the VSM. The remainder of the paper is devoted to exploring and weighing the place of Ashby's Law in the pantheon of management theory.

Ashby's Law of Requisite Variety and Beer's VSM

Ashby's Law

Ashby's Law describes the conditions under which a complex system (in this paper, the organization) can be externally controlled. The chain of propositions that leads from the concept of 'system', through Ashby's Law, to managerial cybernetics and the VSM, is rigorous and detailed. The full argument is found in Beer (1959, 1966) and in Waelchli (1987a). In summary form it runs roughly as follows:

1. There is a way of looking at creation which emphasizes the relationships between things equally with the things themselves. This approach is called the 'system' view.

2. A system is a bounded collection of three types of entities: elements, attributes of elements, and relationships among elements and attributes. Both attributes and relationships are characterized by functions called 'variables', which include the familiar quantifiable variety as well as the

non–numerical types described by Warfield and Christakis (1987). The 'state' of a system at any time is the set of values held by its variables at that time.

3. The values of certain variables of the system must remain within physiologically determined limits for the system to continue in existence as the system; these are called the 'essential' variables (Ashby, 1960, p. 41) of the system; examples are blood pressure and temperature in human systems, and cash flow and net income in the firm.

4. Many system variables display equilibrium; that is, a tendency toward a single or small range of values, and when displaced from these values, a tendency to return. This quality, exhibited by all living systems, is known as teleological or goal-seeking behavior.

5. Within the category of living goal-seeking systems is the class of systems whose goals and reasons for existence are consciously set by man, called 'purposive' (Beer, 1959) or 'purposeful' (Ackoff and Emery, 1972) systems.

6. Most natural systems are 'complex', which means that their possible states are so numerous that they cannot be counted in real time. The unit of complexity is 'variety'. The variety of a dynamic system is the number of distinguishable states it can occupy. The essential quality of a complex system is that its variety is so great that it cannot be controlled or managed by any method that depends on enumerating or dealing sequentially with its states.

7. Ashby's Law of Requisite Variety states that to control a complex system, the controlling system must generate at least as much variety as the system being controlled: 'Only variety in the control mechanism can deal successfully with variety in the system controlled' (Beer, 1959, p. 50).

8. The concept of systemic 'control' operates at two levels. First is physiological control, required to allow a system to continue in existence (see 3 above); the values of all of the essential variables are held within physiologically set tolerances. If physiological control fails, the system dies.

The second level is operational control, or the control of one system by another. This also requires the presence of physiological control, but in addition requires the maintenance of the values of a set of variables (essential or otherwise), chosen by the controlling system, according to its purpose for existence (see 5 above and 9 below), within tolerances set by the controlling system. If operational control fails, the system can still live, but (by definition) it fails to accomplish its purpose. Ashby's Law governs both types of control.

9. An 'organization' is a complex purposive system that man brings into being (or maintains in being) for the purpose of creating some desired change in his environment (i.e. in society). In order to accomplish its

societal purpose the organization must have the ability and power to influence and cause change in other organizations and in the other complex natural systems that make up its environment. It must operationally 'control' some part of its environment, which requires (under Ashby's Law) that it must possess—contrary to normal expectations—at least as much variety as the societal systems it strives to control.

10. In classical cybernetics, there are only three methods that an organization (or any system intent on operationally controlling another complex system) can use to establish the variety surplus it needs: it can amplify its own variety beyond that of the system to be controlled; it can exactly match its variety to that of the system to be controlled (a special case); or it can reduce the variety of the system to be controlled to less than its own.

I will propose below that both of the major methods of variety control are fundamental to management, and that Ashby's Law provides a framework for the understanding of the development of historical management thought and practice from the early days of Fayol and Taylor to the management patterns recently proposed as characteristics of unusually effective organizations (Ouchi, 1981; Peters and Waterman, 1982; Kanter, 1983; Peters and Austin, 1985; Clifford and Cavanagh, 1985).

Additional properties of systems

The last section concluded the chain of logic leading to Ashby's Law, but there are a few additional properties of systems that should be noted. Operational systemic control means establishing and maintaining the controlled system in states dictated by the controller, or, more specifically, adjusting the values of designated variables of the controlled system to values desired by the controller, and maintaining those values.

Cyberneticians assert that changes of value of system variables, and therefore changes of state in all complex purposive systems (such as organizations), are accomplished by flows of 'information' and 'entropy'. Information and entropy are both defined as signals which change the state of the system that receives them. The difference between information (which is also called 'negative entropy', or 'negentropy') and entropy is that information implies purpose, and causes the system to change in ways that favor achievement of system goals, while entropy does not.

Since all organizational goals are set by man, man is the only source of information (negentropy) in an organization. Entropy, which, by contrast, is everywhere present in and around a natural system, can cause changes which actively oppose system goals or can be purely random signals ('noise' in electrical systems). Thus the method an organization must use to control or influence an external system is to transmit information from itself to the other system for the purposes of defeating entropy in the controlled system and moving the values of that system's variables toward the values desired by the controlling organization.

In sum, Ashby's law requires that for a system to control another system (or for an organization to control another organization—or perhaps a market), the controlling system must be able to generate at least as much variety as the system being controlled. The actual mechanism of control, of creating the desired changes in the values of the variables in the controlled system, is the flow of information from the controlling system to the controlled system. Information institutes or maintains control by defeating systemic entropy, moving the value of each designated variable of the controlled system into the range desired by the controller, and holding it in that range.

Algorithms and heuristics

There is another, less conventional, way to think about the two dominant modes of variety control, a way based on the artificial-intelligence research of Douglas Lenat (1984) and others, which, although clearly speculative, has proven useful to me. Decision techniques are traditionally divided into two classes, algorithmic and heuristic. An algorithm is a set of rules for solving a problem. If the rules are followed correctly, a solution (if it exists) is guaranteed. Algorithms work by systematically reducing the variety (size or density) of the solution space, and tend to be effective in solution of relatively simple (i.e. low-variety) problems. It appears that the process of variety reduction is inherent in the algorithmic method. The types of problems to which management appropriately applies variety-reduction techniques may also be algorithmic in form.

Heuristic methods, on the other hand, are rules for enlightened search. They do not guarantee a solution, only improvement—if improvement is possible. The optimum path is not known in advance, nor, in fact, is the method by which one might find such a path. Heuristic methods, in contrast to algorithmic methods, begin with intelligent *expansion* of the potential solution space, thereby initially increasing the situational variety that must be handled, and suggesting the need for compensatory amplification of control variety in the subsequent process of reducing the solution space. It may be then that heuristics, complex systems, and control variety implification are also related. By analogy, the types of managerial problems that are heuristic in form may require the application of control variety amplification. I will look at this possibility in more depth below.

Beer's Viable System Model

The Viable System Model (Beer, 1959, 1966, 1979, 1981, 1984, 1985) is the crown and culmination of Stafford Beer's lifetime of path-breaking work in managerial cybernetics. It is quite obviously redundant, in these pages, to explicate the VSM in any depth, but there are two intertwined aspects of the model, as Beer applies it to the organization, that bear directly on the subject matter to follow, and so require brief discussion. One is Ashby's Law and the central position it occupies in the VSM. The other concerns the triadic relationships of 'management' and 'operations' (within the organization), and the organization's environment. This relationship is sketched in Figure 1 (which represents the VSM in its most elementary form), and amplified below.

For Beer (1985, p. 1), a 'viable' system is one capable of separate existence within a specified environment. The viable system of principal interest to Beer is the organization, embedded in (and ultimately attempting to regulate) its environmental complex of 'markets'. Figure 1 illustrates this VSM, showing the organization divided into 'management' (box) and 'operations' (circle), with management the regulator of operations, and operations the regulator of the environment (blob).

A prime characteristic of viability in a system is the presence of massive flows of information within the system and between the system and its

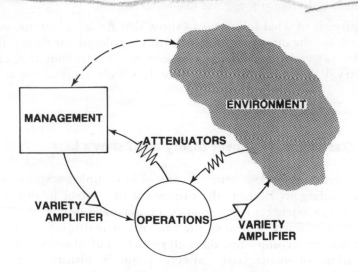

SOURCE: ADAPTED FROM BEER (1985: 27)

Figure 1. The simplified organizational VSM (adapted from Beer, 1985, p. 27)

environment. Much of the detail of the VSM is rooted in the need to design and engineer methods to transmit and manage very large amounts of information in real time. In Figure 1 the information communication channels, represented by arrows, interlock elements of the organization ('management' and 'operations') and join the organization (both management and operations) to the environment.

Ashby's Law requires that the regulator of a system be able to absorb all of the complexity that the system can generate. We normally expect the system under regulation to be intrinsically more complex than the regulator (if not, the situation presents no control problem, and is therefore of minimal interest). Thus we anticipate that the regulator will need to use one or both of the variety management methods described above—situational variety reduction or control variety amplification. Figure 1 illustrates both methods; it shows variety attenuation on the arrow leading from the environment to operations, and on the arrow leading from operations to management. Variety amplification is shown on the arrows leading from management to operations and from operations to the environment.

The primacy of Ashby's Law in the VSM is underscored by Beer's First Principle of Organization (1985, p. 30), which is itself a normative restatement of the law:

> 'Managerial, operational, and environmental varieties, diffusing through an institutional system, tend to equate; they should be designed to do so with minimum damage to people and to cost.'

The purpose of what follows is to show that Ashby's Law has unavoidably—but, for the most part, imperceptibly—colored our formal thinking about the organization and its management. In arguing this case, the elementary VSM, as adduced in Figure 1, will be recalled for additional duty.

Classical management theory and Ashby's Law

While cybernetics is the science of control in complex systems, management, according to Beer, is the profession of control in organizations. Therefore the science of cybernetics in general, and Ashby's Law in particular, are clearly central to the practice of management. Control in organizations is accomplished through reduction of situational variety and amplification of managerial variety. From a historical perspective, methods of situational variety reduction seem to have entered conscious management theory and practice first. I will examine variety reduction in the next section, and then go on to explore what I believe to be the complementary role of control variety amplification.

Management by situational variety reduction

Ashby's first method, situational variety reduction, conjures a vision of the manager struggling (perhaps in structuring the channel from operation to management in Figure 1) to reduce the varieties of a myriad of complex organizational situations to the level of variety personally available to him (or her). What does this vision convey of the genesis of the discipline of management? Here are some candidates: first, that management itself was called into being to tame the unwonted complexity that appeared when men first joined together to do work; second, that management has continued to develop in direct response to the changing patterns of complexity in the world.

From the earliest beginnings, therefore, the practice of management has been the practice of controlling or limiting variety in natural systems, of suppressing entropy so that the remaining natural forces within a system could move it toward man's chosen goals. All of the tools, techniques and theories of management have come about (we hypothesize) as products of the eternal battle against situational variety.

Logically, then, 'management' began when tasks became too complex for one person to master. Wherever people had to work together to accomplish a task, complexity multiplied and what we now call management was required (see Galbraith, 1977). We see this now, of course, in hindsight. At the dawn of commerce, there were no management theorists, or even managers—only owners and workers. Over time, real-world complexity expanded beyond the capacity of the owner to control. By the time the need for management appeared in economic organizations, the problem of organizational complexity had already been faced by the Church and in the military, and hierarchical authority structure had become a recognized response.

As the tremendous power of the technique of specialization (perhaps the first managerial variety-limiting device) became more apparent, some workers' jobs were restructured so that they personally 'did no work', but undertook, as a specialty, supervision of other persons' work. Here, perhaps we find the headwaters of management and, perhaps also, the origin of the idea that management 'is' responsibility for the work of others.

Early managerial themes

By the early 1900s, a few managers had begun to reflect on the implications of these new events in the workplace. Two in particular, Frederick W. Taylor (1911) and Henri Fayol (1916), tried to generalize about the nature of managerial work; Fayol from the top of an organization down, Taylor from the shop-floor up. Fayol suggested that there were universal functions performed by all managers; Taylor that there were certain universal and systematic ways of approaching every type of human labor that led to the most efficient accomplishment of work.

Taylor and Fayol worked within what later came to be called the closed system model of the organization; both adduced principles that are still viable in the appropriate contexts, and that can be analyzed in the light of Ashby's Law.

Fayol's principles of management

The idea that there are 'principles' of management—that this fledgling, empirical profession rests on a theory base—began with Henri Fayol (1916). He postulated a series of general management maxims, many of them formalizations of rules of thumb that had been found, lost, and rediscovered countless times since man first formed purposeful groups. Ashby's first method implies that the Fayolian rules relate directly to complexity and its control. And indeed, Fayol's five elements of management (planning, organizing, commanding, coordinating, and controlling) can be interpreted as variety-limiting or variety-attenuating devices.

Take, for example, planning. The cybernetic interpretation of planning is evident in Ackoff's description:

> '. . . the design of a desired future and of effective ways of bringing it about . . . a process that is directed toward one or more future states which are not expected to occur unless something is done.' (Ackoff, 1970, pp. 1, 3)

Planning, seen cybernetically, is the selection from the unconstrained set of all possible future states, that one, or those few, that the organization or the manager wants to see. The actions necessary to create this future are then deduced, and the appropriate control mechanisms installed.

Organizing (which for Fayol also meant staffing) can be seen as selection from the set of all possible organizational forms the one believed most likely to realize the planned future, and selection from the pool of all possible employees those whose abilities best conduce toward that future. Management then devotes effort and energy to maintenance of the organizational form, supplying the needed human talent, and suppressing those entropic forces (both random and malevolent) that threaten to produce an unwanted future.

And so on for Fayol's other elements. The generic process of variety reduction can be seen to be at work. In each case, a selection process limits the original large set of possible choices to a smaller set. From principles of managerial cybernetics, the key to making the process work is to use policy, experience, judgment, intuition, or analysis to systematically rule out whole classes of potential choices; to eliminate the need to separately consider each choice (cf. Warfield and Christakis, 1987).

In a similar vein, Fayol's principles of Division of Work, Unity of Command and Direction, Centralization, and the Scalar Chain can clearly be seen as devices to limit the operational variety faced by management. The seven famous functions of PODSCORB (Gulick, 1937), pillars of the closed system model of the organization, are either duplications of, or variations on, the Fayol theme, and are variety-limiting techniques, each directed to a different population of variables.

In sum, under the lens of Ashby's Law, Fayol's management principles and processes can be seen as a set of devices whose essential, and common, purpose is the selective reduction of variety within the organization.

F. W. Taylor and scientific management

Taylor (1911) believed that management should precisely define the job and even the exact methodology of the work. Taylor's aim was to have the workman function as nearly as possible like a machine. March and Simon note:

> '. . . the scientific management group was concerned with describing the charac-
> teristics of the human organism as one might describe a relatively simple machine
> for performing a comparatively simple task. The goal was to use the rather
> inefficient human organism in the productive process in the best way possible.
> This was to be accomplished by specifying a detailed program of behavior . . .
> that would transform a general-purpose mechanism, such as a person, into a
> more efficient special-purpose mechanism.' (March and Simon, 1958, p. 13)

Any attempt by the worker to design his own job under this regimen would be anathema. In cybernetic terms, Taylor behaved as if he believed that the workman was a source of entropy in the workplace:

> 'In turn-of-the-century organization theory and its "scientific management"
> legacy, individuals constituted not assets but sources of error. The ideal organiza-
> tion was designed to free itself from human error or human intervention,
> running automatically to turn out predictable products and predictable profits.'
> (Kanter, 1983, p. 18)

If not tightly controlled, man's natural variety would divert himself and the organization from its material goals. Given that Taylor's ideal was the machine, and that his desire was to make the worker as machine-like as possible, this premise is not illogical. Only the legendary John Henry was able to beat the machine at machine-like work—and he not for long.

Where does the Taylor approach succeed? It succeeds where correct execution of a protocol or adherence to what I call above an algorithmic method (March and Simon use the words above, '. . . a detailed program of behavior . . .') produces the desired product or correct outcome; where the correct means guarantees the desired end. It succeeds where work can be simplified to rote; where work is best performed by machines, and where economic motivation dominates. It succeeds where man works alone, or does repetitive tasks with simple machinery. It succeeds, in short, in non-complex systems, where man does work of low variety. The essence of scientific management is the design of low-variety jobs that any man can do. Taylor's method does not appear to succeed as well where the work requires heuristic rather than algorithmic behavior, as we will see below.

A cybernetic summary of classical management theory

In his work called scientific management, F. W. Taylor simplified and depersonalized work in order to standardize jobs. Another way of saying this is that he minimized the variety of work in order to suppress the potential for entropy. This technique of variety reduction worked phenomenally well for him, and has continued to work well where the form of labor is appropriate; even today in the United States, the profitable United Parcel Service appears to use a relatively undiluted form of Taylorism (Machalaba, 1986). Fayol and his followers distilled personal managerial experiences into maxims to guide the general manager. Those principles can now be seen as generic procedures for reducing internal situational variety in the process of managing the organization.

The open system organization, the closed suprasystem, and variety control

The classical methods of Taylor and Fayol, and their followers, implicitly assumed a closed system model of the organization. The organization was considered sufficiently independent of its environment that it could be analyzed separately. The closed system organization produced, as efficiently as possible, a standard product or service, which, in the eyes of the workforce, somehow disappeared outside the boundaries of the workplace and was replaced, equally mysteriously, with new raw materials and wages for the workers (Thompson, 1967). The work was rote and machine-like, with men as the predominate machines, and the goal—in Kanter's words above—to run 'automatically to turn out predictable products and predictable profits'.

In this setting, control, directed toward maximum volume of a standard product, using minimum amounts of labor and materials, and instituted through variety reduction methods, was practical and reasonably effective. The variety that had to be controlled in the closed system organization was normally manageable. There was, of course, the behaviorist objection to the effect of variety reduction methods on the worker. To that we will return.

The open system model

The open system model of the organization was first proposed in the 'forties and is still developing today (Feibleman and Friend, 1945; Parsons, 1956; Chandler, 1962; Emery and Trist, 1965; Thompson, 1967; Lawrence and Lorsch, 1967*a,b*; Berrien, 1968; Buckley, 1968*b*; Baker, 1973; Miller, 1978; Katz and Kahn, 1978; Kanter, 1983; Waelchli, 1987*a,b*). We have only recently recognized that this model carries a significant warning about the older closed system model; not only is the closed system view of the organization philosophically deficient, but, more particularly, the managerial methods that have developed within that model, many still in use today, are also dangerously defective (Waelchli, 1987*a*).

The 'closed' suprasystem

Paradoxically, beginning perhaps with Beer's metasystemic concept of 'completion from without' (1959, p. 81), followed more recently by ideas emerging from the deep intellectual waters of autopoiesis (Maturana and Varela, 1980; Zeleny, 1981) and dissipative structures (Prigogine, 1980), assisted here and there by organization theorists (e.g. Morgan, 1986), the open system model now leads, for some thinkers, directly to a vision of the organization as an element of a larger, effectively *closed*, system. Closed, not in the traditional zero-sum context, nor even in the sense of Beer's 'relatively isolated system' (1966, p. 270), but in the recursive or auto-poietic sense of containing internally the information and structure required for its own self-generation, development, articulation and maintenance.

For Kenneth Boulding (1981) the economic market system is clearly autopoietic; in fact he nominates Adam Smith's 'invisible hand' as the first formulation of the autopoietic concept. At the level of the individual organization, Peters and Austin (1984, p. 28) describe the 'real-world loop'; a system consisting of the firm and its relevant environment, particularly its markets (which are predominantly four: customer, supplier, capital and labor).

For lack of a better title, I call the extended organization–environment complex the 'closed' suprasystem. The suprasystem is defined to contain internally all the forces and factors that measurably affect its states. Its nature implies, as intimated above, that the successful organization not only interacts strongly with its markets, but also exercises some measure of control over them, because the suprasystem operates according to the cybernetic laws of mutually adaptive control. (For earlier thoughts along this line, see Cyert and March, 1963, p. 2.)

From the standpoint of Ashby's Law, the need to deal effectively with multiple and interlocked markets dramatically changes the problem of variety management; it seems unlikely that situational variety reduction, as that technique is practised within the closed system organization, can cope with the complexity of the interactive market environment. A more powerful form of variety management appears to be needed if the organiz- ation is to exercise even partial control over its markets.

Post World War II demand and the lost warning

But because of the pent-up buying power released at the end of the Second World War (concurrent with the first explications of the open system model), the warnings of the open system model and its suprasystem successor were lost in a flood of demand for products ('From 1946 until the early sixties we opened the spigots full bore just to meet domestic demand.' Peters and Austin, 1984, p. xvii). The market effects predicted by the two systemic models appeared only slowly; particularly obscured was the implication that effectiveness of an open system organization, operating within its suprasystem, comes through explicit responsiveness to its markets, especially the customer markets.

Many businesses, and even whole industries, continued to act, with diminishing success, on the implicit closed system assumption of assured, untended markets (Thompson, 1967; Kanter, 1983). Halberstam (1986) pictures the entire US automobile industry suffocating in a cocoon of closed system thinking. The diminishing success was noted, but was blamed on a variety of exogenous factors, until relatively recently, when (among others) the group of observers mentioned above (Ouchi, 1981; Peters and Waterman, 1982; Kanter, 1983; Peters and Austin, 1985; Clifford and Cavanagh, 1985), found, in successful organizations, some of the phenomena implied by Ashby's Law and the suprasystemic model.

So, although an anomaly of demand artificially lengthened its life, the day of organizational effectiveness through closed system behavior and situational variety reduction (alone) gradually passed. We now believe that a firm's markets must be *managed* to some extent by the firm. These markets are, of course, external to the firm and cosmically complex, and so do not seem amenable to the traditional methods of situational variety reduction, except in special cases.

According to Ashby and Beer, there is only one other road to control in complex systems; control variety amplification. If Ashby's Law is valid, and if an organization does manage its environment to any real extent, it appears that it must somehow employ variety amplification. I will argue below that the effective organization does indeed so do—but before we can profitably examine this argument, we must briefly review the organizational 'human behavior' movement, because this movement ultimately revealed to the manager the *means* for effective control variety amplification.

The human behavior movement

Classical organization theory, as outlined above, particularly the ideas of Taylor, provoked instant and intense criticism from a number of sources The definitive behavioral rebuttal to classical organization theory was made in 1957 by Chris Argyris, who distilled 'as much of the existing empirical research . . . on human behavior in organizations . . . as possible' (1957, p. ix). His analysis of the work of scores of behavioral scientists led him to conclude that 'classical' organizations, those derived from Fayolian principles as well as from the Taylor model, were not congruent with the needs of healthy people. The rigidities of the formal organization made the employee feel dependent, submissive and passive. He (or she) was able to use only a few of his or her less important abilities. Because of the constraining effects of the formal organization, the worker was unable to release enough psychic energy to achieve self-actualization. In Argyris' words:

> 'A number of difficulties arise with [the classical] assumptions when properties of human personality are recalled. First, the human personality we have seen is *always* attempting to actualize its *unique organization* of parts resulting from a continuous, emotionally laden ego-involving process of growth. It is difficult, if not impossible, to assume that this process can be choked off . . .' (Argyris, 1957, p. 59)

A cybernetic restatement of Argyris' argument is that man is a complex system, an inherently high-variety entity who cannot function in good health under classical low-variety controls.

Argyris concluded that classical formal organization structure frequently injured both the worker and the organization. This realization caused some behaviorists to attack organizations generally:

'Some human relations researchers have unfortunately given the impression that formal structures are "bad" and that the needs of the individual should be paramount in creating and administering an organization.' (Argyris, 1957, p. 58)

This minority position has proven persistent and dilatory; it seems to have generally delayed and diluted managerial acceptance of the results of behavioral science research, even up to today.

For many managers and theorists, however, the work of Argyris and other behavioral scientists stimulated interest in determining whether the economically productive organization could also be an attractive place to work. Managers and researchers sought ways to implant the goals of the organization in the worker. Important in this task was the idea of 'motivation'. Researchers struggled to understand the factors, particularly non-economic factors, that influenced human behavior in the workplace. A number of theories of motivation (Maslow, 1948, 1954; Argyris, 1957, 1964; McGregor, 1960; Likert, 1961; Hertzberg, 1966) appeared, broadening management's understanding and sharpening its tools. One element of motivation, the idea of 'participation', seems to have been a significant precursor to the effective managerial use of variety amplification. What is the provenance of 'participation' as a practice, as a management challenge and as a manifestation of control variety amplification?

Motivation, participation and Ashby's Law

Man is perceived, for better or for worse, as an infinitely varied being, full of complexity, confusion, change and conundrum. Within this teeming complexity, however, scholars have found some orderly patterns. One in particular interests us here. An early, and still fundamental, human economic duality pits 'man–the–consumer' against 'man–the–producer'. This duality is reflected in the two basic economic concepts of demand and supply: man–the–consumer, demands; man–the–producer, supplies. And both, of course, are the same person. Before the days of group production, man produced for his own consumption (and his family's), and was assumed to balance subjective utility between the pleasures (positive and negative) of production and consumption.

When it became clear that the factory system could produce an immense variety of goods, along with wages that would allow man to materially possess and consume far more than before, it was widely assumed that man would willingly give up the utility of being his own master and producer, sell his time and abilities, and engage in daily work perhaps far less palatable than before, in order to earn the income that permitted him to improve his material standard of living. Philosophically, man–the–consumer was thought to dominate man–the–producer.

For a time, this assumption seemed correct. But at some Maslovian level of material satisfaction, man began to display a distaste for the regimented jobs he performed in factories, and sought more personal utility in work. Man-the-producer emerged, insisting (through researchers and writers on human behavior) that the rote work imposed by the manufacturing system was unacceptably deficient in utility.

Management responded to this development grudgingly and slowly; the worker was permitted limited participation in defining his work, not because management believed that the worker could improve the organization of work (did not F. W. Taylor prove otherwise?), but because limited participation seemed to do no harm, and also seemed to *motivate* the worker to work harder, thus raising output (Kanter, 1983, p. 34). A modern restatement:

> 'We observed, time and time again, extraordinary energy exerted above and beyond the call of duty when the worker (shop floor worker, sales assistant, desk clerk) is given even a modicum of apparent control over his or her destiny.' (Peters and Waterman, 1982, p. xxiii)

As participation was allowed, slowly and painfully, to increase, researchers, and then managers, found evidence that participation, intelligently designed and well managed, could improve the efficiency and quality, as well as the volume, of work (Kanter, 1983).

The process of participation has deepened (fitfully), with the theorists usually in the vanguard of the managers, at least in the West. The pattern has continued to be an initial tolerance of participation in areas where participation offered no harm, followed by a gradually dawning suspicion that the process of participation was improving, beyond expectation, many aspects of the organization's effectiveness.

Is there a theoretical basis for the notion that participation may increase organizational effectiveness? We turn again to Ashby's law.

Qualities of management, markets, and the worker in purposeful systems

Recall that Ashby's Law is originally and fundamentally a law of systems, and, in this paper, purposeful systems. In the domain of the suprasystem embracing the organization and its environment, we focus on three complex, living, high-variety, interlocked systems, each of which, we believe, operates in obedience to Ashby's Law. These three systems are the firm's array of markets, its management, and its workforce (in Figure 1, the blob, the box, and the circle). The last two of these systems are purposeful always, the first, sometimes. Let us consider each system in a bit more detail.

The markets comprise systems that are sometimes purposeful and some-times not. In the first case there exist identifiable persons and organizations whose goals differ from (often oppose) those of the protagonist organiza-tion. In the second case the market systems are made up of unknown humans, acting out of infinitely varied motives, largely oblivious to the organization and its goals. However, owing to the autopoietic nature of the economy, the teleological nature of living systems, and the apparent zero-sum aspect of certain market situations, even this second, non-purposeful, market segment often appears to management as if it pursued goals inconsistent with those of the organization. So, generally, the manager treats markets as purposeful systems whose actions affect his goals and whose goals, diverge from his, therefore requiring some form of active control.

Management, in the highest sense, according to modern theory, tries to create value for its markets (stakeholders); in a more immediate sense it is trying to control these markets, to realize, within those markets, a set of management-defined goals that support the reasons for the existence of the organization, through the notions of control outlined above. As Drucker observes:

> '[The purpose of a business] must lie outside the business itself. In fact it must lie in society since the business enterprise is an organ of society. There is only one valid definition of business purpose: *to create a customer*. Markets are not created by God, nature or economic forces, but by businessmen.' (Drucker, 1954, p. 37; also 1974, p. 61)

The ultimate focus of management is outside the organization, in society, and, more specifically, in the organization's markets.

The worker is an agent for management in the accomplishment of the organization's goals; in this role his immense, diverse, inherent human variety must be focused to support those goals, which are traditionally set by management. This can be done in two very different ways. The worker can be focused as to form, or as to purpose; he can be made to behave in prescribed ways, or he can be asked to pursue, on his own initiative, the societal goals of the organization. Each focus carries an implicit model of management thought, presented, much oversimplified, below.

Algorithmic and heuristic models of management thought

In the first model (focus as to form, a closed system holdover), manage-ment determines, unilaterally, what is valuable to the markets. The organ-ization then produces the good or service according to optimized algorith-mic methods, and presents it to the market on a take-it-or-leave-it basis. The role of the worker in this model is to follow instructions, to execute the algorithm. The worker's inherent variety is here considered entropic; what management wants is a precise, obedient and tireless low-variety machine. Part of management's effort, therefore, is expended on reducing (ideally eliminating) variety in the workforce. Situational variety-reduction methods are management's preferred tools, represented in Figure 1 by variety attenuation between the circle and the box.

The second model takes a suprasystemic, 'close to the customer' form, of the Drucker, Ouchi, Peters, Waterman, Austin, Kanter, Clifford, Cavanagh genre. The firm has the same value-creation goals as before, but 'value' is now determined, not unilaterally by the organization, but jointly, by the customer and the organization, and in near real time. In this model the final form of the good or service may not be known when the work begins. Clearly, this situation cannot be governed by an algorithm; it is, I believe, a heuristic problem.

What is the worker's role in a heuristic organization? We have pictured management as overwhelmed by the variety flowing in from the markets; variety which it cannot materially reduce and still realize 'joint' values. To combat and match this external variety increase, it seems that management must now invoke, focus, and use—rather than suppress—the cosmic variety of the worker. The worker must truly 'participate'. And, in fact, leaders of the 'excellent' firms appear to make this happen; they apparently communicate successfully to the worker a message such as the following:

> 'Only you can make this company a success—we can't do it all. We have common goals, common values, and we have agreed on our strategic approach and target position in the marketplace. Now it is up to you to make it work—and we know you can do it.' (Clifford and Cavanagh, 1985, p. 109)

Management still has a problem with human variety, but now an entirely different one; now management must try to focus and *use* the employee's variety; it does not desire to suppress it. Participation is not only desired, but apparently essential.

In the VSM of Figure 1, the obvious example is the variety amplification link from operation to environment, but this is incomplete. The worker, while remaining in the circle for the accomplishment of daily tasks, has also taken up residence in the box, as he strives to perform the *management* task of realizing corporate goals in the environment (blob). The worker has thus opened, and is part of, a variety amplification channel that connects management with the environment.

This conversion from human variety suppression to variety enhancement is not easily done, as the relevant literature attests; but when it is, it is often remarkably effective. The worker, with a shared corporate value system and corporate goals embedded in his soul, and with freedom, even a charge, to act intelligently on those values, has become, in a sense, an extension of management (Kanter, 1983). He fulfills Drucker's test of a manager (1974, p. 389), one who accepts 'responsibility for contributing to the results of the enterprise'. He is working, directly or indirectly, on the problems and complexity of the markets, adding his considerable variety on the side of management, and thus helping to institute organizational control in those markets. If he is successful, the organization's ability to 'control' its markets has been amplified. The worker has become an engine of managerial control variety amplification. Ashby's second method of variety management now also appears in the manager's toolkit. (For the same conclusion, reached through different logic, see Weick, 1987.)

Summary: Ashby's Law and management

In the dry language of managerial cybernetics, achievement of organizational goals is accomplished in a three-step program: first, selection of those formal variables, inside and outside the organization, that specify (or affect) the reason for being of the organization; second, prescription of acceptable values for each chosen variable; and, finally, expenditure of managerial effort to generate and transmit the information required to move the key values into the desired ranges and keep them there. Ashby's two methods of managing variety in this process take us down two quite different paths of management thought.

The algorithmic model and situational variety reduction

In the first model, which I describe as algorithmic, the governing philosophy is that if one controls the means, the ends are assured. In the factory application of this model, the ends are marketable goods in high volume at low costs. The types of variables that are means to these ends are internal; examples might be output per machine hour, percentage scrappage, time lost through injury, absentee rates, and machine down-time.

In the algorithmic model, management, quite naturally, sees its job as focusing on the means: setting standards for the values of specific internal variables and blocking any forces or actions that threaten to drive the values of these 'means' variables in the wrong direction. The idea of situational variety reduction, particularly reduction of worker variety, is important, since management has devised, at least nominally, an optimized algorithm of work; any deviation from the algorithm threatens to sub-optimize.

The heuristic model and control variety amplification

Heuristics, we recall, imply a search for methods to improve performance against chosen goals. Organizations described by contemporary writers as 'excellent', which I have characterized as heuristic, attempt to control and influence variables that express and measure external (societal) purpose. These goals relate to *ends*, rather than means; examples might be the desire to be the best at something, to provide the maximum in customer–defined value, to help the customer define and solve his problems.

The worker's job now is to devise means that lead to designated results in the organization's markets. In complex market situations, the potential avenues to success are literally endless: '. . . one thousand things done just a bit better' (Peters and Austin, 1984, p. 294). The worker now needs all of his inherent variety to find means that lead toward the desired ends.

Heuristic management's new job is to conceive and clearly articulate the reasons for existence of the organization, in terms of its fundamental values and the changes it desires to make in society (cf. Parsons, 1956). Management must then endlessly wave the banner blazoned with those goals, lead and help in the process of finding means to effect the desired societal changes, and manage the inherent and assumed constraints such as ethical, legal, time, resource, technology, and actions by competitors.

If management does properly the work of value articulation, and if the worker adopts management's vision of societal change and makes it part of his intellectual and emotional fabric, then the worker has become, in a real sense, management. In formal cybernetic terms, he applies his great variety to the task of controlling the complexity of the organization's environment and markets by acting to bring the variables targeted by management into the value ranges specified by management and maintaining them in those ranges. Management has now multiplied itself, and its variety, by the number of workers who strive directly for the organization's societal ends, in countless self-devised and often unseen ways.

The control mechanism just described is control variety amplification, which seems not only to complement situational variety reduction in management practice, but also to give us a new theoretical insight into organizational effectiveness. The modern thesis of 'excellence through people' seems to rest on the mechanism of control variety amplification in compliance with Ashby's Law. As an example, I sense in the 'integrative' management mode described and advocated by Kanter (1983) the themes of heuristic behavior and control variety amplification, while her spiritless 'segmentalist' culture summons echoes of algorithmic thinking and situational variety reduction. A parallel duality seems (to me) to characterize the cited works of Drucker, Ouchi, Peters and Waterman, Peters and Austin,

Clifford and Cavanagh, and other members of what is sometimes called the 'excellence through people' community.

Of law and management

We have seen, briefly, how the two modes of variety management that devolve from managerial cybernetics and Ashby's Law can be related to management theory and practice. We related the method of situational variety reduction to the classical management school as expressed by the principles of Fayol and Taylor, and as applied within the boundaries of the (so-called) closed system organizational model.

We then followed two separate paths to the idea of managerial control variety amplification. On one path we found the highly complex market interaction and control needs implied by the suprasystemic model and, on the other, the needs of healthy men and women in organizations as discovered by the human behavior researchers and articulated by Argyris. Finally, we noted that the phenomenon of control variety amplification seems also to be implied in the managerial patterns found to characterize 'excellent' organizations.

Is Ashby's Law of Requisite Variety a universal—an Iron Law of management? The test condition was that we should be able to find manifestations of Ashby's Law as 'recurrent themes' of management principles, theories and actions. We set out to explore briefly historical and contemporary management thought for evidence relevant to the condition.

As a minimum, it does seem fair to conclude that complexity and complexity control are problems central to all aspects of management. If this is so, and if Ashby's Law is valid, then some pivotal and universal role for Ashby's Law in management seems inescapable. On the other hand, while encouraging and intriguing, the evidence to date is clearly speculative and truant in rigor. We particularly need a translation of our reasonably good theoretical understanding of Ashby's Law, and other principles of managerial cybernetics, into active prescriptions that a manager can use. Beer's prodigious efforts—the Viable System Model in particular—stand as consequential but (so far) lonely monuments to that need. Disciplined research is indicated, perhaps along the 'organizational effectiveness' lines suggested by Lewin and Minton (1986). Based on what we now know, such research seems worth doing.

Finally, a personal note. I recognize that in arguing for the idea of the universality of law, I am swimming against the prevailing current within the social sciences. No apology. To me, Lederman's vision of a 'single and economic law of nature, valid throughout the universe for all time' is compelling. And I can find no evidence that it is a vision reserved for the natural sciences.

References

Ackoff, R.L. (1970). *A Concept of Corporate Planning*. New York: Wiley–Interscience.

Ackoff, R.L. and Emery, F.E. (1972). *On Purposeful Systems*. Chicago: Aldine–Atherton.

Argyris, C. (1957). *Personality and Organization*. New York: Harper & Row.

Argyris, C. (1964). *Integrating the Individual and the Organization*. New York: John Wiley.

Ashby, W.R. (1956). *An Introduction to Cybernetics*. London: Chapman & Hall.

Ashby, W.R. (1960). *Design for a Brain*, 2nd edn. London: Chapman & Hall.

Baker, F. (1973) (ed.). *Organizational Systems: General Systems Approaches to Complex Organizations*. Homewood, Ill.: Richard D. Irwin.

Beer, S. (1959). *Cybernetics and Management*. New York: John Wiley.

Beer, S. (1966). *Decision and Control*. New York: John Wiley.

Beer, S. (1975). *Platform for Change*. New York: John Wiley.

Beer, S. (1979). *The Heart of Enterprise*. New York: John Wiley.

Beer, S. (1981). *Brain of the Firm*. 2nd edn. Chichester: John Wiley.

Beer, S. (1984). 'The Viable System Model: its provenance, development, methodology and pathology', *Journal of the Operational Research Society*, **35**, 7–25.

Beer, S. (1985). *Diagnosing the System For Organizations*. Chichester: John Wiley.

Berrien, F.K. (1968). *General and Social Systems*. New Brunswick: Rutgers University Press.

Boulding, K. (1981). 'Foreword' to Zeleny (1981).

Buckley, W. (1968a) (ed.). *Modern Systems Research for the Behavioral Sciences*. Chicago: Aldine.

Buckley, W. (1968b). 'Society as a complex adaptive system', in Buckley (1968a), pp. 490–513.

Chandler, A.D. (1962). *Strategy and Structure*. Cambridge, Mass.: MIT Press.

Clifford, D.K. and Cavanagh, R.E. (1985). *The Winning Performance*. Toronto: Bantam Books.

Cyert, R. and March, J. (1963). *A Behavioral Theory of the Firm*. Englewood Cliffs, N.J.: Prentice-Hall.

Drucker, P.F. (1954). *The Practice of Management*. New York: Harper & Row.

Drucker, P.F. (1974). *Management: Tasks, Responsibilities, Practices*. New York: Harper & Row.

Drucker, P.F. (1985). *Innovation and Entrepreneurship*. New York: Harper & Row.

Emery, F.E. (1969) (ed.). *Systems Thinking*. Baltimore: Penguin Books.

Emery, F.E. and Trist, E.L. (1965). 'The causal texture of organizational environments', in Emery (1969), pp. 241–57.

Fayol, H. (1916). *General and Industrial Management* (Storrs translation, 1949). London: Pitman.

Feibleman, J. and Friend, J.W. (1945). 'The structure and function of organization', in Emery (1969), pp. 30–55.

Galbraith, J.R. (1977). *Organization Design*. Reading, Mass.: Addison-Wesley.

Gulick, L.H. (1937). 'Notes on the theory of organization', in *Papers on the Science of Administration* (eds. Gulick, L.H. and Urwick, L.F.). New York: Columbia University Press.

Halberstam, D. (1986). *The Reckoning*. New York: William Morrow.

Hertzberg, F. (1966). *Work and the Nature of Man*. Cleveland, Ohio: World Publishing.

Kanter, R.M. (1983). *The Change Masters*. New York: Simon & Schuster.

Kast, F.E. and Rosenzweig, J.E. (1979). *Organization and Management: a Systems and Contingency Approach*, 3rd edn. New York: McGraw-Hill.

Katz, D. and Kahn, R.L. (1978). *The Social Psychology of Organizations*, 2nd edn. New York: John Wiley.

Koontz, H. (1961). 'The management theory jungle', *Academy of Management Journal*, December, 174–88.

Lawrence, P.R. and Lorsch, J.W. (1967a). 'Differentiation and integration in complex organizations', *Administrative Science Quarterly*, **12**, 1–47.

Lawrence, P.R. and Lorsch, J.W. (1967b). *Organization and Environment* (1985 reprint). Boston: Harvard Business School Press.

Lederman, L.M. (1984). 'The value of fundamental science', *Scientific American*, November, 204–13.

Lenat, D. (1984). 'Computer software for intelligent systems', *Scientific American*, September, 40–7.

Lewin, A. and Minton, J. (1986). 'Determining organizational effectiveness: another look and an agenda for research', *Management Science*, **35**, 514–38.

Likert, R. (1961). *New Patterns of Management*. New York: McGraw-Hill.

Luthams, F. (1976). *Introduction to Management: a Contingency Approach*. New York: McGraw-Hill.

Machalaba, D. (1986). Up to speed: United Parcel Service gets deliveries done by driving its workers', *The Wall Street Journal*, 22 April, 1.

March, J. and Herbert Simon, H. (1958). *Organizations*. New York: John Wiley.

Maturana, H. and Varela, F. (1980). *Autopoiesis and Cognition*. Boston: Reidel.

Maslow, A. (1943). 'A theory of human motivation', *Psychological Review*, **50**, 370–96.

Maslow, A. (1954). *Motivation and Personality*. New York: Harper & Row.

Mayo, E. (1933). *The Human Problems of an Industrial Civilization*. New York: Macmillan.

McGregor, D. (1960). *The Human Side of Enterprise*. New York: McGraw-Hill.

Miller, J.G. (1978). *Living Systems*. New York: McGraw-Hill.

Mooney, J.D. and Reiley, A.C. (1931). *Onward Industry!* New York: Harper and Brothers.

Morgan, G. (1986). *Images of Organization*. Beverly Hills: Sage Publications.

Ouchi, W.G. (1981). *Theory Z*. Reading, Mass.: Addison-Wesley.

Parsons, T. (1956). 'Suggestions for a sociological approach to the theory of organizations', *Administrative Science Quarterly*, June, 63–85.

Peters, T. and Austin, N. (1985). *A Passion for Excellence*. New York: Random House.

Peters, T. J. and Waterman, R.H. (1982). *In Search of Excellence*. New York: Harper & Row.

Prigogine, I. (1980). *From Being to Becoming*. New York: W.H. Freeman.

Taylor, F.W. (1911). *The Principles of Scientific Management*. New York: Harper & Row.

Thompson, J.D. (1967). *Organizations in Action*. New York: McGraw-Hill.

Urwick, L. (1937). 'Organization as a technical problem', in *Papers on the Science of Administration* (eds. Gulick, L.H. and Urwick, L.). New York: Columbia University Press.

Urwick, L. (1947). *The Elements of Administration*, 2nd edn. London; Pitman.

Waelchli, F. (1987a). 'A cybernetic model for the proactive organization', in *Decision Support Systems in the Public Sector*, vol. 2 (ed. Shutt, H.J.). Fort Belvoir, Va.: Defense Systems Management College (revised version of 1985 monograph).

Waelchli, F. (1987b). 'Managerial cybernetics: toward a new vision of the organization', in *Proceedings of the 1987 Southeastern Regional Conference of the International Society for General Systems Research*, Norfolk, Va., 10–13 May.

Warfield, J.N. (1986). 'A typology of laws', paper presented to the Annual Meeting of the American Society for Cybernetics, Virginia Beach, Va., February.

Warfield, J.N. and Christakis, A.N. (1987). 'Dimensionality', *Systems Research*, **4**, 127–37.

Weick, K.E. (1987). 'Organizational culture as a source of high reliability', *California Management Review*, **29**, 112–27.

Zeleny, M. (ed.) (1981). *Autopoiesis: a Theory of Living Organizations*. New York: North Holland.

The Viable System Model: Interpretations and Applications of Stafford Beer's VSM
Edited by R. Espejo and R. Harnden
© 1989 John Wiley & Sons Ltd

4

The VSM revisited

Raul Espejo
Aston Business School

This chapter discusses how effective organizations work from the view-point of control and communication processes. These processes are studied from the perspective of the management of complexity. In particular the Law of Requisite Variety is used as a heuristic to develop criteria of effectiveness. The outcome is Beer's model of the organization structure of any viable system (VSM). This paper also offers methodological guidance as to how to study in practice the different regulatory mechanisms entailed by the model.

Introduction

The purpose of this chapter is to discuss how effective organizations work from the point of view of control and communication processes. This discussion should provide criteria to improve the all too natural short-comings of 'real-world' organizations.

Releasing the potentials of people in order to permit them to handle autonomously the multiple problems they confront in their jobs is what gives organizations the flexibility they need to survive in complex and rapidly changing environments. Hence, criteria of effectiveness for an organization emerge both from the need to achieve the cohesion of the whole and the autonomy of the individuals in it.

These are the criteria developed by Stafford Beer in the Viable System Model (Beer, 1979, 1981, 1985). Beer has developed a most comprehensive set of principles and laws of organization; as a whole they permit the establishment of the mechanisms necessary for effective control and communications in organizations. These mechanisms go deeper into the management of complexity, and offer more insights about it, than the traditional mechanisms of feedback and feedforward. This chapter will revisit Beer's mechanisms from first principles using the Law of Requisite Variety as a reference (Ashby, 1964). Broadly speaking this law states that a 'controller' has requisite variety—that is, has capacity to maintain the outcomes of a situation within desirable states (the target set)—if, and only if, it has the capacity to produce responses to all those disturbances that are likely to take the outcomes of the situation out of its target set.

I have suggested elsewhere (Espejo, 1987) that Beer's use of requisite variety in studying control problems has been most impressive, and that nowhere had this use been more apparent than in his Viable System Model. However, I have also made the point that in my opinion he has not always spelt out in detail the actual control mechanisms invoked. Hence this chapter, while revisiting Beer's model, also aims to further our understanding of these mechanisms.

Definitions

Viable systems are those able to maintain a separate existence. Such systems have their own problem-solving capacity. If they are going to survive they need not only a capacity to respond to familiar disturbances, but potential to respond to unexpected, previously unknown disturbances. This latter capacity is the hallmark of viable systems: it gives them the capacity to adapt to changing environments. While a catastrophic event may at any instance fracture the coherence of a viable system, the fact of viability lessens the vulnerability of systems to chance—indeed, it makes them more adaptive to change.

The organization of human activity systems is defined as the set of interpersonal relations which make of the system a whole, independent of the particular individuals involved in these relations who can be any as long as they satisfy these relations (Maturana and Varela, 1980). The emphasis of this definition is in the relations and not in the parts. For instance the organization of a university is defined by, among others, the nature of the relations between students and academic staff and not by the students and staff involved in particular relationships at any given time.

However, the particular social forms, whether formal or informal, taken by these relations at a particular time and context define the structure of the organization. In the above example the concrete teaching departments, committees, groups (formal or informal), services, etc., in existence at a particular time define the structure of the university.

Mechanism is defined as any stable form of communication or interrelation between parts in an organization that permit them (the parts) to work as a whole. With this definition it is possible to redefine the structure of an organization as the set of specific mechanisms defining the interactions between the parts of that organization.

Most importantly, the structure of an organization is defined by the actual parts and actual communication channels in existence and not by the parts and lines of authority formally defined by, for instance, an organization chart.

Viable systems and requisite variety

Figure 1 describes a viable system within its environment and management within the viable system.

A viable system exists within an environment which is beyond the knowledge and control of the people within the system. The complexity of this environment may unfold in a wide range of unexpected forms; not only can people 'see' a limited number of variables among the infinite number of possible environmental variables, but also they can only 'see' some aspects of any particular variable. It is in the nature of human activities that the complexity of the environment is much larger than that of the viable system itself.

Similarly, the management of the viable system is accountable for a situation (i.e. the organization) inherently beyond its own knowledge

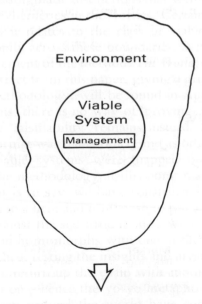

Figure 1. The system in its environment

capacity. In other words, the variety of management is much lower than the variety of the organization itself.

We are thus faced with imbalances in the varieties of management, the organization and the environment. This leaves us with an apparent paradox; if management controls the organization and the organization survives in its environment, then the Law of Requisite Variety implies that their varieties are roughly in balance, at adequate levels of performance. This latter point is captured by Beer's First Principle of Organization:

> 'Managerial, operational and environmental varieties, diffusing through an institutional system, tend to equate; they should be designed to do so with minimal damage to people and to cost.' (Beer, 1979, p. 97)

However, as established above these varieties were recognized to be inherently different. Thus, how can they 'tend to equate'? It seems important to clarify this apparent contradiction.

Of 'all' the environmental variety, only part of it will be relevant to the viable system: namely the part producing the disturbances that the viable system has to respond to in order to maintain viability. However, it is not necessary for the viable system to deal with all of this complexity by itself. It is perfectly possible for an important part of this relevant complexity to be taken up by people or organizations operating in the environment itself. This is the systemic role, for instance, of a network of car dealers *vis-à-vis* the car manufacturer (dealers are in the company's environment both attenuating environmental complexity and amplifying organizational complexity). However—and this is the important point—any residual variety, left unattended by these environmental responses, needs to be met by the organization itself; not to do so would imply a lowering in performance and the risk of becoming non-viable (Figure 2).

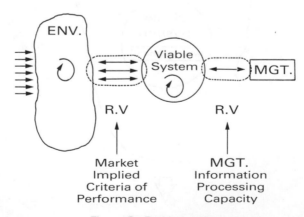

Figure 2. Residual variety

The same argument applies to the relationship between managers and the company. To say that management controls the company does not mean that the varieties of both are the same, but that the residual variety that is left unattended by the processes of self organization and self regulation in the company has to be absorbed, equated, by management. If the information needs implied by the residual variety are beyond the managers' information processing capacity, then control will be inadequate or, in the extreme, not exercised at all (Figure 2).

The above discussion suggests a slightly different formulation of Beer's First Principle of Organization:

> The response varieties of a viable system and its management tend to equate, respectively, the residual varieties of the environment and operations; they should be designed to do so with a minimum damage to people and to cost.

This statement of the principle implies that the matching of varieties is between the organization and its environment and, at a much lower level of complexity, between management and the organization (see Figure 2). This is in contrast with Beer's view which suggests a balance at the same level for the three parts. At the core of this disagreement I see an epistemological problem. Beer is suggesting that the balances in question are those implied by the complexity that is 'seen' in the operations and the environment by the 'viewpoint' of management at that level. In my opinion, the complexity of the environment cannot be meaningfully captured by one viewpoint alone (i.e. management at that level of operations). Since multiple viewpoints are likely to exist within the operations, they and not management alone are responsible for the complexity that is seen in the environment, at that level of operations.

This view has methodological implications; the modified first principle of organization makes apparent that any design of amplifiers and attenuators to match environmental variety should be done with reference to the complexity seen in that environment by the viewpoints within the operations, and not with reference to the complexity seen by the management of these operations alone. Thus, if we turn to Figure 3, it is important to recognize it as a schematic representation of the need for attenuators and amplifiers between the environment, the operations and management. In practice, an in-depth study would have to recognize that most of the attenuators and amplifiers to study are within the three distinguished domains—environment, operations and management.

Though no particular reference is made in this chapter to Beer's Second, Third and Fourth Principles of Organization (which refer to requisite variety from the perspectives of channel capacity, transduction capacity, and sustainability of regulation), they are accepted unchanged in all the following discussions about regulation.

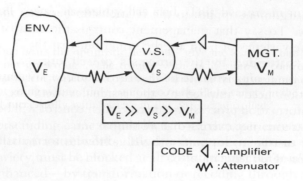

Figure 3. Schematic representation of the need for attenuators and amplifiers

Finally, in several parts of the discussion variety and information will be used as roughly equivalent. However, it is important to keep in mind that they are not at all the same thing. Variety refers to states of a situation, information to representations (e.g. reports) of these states. Since the residual variety relevant to a viewpoint depends on its purposes, and since the information about this variety is produced by viewpoints not necessarily sharing the same purposes, there is always the possibility of mismatches between one and the other. Only if the variety of a situation were completely well-defined and precise, both variety and information could be the same. This is an ideal type situation alone. In practice, as the situation moves away from this ideal the overlap becomes less and less satisfactory and the problem of management is, if it is to avoid the need to see every state by itself, to achieve as close an overlap between one and the other as possible.

Mechanisms for viability: the adaptation mechanism

The problem

To remain viable an organization needs to have the capacity to adapt to new situations. Indeed, an effective organization is one that not only does 'things right' but, most importantly, is one that is able to find the 'right things' to do. This capacity for adaptation is normally associated with the strategic levels of management in an organization.

What can management do if they become aware that the organization is not steering its way in the environment but just aimlessly reacting to external changes?

How can policy-makers increase the likelihood that their visions about the organization's identity will support the organization's long-term viability?

What is the appropriate contribution of policy-makers to policy processes so to make possible an effective use of their limited information processing capacity?

How can policy-makers increase the likelihood that every one in the organization will contribute, to the best of their possibilities, to the decisions necessary for an effective organization?

These are effectiveness questions that need an answer.

Senior managers are confronted with situations that can easily go out of control. For instance, it is not unusual for a board of directors to find out that a new product, in which they have invested large sums, has no market. Equally, it is not uncommon to find boards deciding to invest in the development of new products only to find out much later, after costs have been incurred, that they are technically not feasible, or to find boards approving salaries and wages policies that at a later date trigger damaging industrial relations problems.

Often in such cases managers are aware not only that they had been deciding on issues beyond their own expertise—something that is natural in a complex world—but also that in debating these issues, existing organizational resources, with the necessary knowledge to avoid the problem, had not been used to the best of their possibilities.

Independent of whether the outcomes of a policy have been good or not, the question seems to be what kind of mechanisms are used in the organization to link the so-called 'policy-makers' to the rest of the organization? How sensitive is the organization to the organizational identity and policy issues as seen by policy-makers? Indeed, it is not unusual for people in policy-making positions to feel that they are only rubber-stamping what already has been debated and decided at lower levels in the organization.

These problems seem to be a consequence of ill-structured information processes. A particular casualty of this situation is the espoused theory of stake-holders' participation in policy-making. While democratic organizations espouse the view that their destinies should be in the hands of those representing the stake-holders, in practice there is widespread scepticism about this kind of participation. Such a dilemma may be the outcome of a structural inability to link representatives, managers, policy-makers to the relevant debates taking place in the organization. Often these people feel that their 'briefings' focus their attention on issues for which they are not prepared. The 'residual variety' left for their attention in these briefings is beyond their information processing capacity. In these conditions policy makers may either abdicate their responsibility and just follow the apparent advice of their subordinates, or take decisions hoping for the best. They are in a no-win position. This is frequently the case. Is there any way to make the briefings fall within their information processing capabilities and therefore make more likely that they, and not those under them, will control the related policy process?

The mechanism

The policy function of an organization is discharged by those giving 'closure' to the information loops responsible for defining the organization's identity (i.e. for defining its business areas and their meaning in a particular context). Therefore giving closure to information loops of an operational kind is not, by this definition, part of the policy function.

From the viewpoint of complexity it is a fact that those producing 'identity' closure have a limited information processing capacity; and therefore it is a fact that policy making is inherently a low-variety process. Indeed, in general they are not themselves the ones carrying out the studies of policy concern, but must rely upon the briefings and reports produced for them within the organization. Most of the time policy-makers are in the invidious position of deciding on issues that are beyond their comprehension. If this is so, how can they keep control of these policy processes?

This is a typical case where the residual variety that can be seen by the relevant viewpoints is much smaller than the variety entailed by the situation of concern. In cases like this, the Law of Requisite Variety would suggest the necessity to have effective attenuators of complexity within the high-variety side in order to reduce the residual variety that the low-variety side needs to see in the situation. Ineffective attenuation would imply that the unattended residual variety is larger than the information processing capacity of the low-variety side. In this case, policy-makers would not have capacity to cope with the options and issues as offered by the organizational briefings.

Thus, it is necessary to design effective forms to attenuate the situational complexity. For policy-makers there are two main sources of complexity: these are the organization itself and the organization's environment. On the one hand the states of the organization today define the 'reality' under their control; on the other, the states of the environment define the 'reality' of the threats and opportunities that the organization has to deal with in order to remain viable in the future. Whether we are referring to economic trends, annual accounts, budgets, technological changes, personnel matters, or to any other possible form of information, all fall within one or the other of the above categories; they are either referring to anticipated changes in the environment or to operational problems in the organization (Figure 4).

Quite naturally, since policy-makers cannot access these sources by themselves, they have to rely on filters provided by the organization for these purposes. It is necessary, for instance, to have a finance department to produce the annual accounts or a research and development department in order to keep in touch with technological changes. These two structural filters are the control and intelligence functions of the organization (Figure 5).

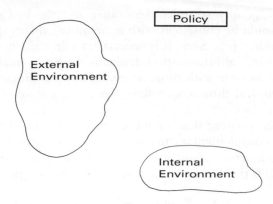

Figure 4. Sources of complexity for policy-makers

They exist in one form or another in any organization. However, they are not necessarily related to well-defined entities in the organization chart: it is perfectly possible that one department performs the two types of filtering functions; or that one person performs in his different roles both intelligence and control functions; or any other combination, including one person performing the three functions of policy, intelligence and control. The problem is how to structure each of these functions and their interactions in order to make policy-making more effective.

Firstly, to oppose technocracy (i.e. the view that policy-making has to be in the hands of those who know about the technologies necessary to carry out the organization's missions), it is necessary to minimize the information needs of policy-makers. However, if this view is taken to its

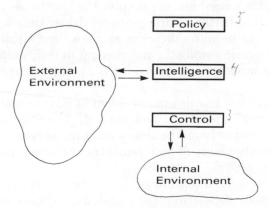

Figure 5. Two structural filters: the intelligence and control functions

extreme it could imply complete ignorance and the loss of control, so 'minimizing' should be consistent with maintaining their capacity to be in control of the policy processes. It is necessary to design a mechanism that, while making minimal information demands on policy-makers, still permits them to be in gear with those assessing opportunities and threats in the environment and those controlling the current state of affairs in the organization.

Secondly, it is apparent that the intelligence and control functions offer alternative, but complementary perspectives for the same problems; those related to the definition, adjustment and implementation of the organization's identity. It is this fact that suggests the need to design the interactions between them.

Policy-making is a process whose outcome is the choice of courses of action for the organization. The issues of policy concern may have their origin in either the policy-makers themselves or in the ideas mooted in the organization. In the former case, there is a need to substantiate these issues if they are going to be more than just ideas; this requires studying these issues from different perspectives, leading to the need to involve structural parts representing the views of both the control and intelligence functions. In the latter case, if the purpose is to reduce the demands on policy-makers, it is necessary that those inside the organization cross-examine and veto their own ideas before sending them to policy-makers.

In either case, policy-making implies the orchestration and monitoring of organizational debates in such a way as to make possible the contribution of people, to the best of their abilities, to organizational adaptation and survival. Extensive debates in the organization including different and opposing viewpoints, should produce informed conclusions and improve the quality of the policy briefings. Policy-makers should only be exposed to issues and alternatives that have been elaborated to the best of the organizational abilities. They need not get involved in the details of the issues of concern; their job is, firstly, to bring into the debate the relevant structural parts, secondly, to monitor these interactions, and finally, to consider alternatives and decide among them according to their preferences, beliefs and values. This model of policy-making is a pointer to avoid information overload.

However, there still remains the problem of how to make effective the interaction between intelligence and control. The effectiveness of their filtering depends not only on the ability and capability of each function in itself, but also on the ability of policy-makers to monitor the interaction of both functions together.

The effectiveness of these filters, from the viewpoint of the policy function, relates to their complexity and the richness of their mutual interactions. Using *reductio ad absurdum*, if the two filters were completely unconnected then, by definition, policy-makers would not only be receiving information independently from both sides, but they would be the only ones responsible for giving closure to each information loop emerging from them. This approach implies, for instance, that there is no chance to disprove, question, or refute at a level different from policy-making, that an issue of environmental concern does not make sense from the viewpoint of the internal organization, and vice versa. Policy makers would be the only communication channel between the two sets of people which in general, as we know, deal with far more complexity than they do. This would be a ludicrous situation and suggests that both sets of filters must be highly interconnected. When this is the case, most of the issues (and related information loops) emerging from each side can get closure (i.e. be cross-checked) with reference to the appreciation of the other filter about the related situations. For instance, while intelligence may suggest options to diversify the company, control may veto some of them on the grounds of operational and coordination difficulties.

Moreover, implicit in the above argument is the need, for specific policy issues, of a balanced interaction between the two filters; otherwise the performance of the policy function will suffer. For instance, if intelligence produces issues of policy relevance at a higher rate than that at which the control function can cope with, then policy-makers will receive unchecked (from the control's point of view) environmental information, for which either they find the likely internal implications or they respond without further inquiries. Either option is ineffective; while the option of policy-makers finding by themselves whether the environmental information makes sense from the viewpoint of the internal state of affairs is bound to slow down decisions, the option of not making further inquiries is bound to produce uninformed, potentially costly, decisions. As suggested above, instances of boards approving investment programmes for products with already declining markets, or of R & D people securing support for a 'beautiful' technology at the same time that the production and sales people are aware of a declining market for the related products, are not uncommon. Decisions over-influenced by one of the filters are likely to be costly and ineffective. On the other hand, if the two filters are interacting effectively, policy proposals with the above pathological characteristics are less likely to emerge. In other words, if the two filters are highly interconnected and the residual complexity that each function has 'to see' in the other is within their response capacity, then the information loops left open for policy attention—that is, the residual variety that this latter function has to deal with directly—is minimized. Such an approach is consistent with the intrinsic limited information processing capacity of policy-makers. Their role in this model is to look after interactions and give closure to issues from the point of view of their preferences, values and beliefs. They should not need to have any technical knowledge about these issues.

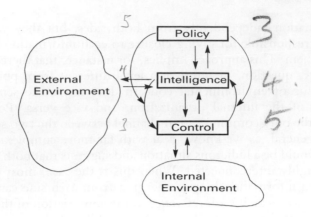

Figure 6. Policy-making: the mechanism for adaptation

It is important to understand that complexity in this case is measured not by the complexity seen from within, by each filter, in the relevant issues, but by the (already filtered) alternatives put forward by each side to the other side.

I call the above arrangement the Mechanism of Organizational Adaptation. Its diagrammatic representation can be seen in Figure 6. It is a means to effectively bridge the variety (and information) gap natural to policy-makers. This mechanism, if well designed, should minimize the residual variety to that necessary for the political discussion of the issues of concern.

An imbalance between intelligence and control in general implies lost opportunities and unnecessary costs in the long run. As for policy-makers, they are likely to find themselves feeling that their information processing capacity is inadequate to cope with the uncertainties of the situations under their attention.

The discussion of this mechanism has made apparent the need for an overall balance between intelligence and control. However, the specific design of this balance will require paying attention to all the relevant issues impinging on the organization's identity. In most cases the description or design of the mechanism will need to take into account a complex web of interactions, far removed from the deceptively simple diagram shown as Figure 6.

This model makes apparent, contrary to the widely held view that the main role of policy-makers is to take decisions, that their key role is to monitor the interactions between the control and intelligence functions in order to amplify their capacity by making as much use of the organizational resources as permitted by the circumstances. Their responsibility is to ascertain that any policy issue either brought to their attention or initiated by them has been adequately cross-examined by the two perspectives. To discharge well this responsibility they need not to be experts or even knowledgeable about the specific policy issues, but they do need a good model of how the organization structure works with reference to their vision of the organization's identity. This model should help in particular to appreciate the need for communication channels to relate the relevant people. This is also residual variety that policy-makers have to deal with.

Summing up, the following three points have been made to support an effective organizational adaptation:

(a) it is necessary to minimize the information requirements of policy-makers; and for this purpose
(b) it is necessary to design control and intelligence functions of roughly similar complexities; and
(c) it is necessary to have highly interconnected control and intelligence functions, as a means to make effective the attenuation of the situational variety.

Finally, the policy, intelligence and control functions correspond with Systems Three, Four and Five in Beer's model of the organizational structure of a viable system, and the mechanism described is no more than an alternative discussion of his Third Axiom of Management (Beer, 1979, p. 298).

Complexity unfolding

Defining the identity for an organization implies defining the primary activities of that organization. In business terms these activities are the products or services implied by the company's 'business areas'. In general they are the services or products offered by the organization to its environment. Though defining these activities is the responsibility of policy-makers, their implementation is not. What we witness is that, as an outcome of self organization, a structure evolves to make possible the implementation of primary activities. Depending on their complexity and control strategies, more or less structural levels with autonomy will emerge to make possible their final implementation. At any stage, if the complexity of a primary activity overloads those responsible for its implementation, it will tend to be broken down into several primary activities at the next lower structural level, and their management passed (devolved) to managers operating at this new managerial level. Such an unfolding of complexity (Figure 7) more often than not is the outcome of a natural, uncontrolled, process of self organization.

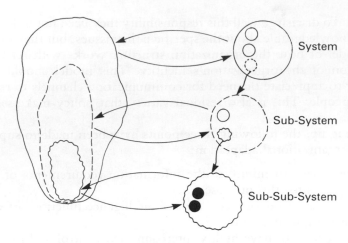

Figure 7. The unfolding of complexity; the complexity of organizational tasks makes necessary autonomous systems within autonomous systems

The modelling implied by Figure 7 is focused only in the products or services implied by the organization's identity. It is not a modelling of any of the other activities that, though necessary to make possible the final tasks of the organization, are only supporting or servicing them. For example, in a university whose identity might be to produce and transmit knowledge, the implied products are those produced by its academic staff in the classroom or in their publications. Their aggregated efforts are consolidated in primary activities like research groups, academic departments, faculties. . . . All other units—like finance, computer services, registry, library—that are not directly producing these results are support activities that facilitate the delivery of the university's 'primary activities'. Therefore, over-simplifying, the unfolding of complexity for a university would have the university itself at the first structural level, the faculties at the second level, the departments at the third level and the lecturers at the fourth level. All of them, at different levels, constitute the primary activities of the organization 'university'.

Primary activities are the 'objects' of management control; the *raison d'etre* of control. Lower-level primary activities are doing what the higher structural levels found they could not do by themselves. Each primary activity is responding by itself to chunks of environmental complexity, and moreover, in general, it is striving for its viability in the same way as the parent primary activity is doing at the more aggregated level. Indeed, if this were not the case, some primary activities of the overall task (i.e. those not responding to environmental challenges) would be endangering the overall viability of the organization.

In this framework control and autonomy are not opposites (Espejo, 1983). Viable subsystems with autonomy are necessary in order to implement the organizational tasks; in other words, it is necessary to have the 'amplification' provided by viable (autonomous) subsystems to make possible the control of the organization's outcomes. The above discussion suggests that in any viable system there is, in one form or another, a complementarity between control and autonomy. Thus the problem is to find criteria to make the most out of it. This is what the discussion of the mechanism of monitoring control should help us to see.

Mechanisms for viability: the mechanism of monitoring-control

The problem

By definition, in order to be an effective filter of the organization's internal variety—that is, in order to have a realistic appreciation of what is going on within the organization—the control function needs to be in control of the organization's primary activities. If there is something that control can contribute to the policy debate it is the accurate appreciation of the capabilities, potentialities and performance of the primary activities. In order to be realistic, such an appreciation needs a control capacity.

However, if managers understand control just as the power or authority to direct, order, or restrain the people under them, then the likelihood is that those managers will suffer 'control dilemmas'.

Two facts underlie these control dilemmas. The first is the unfolding of complexity explained above, unavoidable if the aim is to implement complex tasks. The second is the suggested poor understanding of control. While the first is responsible for imbalances in variety, for managers cannot possibly know everything that is going on within the organization, the second fact is responsible for an apparent inability to accept these imbalances.

Complexity unfolding does not mean managerial abdication of responsibility. It means that while managers cannot know everything that is going on inside the organization, they are still accountable for any loss of control. This is a hallmark of management. There is always the risk for managers to lose touch, be it only temporarily, with their primary activities. Unexpected breaks of control may happen in these periods; but even if no problems emerge immediately, they may emerge at a later stage as a result of their uninformed contributions to policy decisions.

The inherent imbalance between the low variety of management and the high variety of the primary activities they have to control, triggers all kinds of control games. These are interpersonal games where, on the one hand, senior managers control the allocation of resources, and on the other hand junior managers control the information. It is inherent to management that managers operate with an information gap. If junior managers, for whatever reasons, withhold relevant information, the likelihood is that corporate managers will lose actual control of the situation. Most of the time these games may not be the outcome of an intentional behaviour, but simply of poor interpersonal interactions. This situation can be exacerbated by a poor understanding of control processes. How can we minimize the damaging impact of these all too common situations?

The problem seems to be how to avoid losing control of primary activities despite unavoidable information gaps. In terms of requisite variety the problem is how to match, at minimum cost, the residual variety left unattended by the organization of the primary activities with the variety available to management.

The mechanism of monitoring-control gives pointers to achieve this balance effectively.

The mechanism

Complexity unfolding means that the implementation of an organization's missions will always need two or more primary activities; they define its implementation function. Each of these primary activities is autonomous, has its own management and is embedded in its own relevant environment. In Figure 8 the primary activities are called divisions A, B and C.

Depending on the nature of their tasks, the divisions will have stronger or weaker interdependences. All combinations of interactions among them may happen though, for diagrammatic reasons, only some of them are represented in Figure 8; indeed divisions A and C may also be interdependent. They may interact operationally by one providing inputs to another, or through the environment by one affecting the residual environmental variety relevant to the others. But above all they have in common that they

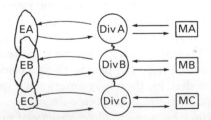

Figure 8. Control of organizational tasks

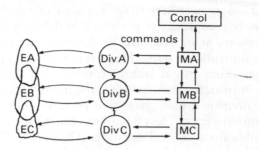

Figure 9. Control of organizational tasks: achieving cohesion by means of the control function

belong to the same organization and, for as long as there is an organization, there is also a degree of cohesion among the parts. Achieving this cohesion is the role of the control function (Figure 9).

If, however, managers in the control function understand control as only commanding or directing, then they are likely to face control problems. It is important to understand their control dilemma.

Often managers make the assumption that complexity is skewed towards higher structural levels; this is the legacy of a positivistic way of thinking where the only relevant complexity was that perceived by those in authority. On the other hand, if we accept that complexity is nothing objective but something that emerges from the interaction of people with a situation, then, naturally, the complexity of the environment is distributed (as people are), and so all primary activities at all structural levels have to operate in complex, sometimes turbulent, environments (Figure 10).

Indeed, this complexification has been prompted by increased competition and sophistication in the products and services offered by modern

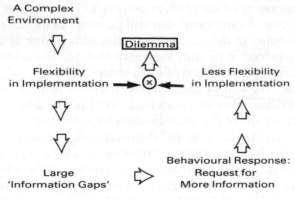

Figure 10. The control dilemma

organizations; it is now necessary to see a complexity that before was easily ignored. The more complexity is perceived in the environment, the more flexibility is necessary at all structural levels: managers have no option but to accept larger information gaps. However, if these gaps are interpreted by them, as they often are, as lack of control, and more commands are issued or more information is requested, the likely outcome is that junior managers (e.g. divisional managers) will perceive more constraints, less room for autonomous action, less flexibility. Such a managerial response is in turn responsible for a control dilemma (Figure 10). While structurally the outcome of this managerial approach is likely to be a larger bureaucracy controlling more and more 'dimensions of control', behaviourally lower-level managers may become increasingly fearful and inflexible, precisely when the need for flexibility is more acute. A proliferation of control games is likely.

Because the imbalance in the varieties of the control and implementation functions is natural, it makes no sense to try to force a balance by increasing the variety of the control function (as is implied by the behavioural response suggested above). What is necessary is to reduce as far as possible the residual variety that the control function needs to take account of in the primary activities. This strategy would permit us simultaneously to increase both the autonomy of the primary activities and the cohesion of the organization.

However, minimizing residual variety is not enough: it is also necessary to ensure that this residual variety is properly communicated to the control function. However small the variety might be, there is always the possibility of corruption in its transmission, and therefore the risk of losing control. Thus arises the need to validate the information used in transmitting such variety.

From earlier discussions it should be apparent that in order to minimize the residual variety relevant to the control function it is necessary to increase the autonomy of the primary activities. The problem is to find the maximum 'degree of autonomy' that still permits organizational cohesion. While the autonomy of the primary activities adds a huge flexibility to the organization—indeed, it permits local responses to environment demands —it also increases the likelihood of inconsistent responses. This is a natural outcome of the freedom of the primary activities in deciding responses. The options to counteract this drawback are either to achieve consistency of responses from above (i.e. coordination by direct supervision) or to let the primary activities do that by themselves (i.e. coordination by self adjustment). The former option is attractive because it permits a comprehensive view of the primary activities in the context of the whole organization. However, in practice it implies not only a potential overload of the

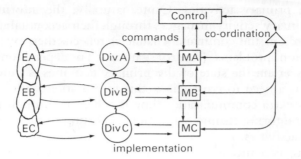

Figure 11. Control of organizational tasks: the coordination function

control function, but a strategy that is based on an undesirable increase in control variety.

The latter option, in which the parts find consistency of responses by themselves, creates the logical necessity of a powerful coordination function (Figure 11). Indeed, the contention is that better interactions between primary activities is more likely to produce consistent responses; this is a natural outcome of a context more supportive of self-regulation. Coordination systems, like those named in the accompanying table, help to damp oscillations among the primary activities, thus reducing the demands on the control function. Thus engineering damping of machine variances not only facilitates communications betwen primary activities, but also permits easier maintenance and support of such machines. More problems remain at the local level permitting a larger acceptable variety (information) gap. Coordination is in any case a very-high-variety function; the stronger it is, the smaller will be the residual variety needing the attention of the control function. Indeed, the more it is developed the more autonomy is possible at lower structural levels.

However, if management is going to be supported effectively by this 'reduced' residual variety, then it needs a capacity to recognize the true

Examples of coordination

Engineering damping of machine variances
Quality control of major raw materials
Damping systems to regulate debtors/creditors/stocks
Damping of idiosyncratic accounting methods
Work procedures
Wages damping across divisions
Production scheduling

states of the primary activities. Quite naturally, the information trans-mitted by primary activity managers through their accountability lines (the flows upwards in the command channels) reflects their own biases and communication problems (control games). To depend only on those reports to ascertain the state of the primary activities is potentially very risky; there is a need to cross-check this information with an alternative source. This extra communication line is achieved through the develop-ment of a monitoring channel with those reporting to the management of the primary activities.

Monitoring is a means to avoid breakdowns in the communications between management operating at successive structural levels in the organization. The control function needs an assurance that the autonomy of the primary activities remains consistent with global policies—that is that the residual variety transmitted by the accountability reports is an adequate reflection of the primary activities variety. There is a need to maintain the integrity of the information flowing between levels. Among other factors, changes in the context of action and the natural changes of people do increase the uncertainty about the meaning of accepted informa-tion procedures, as well as of specific information reports.

Control needs, by exception and sporadically, a high-variety under-standing of those lower-level aspects that are relevant to global cohesion (i.e. of those for which the primary activities management is accountable). Monitoring is a low-variety channel that carries high variety about a few, specific issues.

Summing up, the following three points have been made to support an effective control of primary activities:

(a) it is necessary for the control function to minimize issuing commands and directives to the primary activities;
(b) it is necessary to develop as much as possible coordination by mutual adjustment, rather than by direct supervision, among the primary activities; and
(c) the control function needs to develop a capacity to monitor the primary activities in order to minimize breakdowns in their communi-cations.

Indeed, a better developed coordination function permits the achievement of cohesion with fewer commands and more general policies. However, if cohesion is going to be achieved, an adequate monitoring of primary activities is necessary as well. Thus quite contrary to the widely held theory-in-use that the communications between two successive structural levels take place mainly through a command channel, the above conclusions suggest that there are two other channels for this purpose: the coordination and monitoring channels (Figure 12). While the coordination channel can

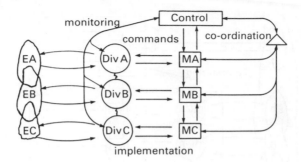

Figure 12. Control of organizational tasks: mechanism of monitoring-control

be used to induce self regulation, the monitoring channel helps to guard against communication breakdowns.

From the viewpoint of information processing, the capacity of managers carrying out the control function needs to be in balance with the actual information flowing through the three incoming channels. If the information reaching control managers is beyond their capacity, one of the options is to design a stronger coordination function in order to induce autonomy, reduce residual variety, and thus reduce the amount of information aiming for their attention. Provided it is backed by a parallel strengthening of monitoring, this approach permits a larger information gap without a loss of contact with implementation activities.

The links provided by the three channels among the control, coordination and implementation functions define the mechanism of monitoring-control (Figure 12). These three functions correspond to Systems Three, Two and One of Beer's model of the organizational structure of a viable system.

Model of the organization structure of any viable system

The discussion of the above three sections led us to the 'discovery' of what we may call Beer's regulatory mechanisms. These mechanisms are extremely general. For instance, the mechanism of monitoring-control is applicable to any organizational entity with structural levels, whether or not it is a primary activity. However, the focus of the discussion has been on primary activities and their capacity to remain viable. Together, the adaptation and monitoring-control mechanisms define the set of functions and relations necessary for effective viability. According to our earlier definition of structure, these are the mechanisms defining the organization structure of a viable system (Figure 13).

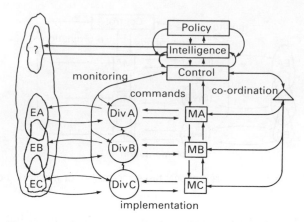

Figure 13. The two mechanisms defining the organization structure of a viable system

However, perhaps the most powerful insight into the management of complexity is made apparent when the mechanisms of adaptation and monitoring-control are related to complexity unfolding. Since, in general all primary activities, at all structural levels, have the same problems of viability, the same two mechanisms are valid for all of them. The same model of the organization structure at the global level is repeated in each of the primary activities at all structural levels; this is Beer's Principle of Structural Recursion (Beer, 1979, p. 118). Figure 14 permits us to appreciate the idea of structural recursion.

Then, the same criteria of effectiveness as discussed above apply to the subsystems within a viable system, and to the sub-subsystems within a subsystem, and so forth. This proposition has deep consequences. It is saying that in truly effective organizations, policy, intelligence, control, coordination and implementation are distributed at all levels. Autonomy should exist at all levels. Contrary to the established knowledge all structural levels, from the higher to the lower, should be concerned with the short, middle and long term. This proposition makes it apparent that reducing people working at lower structural levels to the status of operators alone not only reduces unnecessarily their individual freedom, but also reduces the effectiveness of the organization as a whole.

Conclusion

The Viable System Model offers a paradigm for problem-solving. Its understanding offers a mental tool to approach the creation and design of effective contexts for the participation of people in human activities.

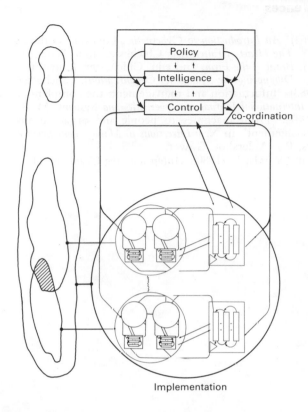

Figure 14. The principle of recursion (adapted from Beer, 1985)

The mechanisms of adaptation and monitoring-control are particular instances of mechanisms that emerge when there is a need to study the interactions between high- and low-variety parts. In either case the problem was to find effective means to attenuate the variety of the high-variety side, in order to reduce the residual variety relevant to the low-variety side. The lower this residual variety, the more feasible it is for the low-variety side to control the situation. Beer's model gave us the direction to discuss these mechanisms. In fact, as suggested in the introduction to this chapter, this chapter works out in detail, from the viewpoint of the Law of Requisite Variety, the mechanisms invoked by Beer when he develops his model. The most powerful idea behind the discussions of this chapter is the concept of residual variety, which highlights the important heuristic value of Ashby's Law.

References

Ashby, R. (1964). *An Introduction to Cybernetics*. London: Methuen.
Beer, S. (1979). *The Heart of Enterprise*. Chichester: John Wiley.
Beer, S. (1981). *Brain of the Firm*, 2nd edn. Chichester: John Wiley.
Beer, S. (1985). *Diagnosing the System for Organizations*. Chichester: John Wiley.
Espejo, R. (1983). 'Information and management: the complementarity control–autonomy', *International Journal of Cybernetics and Systems*, **14**, 85–102.
Espejo, R. (1987). 'From machines to people and organisations: a cybernetic insight of management', in *New Directions in Management Science* (eds. Jackson, M. and Keys, P.). Aldershot: Gower.
Maturana, H. and Varela, F. (1980). *Autopoiesis and Cognition*. Dordrecht: Reidel.

Part Two
Applications of the VSM

The Viable System Model: Interpretations and Applications of Stafford Beer's VSM
Edited by R. Espejo and R. Harnden
(c) 1989 John Wiley & Sons Ltd

5

P.M. Manufacturers: the VSM as a diagnostic tool

Raul Espejo

Aston Business School

This chapter illustrates the use of the Viable System Model as a diagnostic tool. The emphasis is on the practical use of the model rather than on the complications of the process of intervention itself. The chapter illustrates how to define levels of recursion and how to study the mechanisms of control and adaptation in a business organization. At the end it offers a set of 'diagnostic points'.

Introduction

The purpose of this chapter is to illustrate the use of the Viable System Model (VSM) as a diagnostic tool. The model is applied to a small British company as it was observed in 1978 (Espejo, 1980). The emphasis of the discussion is upon general problems like 'How apt is the organization in reflecting and deciding about its policy?' or 'How likely is it that people in the organization will discover imaginative answers to cope with environmental threats and opportunities?' or 'How likely is it that management will keep the organizational activities under control?' Indeed, these are effectiveness questions and the contention of this chapter is that organizations with 'good cybernetics' are more likely to score higher in answering them.

Good cybernetics suggests effective mechanisms both to discover problem situations and to regulate relevant organizational tasks. A detailed study of these mechanisms and related criteria of effectiveness is discussed in Chapter 4.

103

The first part of this chapter describes P.M. Manufacturers, the subject of this study, and the problems facing the company in 1978. Then the question is raised about the company's ability to discover solutions to its apparent problems by 'itself'. Studying the cybernetics of its organization makes apparent several weaknesses directly impinging upon the company's ability to formulate effective policies. This study is at the core of the chapter. Finally some conclusions are drawn for the company.

The company: P.M. Manufacturers

P.M. is a small company in the electrical engineering sector. Its main business is the manufacturing, or more accurately, assembly of engine-driven electrical generating sets. However, the company has also been, for the past two years (i.e. since 1976), in the business of 'procuring spares' for third parties and 'servicing generators' on-site. These two activities are referred to in the company as 'non-manufacturing activities'.

The company has around 40 employees and an estimated sales turnover of £2 million for the current financial year (i.e. 1978). It is part of a larger engineering group with two more operating companies, one related to land development and the other to civil engineering. The operating companies enjoy a fair amount of discretion. The group's corporate structure is formed by a chief executive, supported by a finance department and a

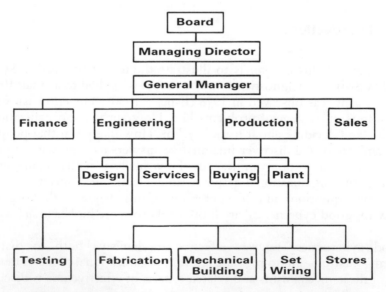

Figure 1. The organization chart for P.M. Manufacturers in 1978

secretariat with planning and administrative responsibilities. Altogether the group has 600 employees.

P.M. Manufacturers sells standard and non-standard generating sets to order. It is not its policy to stock products. The market is mainly abroad, in particular Nigeria and the Middle East. The company's organization chart is shown in Figure 1.

The production department

The 'resources manager' is responsible for the production department. He controls the 'buying manager', responsible for all procurement, except engines, and the 'senior foreman', responsible for the day-to-day running of the plant.

It is in the plant that all the manufacturing activities take place. A flowchart of these activities is presented in Figure 2. Manufacturing are chiefly responsible for the assembly of the generators. At present only four of the five operations implied by the flowchart are active, these being:

fabrication: manufacturing bed frames for generators
building: mechanical assembly of engines and alternators (the two major components of a generator)
set wiring: installing the electrical system for the generator (this includes an electronic control panel)

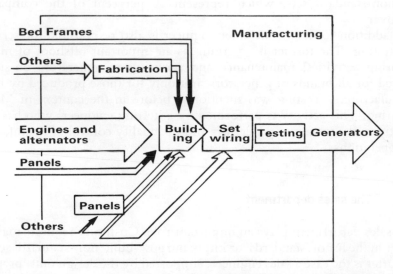

Figure 2. Manufacturing flowchart

testing: finishing and testing generators under several loading conditions.

Although the factory has capacity to produce control panels, their production is at present subcontracted.

In the formal organization structure the testing unit, manned by the 'testing engineer', is under the control of the 'chief engineer'. Each of the other activities is under the control of a 'leading hand'; the three leading hands formally report to the senior foreman.

The senior foreman also supervizes the 'stores manager'. The store is physically in the shop-floor itself.

The engineering department

Most of production is standard sets, but the company also offers non-standard sets on request. Over time production is becoming more and more restricted to standards. In this trend the design unit of the engineering department has a major role.

Historically the company has evolved from producing a large variety of models in small quantities to larger quantities of a few standard sets. Today standard sets are responsible for 70 per cent of the company's turnover. This has simplified not only manufacturing but also sales and financial activities. For instance, the production of standard sets permits a 'price list' that simplifies the interactions with customers. Quotations are done only for non-standard sets, which represent 20 per cent of the company's turnover.

In addition to design the chief engineer is also responsible for services and testing. The former is emerging as an important offshoot of manufacturing activities; maintenance and servicing of generators on-site is offered for all brands of generators, not only for those produced by P.M. Manufacturers. Testing was mentioned before in the context of Manufacturing. This activity is performed by a 'testing engineer' who has two functions: the testing of generators and the quality control of manufacturing operations.

The sales department

The sales department plays an important role. Customer requests may fall either in the list of 'standards' or imply purpose-built 'non-standard' sets. If the latter is the case a sales engineer, supported by the design unit, prepares a quotation. This activity consumes most of the sales engineer's time, but only one in ten of the quotations is likely to be successful. Manufacturing orders are accepted by the sales manager after consultations with the general manager and managing director. The resources manager, in fact, is not consulted; he receives manufacturing orders in a way that practically implies that production scheduling is done by the 'sales manager'.

The company also sell its procurement capabilities, this mainly to foreign customers. One of the sales engineers is devoted full time to coordinating procurement, storing and despatching of spares and parts, but he also oversees the 'services' on-site performed by the service engineers.

The finance department

The finance department discharges both accounting and administrative duties. It is headed by a 'chief accountant'. There is limited financial discretion at this level since the group controls this business dimension. Cost accounting of manufacturing orders and company accounts are the main outputs of this department. Cost analysis is done for each manufacturing order, both periodically and after completion. There is no cost analysis for product lines, so company finance reports are fairly detailed in nature. Wages, salaries, overtime and other administrative duties are also the responsibility of this department.

Corporate management

The heads of engineering, sales, finance and production report to the 'general manager'; and all of them constitute the 'management team' of P.M. Manufacturers. The general manager is responsible for the day-to-day running of the company. Among his responsibilities is the acceptance of manufacturing orders and the procurement of engines for generators.

The 'managing director' is the main shareholder of the group. Since P.M. is the most recent acquisition he wants to keep his fingers close to the operations. However, in practice the group consumes most of his time. He is also the chairman of the board.

The company's problems

The managing director does not make a secret of the fact that the group is prepared to give an umbrella to P.M. only as long as it remains financially viable. At the first turn in luck it will go.

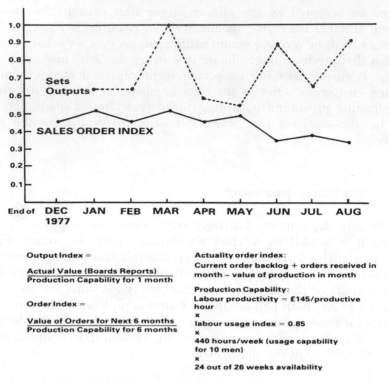

Figure 3. Indices of manufacturing output and sales

The company has been marginally profitable since it was absorbed by the group three years ago. However, there are several warning signs of difficulties in the near future. The company performance in the past year or so is worrying. Indices of production, sales and finance suggest that the situation is not good (see Figures 3 and 4).

Figure 3 gives information about manufacturing and sales performance. Even using fairly conservative labour productivity figures it is possible to appreciate important slack on the shop-floor; in some months more than 40 per cent of the available productive capacity was not used in manufacturing. The sales order index implies that at any given time the firm orders for the next six months do not exceed 50 per cent of the production capability for that period, and that in the last three months the situation has deteriorated even further, so that the order book covers only 40 per cent of the production capability for the next six months. The company is finding it difficult to have its products accepted in the market. An enquiry on 'sales' makes even more apparent the company's fragility. P.M. is extremely dependent on the orders of only one foreign customer. Around 60 per cent

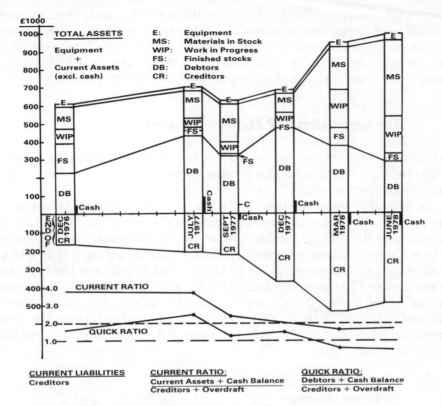

Figure 4. Liquidity indices

of the company's turnover comes from a major distributor in Nigeria. Everyone is conscious that he is holding the company's future in his hands. Just three months ago the Nigerian Government decided on new import restrictions. The distributor had to readjust his order programme with drastic effects in the company. All efforts to increase sales in the UK market have failed so far, suggesting that P.M.'s competitive position is not good enough.

These problems are compounded by a worsening liquidity situation. This can be appreciated from Figure 4. In an effort to keep production costs down the company accepted deals which have increased both the money they owe to creditors and the materials in stock and work in progress. At the same time, to improve sales P.M. is giving more generous credit. Not surprisingly the company has an acute cash problem. However, this recent problem appears only to confirm the trend of the last eighteen months.

On a more positive note, for the past two years P.M. has been selling its procurement expertise to foreign customers, making high profits. However, while the perceived significance of this activity is high, the company has failed to increase its contribution to the sales turnover: it is still under 10 per cent. Additionally, in an effort to reduce costs and achieve a competitive position in the UK markets, P.M. is trying new technology, namely the incorporation of microprocessors to control panels.

The cybernetics of P.M. Manufacturers

Are the above problems the outcome of hard luck or perhaps the natural outcome of an organization that permits nothing better? If so, how can the situation be improved?

An answer to these questions is the aim of a study of the cybernetics of the organization. It has already been said that the essence of this analysis is to elucidate the 'actual' mechanisms regulating organizational activities. The study will be done in two stages. First, I shall discuss the way in which autonomy and discretion appear to be allocated at different levels in the organization. The outcome is a model of the necessary changes in structural recursion for more effective organizational responses to environmental demands. Second, I shall study the effectiveness of regulation for two structural levels: that is, the divisions within P.M. Manufacturers, and the company itself.

Structural recursion

Figure 5 postulates the structural levels which appear necessary to implement the group's policies.

While in the context of the group, P.M. is clearly one of three implementation subsystems, with responsibility over a particular policy area, within P.M. the situation is less clear. Only manufacturing activities respond to the image that the company has of itself. Non-manufacturing activities have no identity of their own: they are perceived as byproducts of manufacturing. However, as explained below, both are business areas which imply complex activities that want managerial autonomy.

Company's structural level

The view that each of these two business areas should have structural recognition, in the form of autonomous divisions or subsystems, is a cybernetic conclusion that is not apparent in the company's formal organization structure. While manufacturing activities are clearly the responsibility of the resources manager, the management of 'non-manufacturing' is

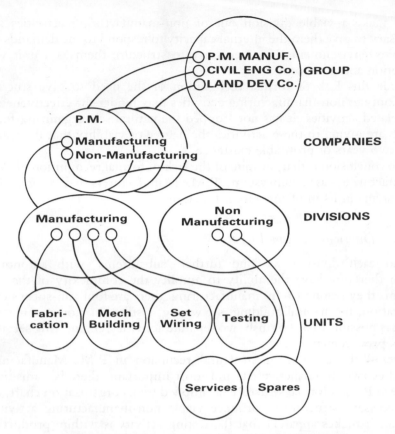

Figure 5. A structural recursion for P.M. Manufacturers

disseminated in several organizational parts: services, buying, sales, stores. There is no structural recognition of the contribution of non-manufacturing activities to the company's viability. Yet, these are important current activities of the company.

This assertion arises from the fact that services and procurement are activities that the company wants and needs to make viable. Their 10 per cent contribution to sales turnover disguises the fact that their contribution to the company's profits is much higher. While the production department (i.e. manufacturing) adds little to the value of the final product, and has negligible control over its price (because of the company's uncompetitive position), the value added by non-manufacturing activities is larger, mainly because the company controls their prices. In the end, their contribution to the company's profits is much greater than that suggested by their small size.

To make a viable division out of non-manufacturing activities it is necessary to give them the internal capacity to respond to the demands of a complex environment. It is necessary to structure them as a unit with discretion and autonomy.

While this lack of formal recognition of the need to give stuctural autonomy to non-manufacturing activities may hinder the effectiveness of the related activities, it has not stopped the natural self-organizing forces which are giving to these activities the degree of viability that they show after two years of profitable existence.

The conclusion is that, in spite of the lack of formal recognition of 'non-manufacturing' as a subsystem, P.M. Manufacturers has *de facto* two subsystems in its implementation function.

Divisions' structural level

Within each division we find further 'subsystems' with autonomy, which therefore have the ability to amplify the complexity of the sub-systems they belong to. In 'manufacturing' there are four sub-subsystems: fabrication, mechanical building, set wiring, testing. In 'non-manufacturing' it is possible to distinguish two separate activities: engineering services, spares procurement.

Overall the analysis of structural recursion in P.M. Manufacturers identifies two anomalies. First, and most important, there is a mismatch between P.M.'s formal structure, as implied by its organization chart, and the necessary structure to produce viable non-manufacturing activities. Second, it makes apparent that the testing activity is within 'production' and not within 'engineering'. These mismatches are further explored below while discussing regulatory problems in the company.

Discretion at each structural level

Each unit, at its own structural level, is in interaction with a relevant complex environment. By definition these environments are encapsulated in the environment relevant to the higher structural level. Hence, the group's environment encapsulates the environments relevant to the three operating companies. Particular to the group is an overall financial discretion; in particular the group controls investment decisions. While P.M. corporate management retains discretion in sales, engineering, costs and the procurement of engines, the divisions have discretion in product development and procurement. Indeed, technological and operational discretion are seen as necessary to make possible the activities on the shop-floor.

Effectiveness of regulation

I shall study the effectiveness of regulation with reference to Figure 6. This figure focuses attention on the mechanisms of monitoring control and adaptation for both P.M. Manufacturers and each of the two divisions, manufacturing and non-manufacturing. I shall start the analysis in the divisions and then discuss in general the company.

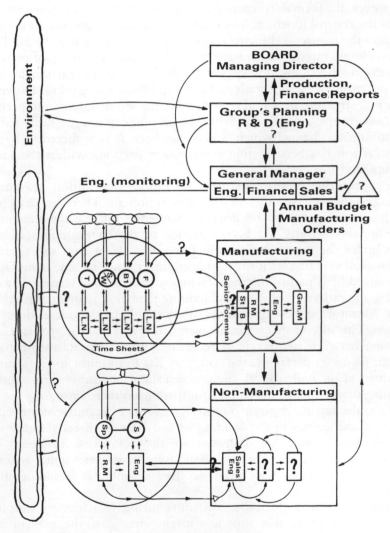

Figure 6. The postulated VSM for the company

In the divisions: manufacturing

The resources manager is responsible for the control of manufacturing activities. In this task he is supported by the senior foreman and the buying and stores managers. Control in this case means to produce good-quality generators, on time. (Note that cost is not a concern at this level of management. Even if costs are comparatively high, still production could be under control!)

However, the formal structure creates problems in the effective performance of the control function. The conventional organization chart does not recognize the nature of the interactions between buying and stores. These are two closely interdependent activities. The effective supply of materials and parts to the shop-floor suggests a high-variety interaction between those procuring these materials and parts and those using them. However, the formal structure in manufacturing puts the senior foreman in between these two groups, a position in which he becomes an unnecessarily narrow communication channel which 'stores' has been *de facto* forced to bypass. This situation has been fueling unnecessary frictions within the manufacturing division.

The monitoring of manufacturing operations (fabrication, mechanical building, set wiring, testing) is fairly hazy at present. There is a perception that this auditing has to be done by someone outside manufacturing. Therefore this activity has been allocated to the testing engineer, who works under the control of the chief engineer. However, in cybernetic terms we can say, first, that the testing engineer is not outside manufacturing—indeed he is part of the manufacturing process; and second that he is auditing operations within manufacturing and not manufacturing as a whole. Monitoring these operations surely should be the role of the senior foreman. This obviously creates problems. It seems that the problem is the confusion between two levels of recursion. While manufacturing has to monitor its own operations, the company has to monitor manufacturing. Therefore, at the higher level, the concern should be subsystems, and not sub-subsystems; this seems to be an intrusion on their autonomy.

In fact, the testing engineer reports problems in manufacturing operations (e.g. difficulties in the welding of bed frames) directly to the chief engineer before they are reported to the resources manager. This has created ill-feeling between these two managers: quite naturally the resources manager feels that the chief engineer is intruding in his territory.

Despite the above difficulties, manufacturing is achieving output as necessary. However, this situation might change if the present slack resources are removed by a tighter production programme. The analysis of regulation in the division says nothing about the control of 'production costs'; this is not an aspect under the discretion of the resources manager.

The discussion of the manufacturing division's adaptation to changing environmental conditions is limited to those dimensions that are not at the discretion of higher structural levels. Indeed, this level is unlikely to question whether it is operating in the right markets, or whether it is producing the right products. Quite naturally, it focuses its adaptation concerns on changes in technology, in production procedures and in procurement. In fact it is possible to appreciate an ongoing interaction between manufacturing and engineering, with useful outcomes. New types of generators are constantly under development, and efforts to introduce a new technology (a microprocessor in the generator control panel) are in progress. In this form engineering is providing 'intelligence' capacity to manufacturing. The management of the interactions between these two departments is a continuous concern of the general manager, who is therefore performing the policy function at this structural level.

In the divisions: non-manufacturing

This subsystem is not formally recognized by the present structure. In other words, there is no unifying managerial capacity related to this task. As suggested before, this may imply the non-viability of the task in the long-run. In Figure 6 we can see that while services and spares, the two subsystems of non-manufacturing, have their instances of control in engineering and production, their overall coordination is in sales. This coordination occupies one of the sales engineers full time. He does all the administration and is not actively in sales. Moreover, and most importantly, he is not operating within a policy framework, something which would be necessary to give him discretion to exercise control. The fact that monitoring control in this subsystem is not performed effectively is reflected by the number of operational problems that regularly require the attention of senior managers. Indeed, the sales engineer cannot take freely operational decisions: he feels the need to involve senior managers in these problems. The result is that senior managers are overloaded by trivial operational problems.

The adaptation of this subsystem to market threats and opportunities may be related to the hazy concern of the company's senior management to developing these activities. This is indeed limiting its viable expansion.

The lack of a viable organization for non-manufacturing suggests that corporate management is absorbing a good deal of its implied complexity. In fact they are overloaded with the details of an activity which, important as it is, represents no more than 10 per cent of the company's turnover.

In the company

Not surprisingly the weaknesses of the two subsystems have implications in the overall organizational structure of the company. Figure 7 is perhaps a more accurate description of the way P.M. absorbs the complexity of its activities.

Corporate management in P.M. is mainly responsible for selling the company's products at a financially viable price. While discharging this task the main parameters for production are set. In this, the general managers operate with the support of the sales, finance and engineering departments. In particular, sales performance defines the level of activity in each of the divisions. However, this performance is not independent of production costs, a dimension under corporate control. In fact most of the references used in the company to assess performance are in money terms: particular attention is paid to sales turnover targets and gross margins in manufacturing and non-manufacturing activities. These references are the only explicit recognition of non-manufacturing as an activity in its own right.

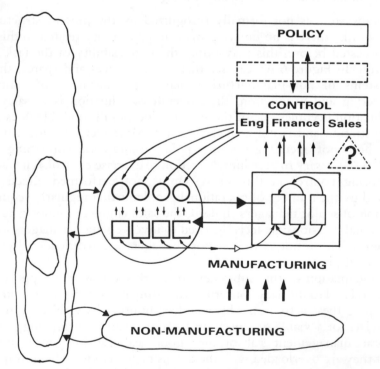

Figure 7. How P.M. Manufacturers absorbs the complexity of its activities

In practice the control of subsystemic activities is hindered by this lack of recognition of the non-manufacturing subsystem. On the one hand there is no effective organizational sponge with discretion to absorb the complexity of the multiple non-manufacturing transactions, thus increasing unnecessarily the demands upon corporate management. On the other hand, as explained below, this defect limits the opportunities for self-regulation in the implementation function and blurs the monitoring of manufacturing activities.

In the context of P.M. Manufacturers, there is a risk of confusing the need for coordination between the engineering, sales and finance departments, and the coordination between the two subsystemic operations, manufacturing and non-manufacturing. In systemic terms there is a difference between these two coordinations. While interdepartmental coordination is a mechanism to increase or amplify the variety of the control function, the self-regulation of subsystems is a mechanism to decrease (filter) the variety reaching that function. Because the variety of the implementation function is far larger than that of the control function, the latter mechanism has a far more relevant systemic meaning.

Because the organizational structure of P.M. Manufacturers fails to recognize non-manufacturing as a subsystem, the company does not benefit from this damping mechanism, and this is another way of understanding why senior managers operate at a fairly detailed level—something that is limiting their capacity to deal with more strategic issues, thus reducing their effectiveness.

Monitoring the implementation function is necessary for effective regulation, and quite rightly there is a perception of this need in the company. However, it is perhaps because there is no recognition of two operational subsystems that the engineering department confuses two levels of recursion and monitors at the wrong level. This point was explained before and the suggestion was made that the formal position of the testing engineers creates frictions and misunderstanding. Monitoring the 'right' level would imply, for instance, the assessment over time of whether the level of manufacturing activities was well matched to the factory capabilities, or whether the resources allocated to the non-manufacturing operations were reasonable for the magnitude of the task. Neither of these assessments is performed by the control function.

Overall, the way in which control is carried out appears to be fairly ineffective, and no doubt this is an important contributor to the present level of uneasiness in the company. This situation is made worse by the lack of an effective mechanism for adaptation.

To say that there is a need for functional capacity to 'create' the company's future is just to recognize the need for a mechanism to make less painful its learning and adaptation processes. Unfortunately, P.M. Manufacturers, at its corporate level, appears not to be aware of this.

Planning the long term is indeed a very limited activity in this company. Its corporate plan is produced once a year by a planning team in the group's secretariat, with a limited participation of P.M.'s senior management. In this scenario what is seen as the most important payoff of planning—that is the process itself—happens outside the boundaries of the company (i.e. in the group). Planning to be effective has to be done continuously by the whole organization and not only by a few of its members, let alone by people outside its boundaries.

In the end, the balance between the intelligence and control functions necessary for effective adaptation is strongly dominated by the latter function. Supporting this view is the information normally prepared for board meetings: by and large, this is information concerning manufacturing orders, production problems and financial details.

Thus the policy function is plunged into operational details, far from the normative role that is suggested by our cybernetic model. This is a typical case in which policy-making is hindered by a structural bias towards the control function.

This weakness in the policy function is reflected by lack of a well-defined, and insightful, identity for P.M. Manufacturers. Simply put, discussions about the company's identity are not perceived as relevant. The directors, as a corporate team, have not recognized that, whether they like it or not, the company's identity has been changing over the past two years from the original manufacturing identity towards a 'service oriented' identity.

Conclusions

Studying the cybernetics of P.M. Manufacturers has permitted an appreciation of the company's weaknesses. While no attempt has been made to find specific policies to overcome the present fall in sales and liquidity, the point is that the structure of the company does not help people in the organization itself to find solutions for their problems.

One of the most important conclusions of this study is, perhaps, the effect that P.M.'s weak identity is having on its ability to respond to unexpected changes in the market. While there have been some efforts to make the manufacturing division viable, albeit with very limited resources, it is not possible to recognize parallel efforts at the corporate level. The development of non-manufacturing activities—the result of a free opportunity in the market—unfortunately has not been well structured. In general, there is not sufficient organizational capacity to develop the several good ideas mooted. All this accounts for the present fragility of the company which is extremely dependent on the demand of only one customer. In cybernetic terms this situation could be explained both by the lack of adaptive and control capabilities at the corporate level, both problems, very likely, being the consequence of the company's hazy identity.

Apparently corporate managers have not absorbed the implications of a changed environment and the structure is lagging behind events.

A number of diagnostic points were produced by the cybernetic analysis of the company. These points are summarized in what follows from the perspective of the manufacturing and non-manufacturing subsystems and of the company as a whole.

From the viewpoint of the manufacturing division:

(a) The testing engineer is systemically within manufacturing, and not within engineering as suggested by the organization chart.
(b) Monitoring of the manufacturing primary activities is done by the corporate control function and not by its own control function.
(c) The buying and stores units are linked by a too narrow communication channel (the senior foreman).
(d) Despite the above problems, manufacturing is well under control. It is producing good-quality generators, on time.

From the viewpoint of the non-manufacturing division:

(e) Non-manufacturing activities are not recognized as a 'business area'. This is limiting its viable expansion.
(f) There are signs that a 'division' is emerging as an outcome of self-organizing forces.
(g) However, this is hindered by the fact that the sales engineer responsible for coordinating these activities is operating without a policy framework.

From the viewpoint of the company as a whole (i.e. at the corporate level):

(h) The lack of a viable organization structure for non-manufacturing implies overloading managers with trivia.
(i) The inadequate monitoring of manufacturing implies overloading the chief engineer with unnecessary detail.
(j) The non-recognition of non-manufacturing as a viable system hinders the development of effective coordination and monitoring mechanisms.
(k) There is no 'intelligence' capacity within this level. This is limiting the viable development of the whole company.
(l) The company has a weak identity.

Summing up, in terms of the organization chart, the study supports the need, first, to move the testing engineer from the engineering department to the production department; second, to have buying and stores operating one within the other, or at least at the same structural level; and third, most importantly, to hive off non-manufacturing activities in one subsystem with discretion and autonomy. A decision to do this should have three implications.

First, it should improve the control mechanisms at two levels of recursion. In particular, it should clarify the monitoring of manufacturing activities.

Second, the creation of a non-manufacturing division should not only permit an improvement in the viability of the related activities; it should also have spinoffs that help to improve the effectiveness of the company's control and coordination functions.

Third, corporate managers should find that they have more time and opportunities to discover problems and carry out debates about the development and adaptation of the company to its market. Indeed, this is the suggested approach to achieve a balance between the control and intelligence functions at this level.

Reference

Espejo, R. (1980). 'Information and management: the cybernetics of a small company', in *The Information Systems Environment* (eds. Land, F., Lucas, H., Mumford, E. and Supper, C.). Amsterdam: North Holland.

The Viable System Model: Interpretations and Applications of Stafford Beer's VSM
Edited by R. Espejo and R. Harnden
© 1989 John Wiley & Sons Ltd

6

The organization of a Fortress Factory

R.A. Foss
Independent research worker

This chapter describes an application of the VSM to a very old but successful system which may, nevertheless, be unfamiliar to management theorists. The subject is a complex social system which is ultimately dedicated to making and maintaining itself in a true autopoietic sense. The performance of the system is described in detail.

We would surely be deluding ourselves if we were to imagine that we could obtain an adequate understanding of organization from studies of human systems alone. We would be examining the twigs of a tree, with no understanding of the roots, trunk and branches which give rise to them. But it would not just be the ignorance of preceding and underlying systems that would prevent us from obtaining a higher-level understanding of organization. We would still not understand viability. Natural systems have undergone millions of years of 'experimentation in organizational design' and have proved their viability, whereas artificial systems may have been insufficiently tested by selective forces. It is probable, therefore, that artificial systems contain 'design faults' which may become selectively disadvantageous as the complexity of their environment increases.

Our search for a higher-level understanding of organization would also require us to study a diverse range of viable systems, so that any organizational invariances which might be found would be of general validity to all viable systems. If this range included an increasing number of levels of recursion we might also learn whether new whole-system properties emerged at the higher levels. Comparing and contrasting the organizational topologies of diverse viable systems in this way should provide insights helpful to the construction of artificial systems having all the effective organizational characteristics of natural viable systems.

Such an approach would have been unthinkable twenty years ago, because neither language nor models existed that were capable of discussing or representing such diverse systems in terms of their regulatory subsystems. The Viable System Model (Beer, 1972, 1979) was the first to offer such a prospect because it was written in a metalanguage powerful enough to encompass the diversity. The following pages describe an application of the model to a very old but successful system which will nevertheless be unfamiliar to most management theorists. In this very unfamiliarity, however, may lie an opportunity to escape from the confines of conventional thinking.

The viable system of our concern can be best described as a Fortress Factory. This is an apt description for its walls are so fiercely defended that many outsiders dare not approach it and yet, once inside, the observer will find a level of industry and hard work unequalled anywhere on Earth. It is a complex social system (Butler, 1974; Free, 1977; Wilson, 1971) which is ultimately dedicated to making and maintaining itself in a true autopoietic sense (Maturana and Varela, 1975). The organization required to do this is really quite remarkable and has fascinated the observer community for centuries (Butler, 1609).

Figure 1 shows a VSM mapping of the three outermost levels of recursion (Beer, 1979). Recursion 3 is not strictly speaking a viable system, but a Management Development Centre containing several thousand junior managers who are initially dependent on the rest of the system for all their needs (Butler, 1974; Free, 1977; Michener, 1974). The fact that nearly all managers are 'home grown' and not imported from outside in autopoietic systems of this sort means that the present Recursion 1 and Recursion 2 metasystems have considerable power over the determination of future Recursion 1 and Recursion 2 functions and their variety balances. The first step in determination is the differential allocation of sets of programmes (Butler, 1974; Free, 1977; Koeniger, 1970; Michener, 1974) to young managers by Systems 3 and 4 of Recursion 1, within certain System 5 constraints (Free, 1977; Winston, Orley and Gard, 1983). Subsequently Recursion 2 managers have the power to select and develop those junior managers who have the required programme content, by a process of differential resource allocation (Beetsma, 1979; Free, 1977; Rembold, 1969). By these means, the requisite balance of future Recursion 1 and Recursion 2 managerial functions is determined within Recursion 3.

We see here an interesting division of power between the recursions, with Recursion 1 firstly allocating the programmes inter-recursively and Recursion 2 subsequently allocating the resources intra-recursively. At first sight it might seem that Recursion 1 is all-powerful because it plays first, leaving Recursion 2 seemingly no option but to provide resources accordingly (Dawkins, 1976). This is actually not really the case, for Recursion 2 can construct the Management Development Centre in such a way as to accommodate only certain types of junior manager (Butler, 1974; Fell, 1979; Free, 1977) and the construction is carried out before the programmes are allocated (Koeniger, 1970). Also, Recursion 2 does have the final say in the matter, because it allocates resources selectively only to junior managers who have the required programme content (Dawkins, 1976; Free, 1977; Trivers and Hare, 1976). If a certain type of manager is not required by the Fortress Factory in the medium term, then not only will resources not be provided for them, but in times of great scarcity previously allocated resources may actually be withdrawn and recycled into the required areas (Free and Williams, 1975; Free, 1977; Meyer, 1982). In order to be able to make decisions of this importance, Recursion 2 managers must in some way be aware of the requirements of the whole system. It is this awareness, as we shall see later, that makes closure for the whole system possible.

Recursion 2 is in some ways rather unusual, as Figure 1 suggests. The managerial domain consists of 15–60,000 middle managers, some of whom run the internal milieu and are called House Managers, and others who deal with the outside world and are called Field Managers. They form System 3 and System 4 respectively, at this level of recursion. They are a very cohesive group, in spite of their great numbers, and this is no doubt due to the fact that they share a very high proportion of both their genetic and cultural information content (Getz and Smith, 1983; Hamilton, 1964; Wenner, 1974). It seems that in viable systems, cohesiveness is directly related to the proportion of shared information (Beer, 1979; Hamilton, 1964).

Turning first to Recursion 2 System 3, the House Managers, careful observations have shown that they generate exactly the same amount of variety horizontally while working within operations as they do vertically while working between operations (Lindauer, 1953; Seeley, 1982). A balance between independence and cohesion within System 1 was of course a major prediction of the VSM (Beer, 1979). The important consequence of this is that very little vertical variety escapes from Recursion 1 System 1, and so Recursion 1 System 3 has a relatively easy task in balancing it. This is partly the reason why a single Chief Executive, without help from any other directors, is able to control the activity of up to 60,000 middle managers! It is a superb example of how balancing variety in System 1 can be used to reduce metasystem size.

Examination of Figure 1 shows that the operations and environments of Recursion 2 are actually nested one within another. The reason for this most unusual state of affairs lies within System 2, which schedules the

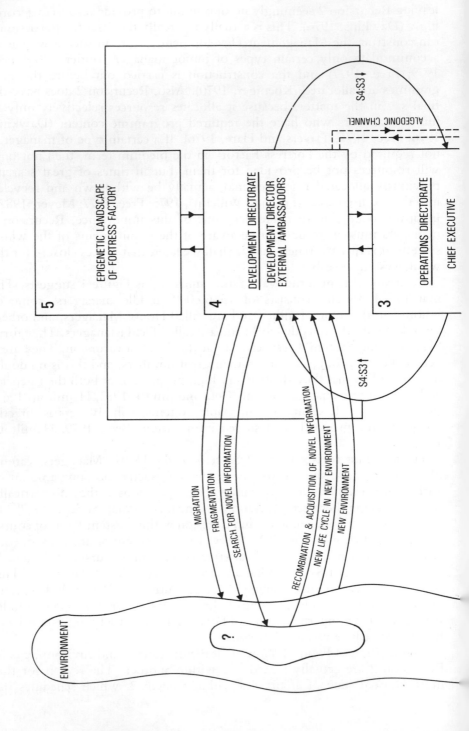

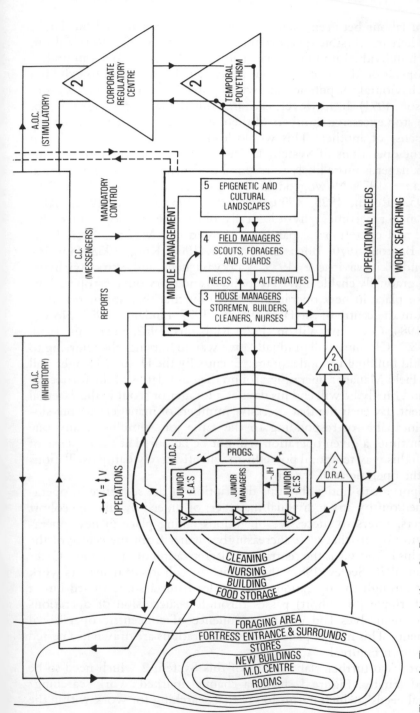

Figure 1. The organization of a Fortress Factory. ↕ *V*, horizontal variety; ↕ *V*, vertical variety; M.D.C., Management Development Centre; E.A.'s, External Ambassadors; C.E.'s, Chief Executives; PROGS, programmes; C, specific types of accommodation; —JH, insufficient information for full development of potential; D.R.A., differential resource allocation; C.D., centrifugal discretization; C.C., command channel; A.O.C., anti-oscillation channel; O.A.C., operational audit channel; S4:S3 ↑, the ratio of System 4 to System 3 influences increases; S4:S3 ↓, the ratio of System 4 to System 3 influences decreases

division of labour between them. In contrast to some related but larger systems where irreversible specialization is the norm (Sands, 1981; Wilson, 1971), each individual middle manager in the Fortress Factory can undertake any operation that is carried out in System 1. He is said to exhibit a type of behavioural totipotency known as polyethism (Oster and Wilson, 1978; Wilson, 1971). If, however, each individual manager can do any job, what is to stop excessive numbers of them carrying out one particular job at the expense of another? This would lead to uncontrollable oscillation between the operations of System 1. The answer is that the job stimulus that the manager is more likely to respond to depends on his physiological age (Lindauer, 1953; Nowogrodzki, 1984; Oster and Wilson, 1978; Ribbands, 1952; Rosch, 1925, 1930; Sakagami, 1953; Seeley, 1982). As a young manager becomes physiologically older, the development of his perceptual and operational capabilities (Arnold and Masson, 1981; Berthold and Benton, 1970; Boch and Shearer, 1966; King, 1933; Skirkyavichyus and Skirkyavichene, 1979; Vaitkyavichene and Skirkyavichyus, 1979) will gradually enable him to respond to and carry out new operations which take place in new operational environments that are increasingly further from the centre of the factory (Free, 1977; Lindauer, 1953; Nowogrodzki, 1984; Oster and Wilson, 1978; Ribbands, 1952; Sakagami, 1953; Seeley, 1982). Cleaning will gradually give way to nursing, then nursing to building and building to food storage. Eventually the House Manager will become a Field Manager operating around the boundaries of the fortress as a guard, and finally he will end his days as a forager or scout in the external environment. By ensuring that excessive numbers of managers are physiologically incapable of responding to the same job stimulus at any one moment in time, a given operation will not be favoured at the expense of another; in this way temporal polyethism will effectively damp oscillations between the operations of System 1.

It is important to notice that this description of the sequence of operations carried out by any individual manager, or indeed by any age cohort of managers, refers to successive operations taking place in new operational environments that were increasingly further from the centre of the factory. This is known as the centrifugal discretization of operations (Oster and Wilson, 1978; Seeley, 1982); the various age cohorts of managers work their way through the operations like ripples from a stone tossed into a lake. Each ripple (or cohort) passes through a succession of operations which are themselves laid out in a sequence of concentric operational environments. This now explains why the operations are drawn in a nested arrangement in Figure 1.

There are two further elaborations of this System 2 which need to be mentioned. Firstly, Figure 1 shows a channel marked 'work searching/operational needs', which means to say that searching for work within the operations can provide information on operational needs, which can then modify, in the short term, the rate at which managers progress through their own intrinsic schedules (Free, 1977; Lindauer, 1953; Nowogrodzki, 1984). For instance, a shortage of storers will be sensed by the nearby builders, and many of them will change to food storage tasks at a younger age than usual (Seeley, 1982; Lindauer, 1971). Secondly, long-term modification of these working schedules is under metasystemic control by the Corporate Regulatory Centre. This is especially important when the Fortress Factory is coming out of a period of recession. Sensing that the system has reached its lowest ebb and that environmental prospects are just starting to improve (Kefuss, 1974), the Chief Executive will signal the Corporate Regulatory Centre to stimulate the activity of System 1 managers (Pain, Roger and Theurkauff, 1974). The rate of operational activity will then begin to increase and the Fortress Factory will enter a new growth phase.

The consequences of this amazing System 2 are extremely interesting. The fact that the temporal schedule of jobs undertaken by any one age cohort of managers followed the same sequence as the spatial layout of operations in a centrifugal direction means:

(1) The factory layout is specifically tailored to the changing physiology and developing abilities of the individuals who have to work within it. During the course of evolution, layouts became adapted to the workforce and not vice versa, as was expected of many human systems.

(2) Every manager learns the business from inside out, so that by the time he reaches a System 4 position, he will understand completely the needs of System 3. The System 4 dealings with the external environment are therefore always relevant to the needs of System 3. There is no ivory tower research here.

(3) The factory layout ensures that no manager has to waste time and energy crossing from one side of the factory to the other in order to carry out his next job. His next job is conveniently located right beside his present one.

(4) The juxtaposition of successive jobs also has two other advantages. Firstly, if there is a shortage of managers in his next job, he can easily substitute for them while they are absent. Secondly, flexible manning of this sort allows him to learn about the next job while finishing his previous one.

(5) There must be very great opportunities for the promotion of synergy within a System 1 where nested operations overlap (Beer, 1979) both in time and in space.

Now System 4 at this level of recursion consists of several thousand Field Managers who undertake the jobs of guarding the fortress, foraging for food and raw materials in the external environment and scouting for new sources of these requirements. The changing needs of System 3 are continually being communicated to these System 4 managers (Butler, 1974; Free, 1969, 1977; Lindauer, 1971; Michener, 1974), who of course can immediately transduce the meaning of such requests because of their past experience in System 3 roles. Now System 4 as a whole contains a spatiotemporal model of the surrounding environment (Dyer and Gould, 1981; Edrich, 1969, 1981; Lindauer, 1971) made up from the communicated models of particular environmental niches which are contained within the brains of members of about nine (Visscher, 1982) foraging groups. As the various foraging groups return from their expeditions they communicate (Lindauer, 1971; Von Frisch, 1967) the profitability of foraging in their particular niche on an energy cost–benefit basis (Waddington, 1982). Since foragers from all groups listen in to each other's reports, the word soon gets around if just one or two niches are presently providing the latest System 3 needs on the most profitable terms. Gradually foragers to less profitable niches are persuaded to visit the more profitable niches (Lindauer, 1971; Von Frisch, 1967; Waddington, 1982), and so System 4 as a whole adapts itself to a changing environment within an accepted range of alternatives. Scouts, who form about one-quarter of the foraging force (Seeley, 1983), search outside this presently accepted range of niches for possible new sources of food and raw materials. This is necessary because the environment is a time:space patchwork, not just varying from place to place at any one time but also varying over time in any one place.

System 4 usually invests most of its efforts in foraging within niches of known profitability and relatively less effort in finding new niches. The balance of investment between the accepted and problematic environments will vary, however, according to changes in the environment and in the needs of System 3 (Seeley, 1983). Another interesting aspect of System 4's interaction with the environment is that it is highly selective and beneficial to both the Fortress Factory and the viable systems with which it trades (Wells and Wells, 1985). The long-term consequence of this symbiotic relationship is the coevolution of both systems.

It is the responsibility of the youngest members of System 4 to guard the Fortress Factory against attack by its enemies and to defend its stores from pilfering by competitors (Butler, 1974; Free, 1977; Michener, 1974). Their problem is not only to fight off individuals or even large armies of marauders who threaten the fortress, but also to distinguish friend from foe in the first place. Any viable system which fights off all outsiders may deny itself the acquisition of novel information (Lumsden and Wilson, 1981; MacArthur and Wilson, 1967; Michener, 1974) or functions; on the other hand some outsiders will be of evil intent. How are they to distinguish one from the other? It seems that they first examine the outward appearance of the outsider, using two levels of screening, and if he is not too unlike themselves he is allowed to enter (Butler, 1974; Free, 1977). Once inside the system they then examine his purpose. If he behaves submissively when challenged, he is considered to have entered the system accidently but innocently, and in such cases he will probably be accepted. If, however, he is dominant and 'confident' when challenged, he will only be accepted if he is bringing useful gifts into the system (Butler, 1974; Free, 1977; Michener, 1974). These security screening procedures serve to filter out more probable danger at an earlier stage and less probable danger at a later stage, while still enabling the system to benefit from the acquisition of variety from harmless outsiders.

Recursion 2 System 5 contains an interesting mixture of intra-recursively and inter-recursively derived variety. To some extent, the all-pervading ethos of middle management is determined by information transmitted from brain to brain and is therefore of cultural origin (Ambrose, Morse and Boch, 1979; Free, 1969; Lindauer and Medugorac, 1967). This much is acquired intra-recursively, as behavioural norms learned from other individuals. But the brain size (Witthoft, 1967) of these managers is such that it can be largely programmed in advance by the genome (Lumsden and Wilson, 1981) via inter-recursive channels. It would be wrong to conclude, however, that behaviour is entirely controlled by genetic influences, for this would cast the genes as 'puppeteers' who continually intervene during the performance of a behaviour, something which is clearly impossible even in a moderately complex world (Dawkins, 1976). Rather the genes are 'programmers' who guide the development of the brain in such a way as to provide it with three main capacities. Firstly, genes can program the brain in advance with rules and advice on how to deal with predictable eventualities (Collins, 1979; Dawkins, 1976; Mobbs, 1980; Rothenbuhler, 1964). Much of middle management behaviour would be hard-wired in this way. Secondly, genes build into the brain a capacity for learning, which enables the manager to deal with unpredictable eventualities by learning which of his inherited repertoire of behaviour patterns is best for the given situation (Lindauer, 1971; Wilson, 1971). Thirdly, genes also build in a capacity for simulation, which is required in many System 4 situations (Lindauer, 1971; Lopatina, 1979; Von Frisch, 1967). So the ethos of middle management is influenced slightly by culture but mainly by epigenetic rules that have proved successful during the past evolutionary experience of the line.

Recursion 1 System 3 consists of a single Chief Executive who is capable of controlling the behaviour of up to 60,000 System 1 managers! The main sources of this amazing power are as follows:

(1) He has the ability to control the allocation of sets of programmes to junior managers via inter-recursive channels.

(2) Horizontal and vertical variety within System 1 is largely balanced already, so little vertical variety escapes.

(3) He has the ability to speak a powerful metalanguage (Callow, Chapman and Paton, 1964; Crewe, 1982; Law and Regnier, 1971; Williams, 1981) to System 1 managers via three main channels: (a) the predominantly stimulatory Anti-Oscillation Channel (see Figure 1) for initiating long-term action (Butler, 1973; Chauvin, Darchen and Pain, 1961; Jaycox, 1970); (b) the Command Channel, which utilizes managers as messengers of mandatory control signals in a type of 'contagious diffusion' process (Ferguson and Free, 1980; Hagerstrand, 1967; Seeley, 1979); and (c) the mainly inhibitory (Butler and Callow, 1968; Butler and Fairey, 1963) Operational Monitoring Channel, which requires a regular physical metasystemic presence throughout the entire operational area (Lensky and Slabezki, 1981).

(4) He has a synoptic view of System 1 operations obtained both on audit and also from regular managerial reports on the state of their operations (Engels and Fahrenhorst, 1974; Fournier, Darchen and Delage-Darchen, 1983; Halberstadt, 1980).

(5) He has received all the necessary information from Recursion 1 System 4 about new but workable ways in which the entire system can be adapted to its environment (Adams, Rothman, Kerr and Paulino, 1977; Free, 1977; Page and Metcalf, 1982; Taber and Wendel, 1958).

One very interesting aspect of the Chief Executive's power-base is that it has gradually changed during evolution from a dependence on physical bullying to a subtle manipulation of System 1 managers by a complex language of signals (Michener, 1974; Wilson, 1971). This is a good example of the evolution of a metalanguage and suggests that metasystems which rely on bullying do not yet speak a metalanguage.

The effect of a metasystem on its System 1 is not always obvious until it suddenly ceases to function. When the Chief Executive of this system dies or is lost, all 60,000 managers know about it within the hour (Butler, 1974; Fell and Morse, 1984; Yushka and Skirkyavichyus, 1976), thanks to the rapid rate of signal fade (Johnston, Law and Weaver, 1965; Juska, Seeley and Velthius, 1981; Pain and Barbier, 1981) which ensures the sensitivity of its communication system. This shows that viable systems respond to change, not to the level of the message. The whole system then takes on an air of disorganized restlessness (Butler, 1974; Free, 1977), operational activity slows (Tsibul'skii, 1975), and steps are taken to develop the youngest junior managers as potential replacements (Butler, 1974; Free,

1977; Fell and Morse, 1984; Punnett and Winston, 1983). We may conclude from this that the Chief Executive is normally responsible for procuring cohesion (Butler, Callow and Chapman, 1964; Ferguson, Free, Pickett and Winder, 1979; Winston, Slessor, Smirle and Kandil, 1982) and stimulating operational activity within System 1 (Chauvin, Darchen and Pain, 1961; Free, 1967; Jaycox, 1970), and also for inhibiting the development of competing metasystems (Butler, Callow and Johnston, 1961; Johnston, Law and Weaver, 1965; Lensky and Slabezki, 1981). If no new Chief Executives are produced by System 1, the internal milieu can be restabilized by some leading System 1 managers who act as a substitute metasystem (Crewe and Velthuis, 1980; Free, 1977; Michener, 1974; Sato, 1982). System 1 is therefore itself a viable system, although it cannot grow any more under these conditions. It can, however, generate variety *vis-à-vis* the external environment by enlarging Recursion 1 System 4 (Butler, 1974; Free, 1977; Michener, 1974).

Systems 3–2–1 of Recursion 1 between them stabilize the internal milieu to a remarkable degree. All members of the Fortress Factory live in an internal environment where temperature, humidity, gas exchange and other requirements are held between critical physiological limits purely by cooperative effort. No individual member is capable of controlling any of these parameters for himself over anything like the same range of values, and yet the system as a whole is able to do so. We see here the emergence at a higher level of recursion of new whole-system properties, such as homeothermy (Southwick and Mugaas, 1971; Southwick, 1983), which are not found in the contained viable systems themselves. The very important consequence of this is that the Fortress Factory is able to colonize new environmental niches in which single individuals would perish within hours.

Recursion 1 System 4 consists of the Chief Executive designate, who acts as the Development Director until such time as he has obtained and recombined all the necessary information that will be required to adapt the Fortress Factory to its environment (Berthold and Benton, 1970; Free, 1977; Page and Metcalf, 1982; Taber and Wendel, 1958). He then switches into his System 3 role (Eid, Ewies and Nasr, 1980; Free, 1977; Berthold and Benton, 1970), relying thereafter on a Development Directorate formed from special ambassadors to generate his variety *vis-à-vis* the external environment. Later he may actually switch back again into a System 4 role (Berthold and Benton, 1970; Free, 1977), if the viable system needs to adapt itself to environmental change. Figure 1 shows that such adaptation can take place in three main ways, namely by migration, fragmentation and/or by the acquisition and recombination of existing or novel information (Butler, 1974; Free, 1977).

Migration is the simplest of the three adaptive strategies employed by the Fortress Factory. If its microenvironment changes in such a way as to present a variety state which it cannot match, the Fortress Factory can quite easily abandon its present home and move to another place where a match is once more possible (Butler, 1974; Free, 1977). Recombination is more complex, involving the selection of adaptively successful whole solutions followed by a reshuffling of the sub-solutions which comprise them. New combinations of these sub-solutions are then used to construct new whole solutions, some of which will be better adapted to the changing environment in which they live (Brady, 1985). This process is repeated at frequent intervals, so that at least some members of the line will always be adapted to the prevailing environment.

Fragmentation is an interesting solution to the problem of maintaining the average size of a growing system close to the optimum both for internal control purposes and for the requirements of the particular environmental niche. As System 1 grows past a certain critical size (Simpson, 1973, 1974; Winston and Taylor, 1980), it becomes increasingly difficult for System 3 to balance its variety. At least to start with, this is not because System 3 is incapable of generating sufficient variety (Seeley and Fell, 1981), but because problems arise within the three channels of communication (Baird and Seeley, 1983; Butler, 1974; Caron, 1981; Lensky and Slabezki, 1981; Simpson, 1974). Channel capacity is limited here both by the surface area across which transduction can take place and by the rate of signal breakdown. The combined effects of these two limitations leads to a decrease in the signal:noise ratio, with the result that metasystemic influence on System 1 begins to wane. System 1 becomes disorganized and starts to develop a number of young junior managers in Recursion 3 as potential new Chief Executives (Butler, 1974; Fell, 1979; Free, 1977; Winston and Taylor, 1980).

Now this poses a number of problems for the Fortress Factory. Chief Executives are a competitive breed (Lensky, Darchen and Levy, 1970), and when there are anything up to twenty (Fell and Morse, 1984) of them waiting to head just three or four viable system fragments (Otis, 1980) there are all the makings of an ugly situation. It is enough to set the alarm bells ringing along the algedonic circuits (see Figure 1), whereupon System 1 managers do everything they can to prevent these power-hungry individuals from meeting, including locking them into their rooms in the Management Centre (Bruinsma, Kruijt and Van Dusseldorp, 1981; Fletcher, 1978; Simpson and Cherry, 1969). Such segregation is necessary to avert almost certain murder, but eventually the first Chief Executive designate announces that he is ready to leave. In an amazing reversal of normal behaviour, the System 1 managers then start to destabilize the system (Lindauer, 1955), with the result that a group of about 15,000 of them (Burgett and Morse, 1974; Avitabile and Kasinskas, 1977), plus their Chief Executive designate, pack their bags and leave home. Meanwhile further careful segregation of young metasystems and controlled fragmentation of System 1 may allow several more viable fragments to leave for new homes.

A viable fragment (Wille, 1974) now sets up a self-organizing network (Beer, 1979) to search for a new home, as shown in Figure 2. First of all it forms into a homomorphic model of a Fortress Factory (Avitabile and Kasinskas, 1977; Burgett, 1971; Lindauer, 1971) and then System 4 scouts set out to interrogate the problematic environment for suitable sites (Ambrose, 1976; Lindauer, 1971; Seeley, Morse and Visscher, 1979). Many such sites are investigated by the scouts, who then return to report on the location and suitability of each one (Lindauer, 1971). The valuation of a site is relative to the characteristics of the model. Scouts listen in to reports of sites that they have not visited, and if these sound better they will go off to check it out for themselves, so there is no face-value acceptance. Neither is there any filtering of reports; everybody has their say. Reports of better sites recruit more scouts to investigate them and scouts reporting on poorer sites are persuaded to visit better ones. In this way, initial reports of up to twenty possible sites are gradually reduced to one preferred site, which must remain the best over an adequate period of time (Lindauer, 1971). Only then will the viable fragment move off and take up residence there. Note that during this search for a new home, consensus is reached using a non-hierarchical, iterative and dynamic decision-making process in which unfiltered reports are focused on to a model of 'self'.

Fragmentation is normally followed by recombination (Free, 1977) and then by a new period of development in which the Fortress Factory grows to its former size using recombined information. The advantages of fragmentation and recombination have already been mentioned, but those of development are not so obvious here as in some other viable systems. There must be some other reason, apart from optimum size, why so many viable systems do not just keep on growing to some enormous size but instead opt to create new individuals by the apparently laborious process of passing through cyclically repeating developmental periods. The answer would seem to be that in order to put together new complexity, new developmental beginnings are required (Bonner, 1974; Dawkins, 1976). A new viable system must start from scratch, so that revolutionary change can be allowed to act on sensitive key points in early development to bring about a new fundamental restructuring of the system. Developmental recycling allows a return back to the drawing board in every generation. Continued growth merely allows superficial tinkering.

There may be lessons here for the designer of organizations. These observations suggest that in mature viable systems, which are of course

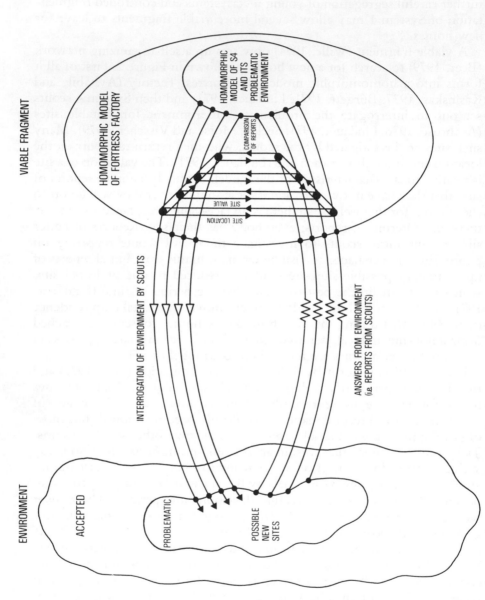

Figure 2. How a viable fragment uses a self-organizing network to search for a new home

autopoietic, no fundamental restructuring of organization is really possible. Instead one may need to return to the drawing board by creating a new embryonic system, which can then restructure itself by altering the developmental timing and hence balance of functions during a new period of development. This will require the designer to have a comprehensive understanding of organization, because an act of creation rather than acquisition or growth is involved.

System 5 of Recursion 1 is the last system to be considered in this organizational model of the Fortress Factory and in some ways it is the most mysterious. It has been described how a single individual has the ability to switch between the most senior System 3 and System 4 roles on behalf of the whole system and so there is here an approximate inbuilt variety balance. All managers are able to monitor the balance between System 3 and System 4 influences (Adler, Doolittle, Shimanuki and Jacobson, 1973; Vaitkyavichene, 1982), but only System 1 managers can make the necessary adjustments (Butler, 1974; Free, 1977; Michener, 1974). The actual 3:4 balance which is required at any point in time will depend on the stage of the Fortress Factory's life cycle (Allen, 1965; Free, 1977; Free and Williams, 1975). Since this cycle is some ten times longer than the average life-span of any individual System 1 manager, it is most unlikely that this knowledge can be acquired culturally. It must be hard-wired into their brains epigenetically, as inherited norms for a particular balance between System 3 and System 4 influences (Free, 1977; Louveaux, Mesquida and Fresnaye, 1972; Winston, Orley and Gard, 1983). On the surface, then, System 1 managers close the system, but underneath it all their behaviour has been strongly influenced by their inherited set of genes. This indicates that closure within the Fortress Factory is both an inter- and intra-recursive process.

The final word should be on performance. The Fortress Factory's most limited resource is energy, and if one believes that efficiency should be measured in terms of the return on the most limited resource, then for every one kilocalorie of energy spent on foraging, 29 kcal of food energy are harvested (Southwick and Pimental, 1981). The Kung bushman, an expert forager with similar priorities, can only manage 7.8 kcal, and modern agriculture, although presently having different limiting resources, creeps in at 0.32 kcal (Green, 1978). Now, although the communal 'brain' of this Fortress Factory does contain some 50 billion neurons (Witthoft, 1967), behaviour is largely hard-wired, and yet the system is energetically four times more efficient than a man and one hundred times more efficient than modern agriculture. This must speak volumes for the effectiveness of its organization, of which the foregoing account is a mere caricature.

And finally, what is the system described in these pages? Figure 3 is another model of it. It is in fact the bee colony.

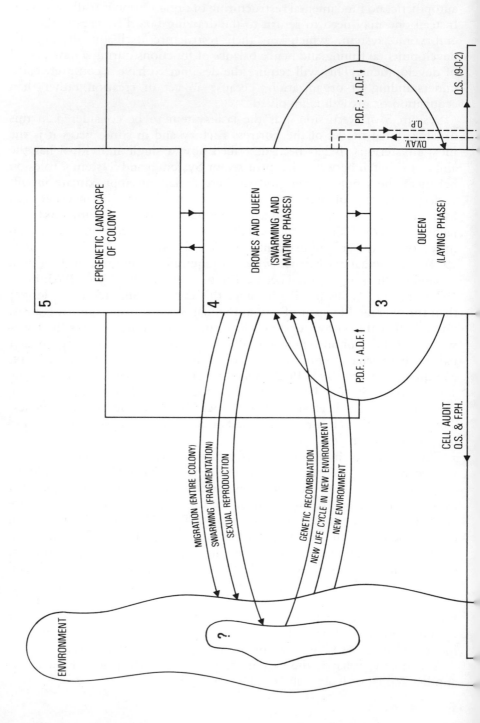

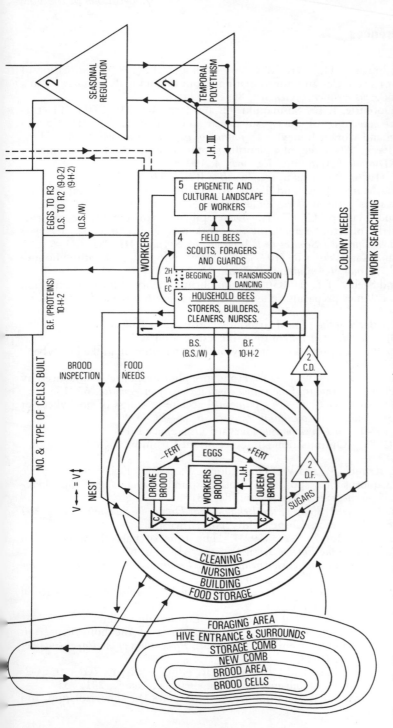

Figure 3. The organization of a bee colony. ↔ V, horizontal variety; ↕ V, vertical variety; ↑ +FERT, eggs fertilized; −FERT, eggs not fertilized; +FERT, eggs fertilized; −J.H., reduced juvenile hormone production; C, specific cell type; D.F., differential feeding; C.D., centrifugal discretization; B.S., brood substance (glyceryl-1,2-dioleate-3-palmitate); B.S./W, amount of brood substance per worker; B.F., brood food; 10-H-2, 10-hydroxy-trans-2-decenoic acid; 2.H, 2-heptanone; I.A., isopentyl acetate; E.C., (E)-citral; J.H.III, juvenile hormone III; Q.S, queen substance; 9-0-2, 9-oxo-trans-2-decenoic acid; 9-H-2, 9-hydroxy-trans-2-decenoic acid; Q.S./W, amount of queen substance per worker; F.P.H., footprint hormone; D.V.A.V., dorsoventral abdominal vibration; Q.P., queen piping; P.D.F.:A.D.F. ↑, the ratio of pro-drone factors to anti-drone factors increases; P.D.F.:A.D.F. ↓, the ratio of pro-drone factors to anti-drone factors decreases

References

Adams, J., Rothman, E.D., Kerr, W.E. and Paulino, Z.L. (1977). 'Estimation of the number of sex alleles and queen matings from diploid male frequencies in a population of *Apis mellifera*', *Genetics*, **86**, 583–96.

Adler, V.E., Doolittle, R.E., Shimanuki, H. and Jacobson, M. (1973). 'Electrophysiological screening of queen substance and analogues for attraction to drone, queen and worker honeybees', *Journal of Economic Entomology*, **66**, 33–6.

Allen, M.D. (1965). 'The effect of a plentiful supply of drone comb on colonies of honeybees', *Journal of Apicultural Research*, **4**, 109–19.

Ambrose, J.T. (1976). 'Swarms in transit', *Bee World*, **57**, 101–9.

Ambrose, J.T., Morse, R.A. and Boch, R. (1979). 'Queen discrimination by honeybee swarms', *Annals of the Entomological Society of America*, **72**, 673–5.

Arnold, G. and Masson, C. (1981). 'Development with age of the external structure of the olfactory sensilla of the worker honeybee antenna', *Comptes Rendus Hebdomadaires des Seances de l'Academie des Sciences*, **III**, 292, 681–6.

Avitabile, A. and Kasinskas, J.R. (1977). 'The drone population of natural honeybee swarms', *Journal of Apicultural Research*, **16**, 145–9.

Baird, D.H. and Seeley, T.D. (1983). 'An equilibrium theory of queen production in honeybee colonies preparing to swarm', *Behavioural Ecology and Sociobiology*, **13**, 221–8.

Beer, S. (1972). *Brain of the Firm*. Harmondsworth: Allen Lane.

Beer, S. (1979). *The Heart of Enterprise*. Chichester: John Wiley.

Beetsma, J. (1979). 'The process of queen-worker differentiation in the honeybee', *Bee World*, **60**, 24–39.

Berthold, R. and Benton, A.W. (1970). 'Honeybee photoresponse as influenced by age. Pt 1: Workers', *Annals of the Entomological Society of America*, **63**, 136–9.

Boch, R. and Shearer, D.A. (1966). 'Isopentyl acetate in stings of honeybees of different ages', *Journal of Apicultural Research*, **5**, 65–70.

Bonner, J.T. (1974). *On Development*. Harvard University Press.

Brady, R.M. (1985). 'Optimisation strategies gleaned from biological evolution', *Nature*, **317**, 804–6.

Bruinsma, O., Kruijt, J.P. and Van Dusseldorp, W. (1981). 'Delay of emergence of honeybee queens in response to tooting sounds', *Proceedings, Koninklijke Nederlandse Akademie van Wetenschappen*, **84**, 381–8.

Burgett, D.M. (1971). *A Study of the Behaviour of Drones in Swarming Honeybees Apis mellifera*. MSc Thesis, Cornell University, Ithaca.

Burgett, D.M. and Morse, R.A. (1974). 'The time of natural swarming in honeybees', *Annals of the Entomological Society of America*, **67**, 719–20.

Butler, C. (1609). *The Feminine Monarchie: On a Treatise Concerning Bees, and the Due Ordering of Them*. Oxford: Joseph Barnes.

Butler, C.G. (1973). 'The queen and the "spirit of the hive'', *Proceedings of the Royal Entomological Society*, A, **48**, 59–65.

Butler, C.G. (1974). *The World of the Honeybee*. Edinburgh: Collins.

Butler, C.G. and Callow, R.K. (1968). 'Pheromones of the honeybee (*Apis mellifera L*): the "inhibitory scent" of the queen', *Proceedings of the Royal Entomological Society, London*, B, **43**, 62–5.

Butler, C.G., Callow, R.K. and Chapman, J.R. (1964). '9-hydroxydec-trans-2-enoic acid: a pheromone stabilising honeybee swarms', *Nature*, **201**, 733.

Butler, C.G., Callow, R.K. and Johnston, N.C. (1961). 'The isolation and synthesis of queen substance, 9-oxodec-trans-2-enoic acid, a honeybee phero-mone', *Proceedings of the Royal Society*, B, **155**, 417–32.

Butler, C.G. and Fairey, E.M. (1963). 'The role of the queen in preventing oogenesis in worker honeybees', *Journal of Apicultural Research*, **2**, 14–18.

Callow, R.K., Chapman, J.R. and Paton, P.N. (1964). 'Pheromones of the honeybee: chemical studies of the mandibular gland secretion of the queen', *Journal of Apicultural Research*, **3**, 77–89.

Caron, D.M. (1981). 'Congestion, seasonal cycle and queen rearing as they relate to swarming in *Apis mellifera*', *Annals of the Entomolgical Society of America*, **74**, 134–7.

Chauvin, R., Darchen, R. and Pain, J. (1961). 'Sur l'existence d'une hormone de construction chez les abeilles', *Comptes Rendus Academie Sciences, Paris*, **253**, 1135–6.

Collins, A.M. (1979). 'Genetics of the response of the honeybee to an alarm chemical, isopentyl acetate', *Journal of Apicultural Research*, **18**, 285–91.

Crewe, R.M. (1982). 'Compositional variability: the key to the social signals produced by honeybee mandibular glands', in *The Biology of Social Insects* (eds. Breed, M.D., Michener, C.D. and Evans, H.E.). Westview Press.

Crewe, R.M. and Velthuis, H.H.W. (1980). 'False queens: a consequence of mandibular gland signals in worker honeybees', *Naturwissenschaften*, **67**, 467–9.

Dawkins, R. (1976). *The Selfish Gene*. Oxford University Press.

Dyer, F.C. and Gould, J.L. (1981). 'Honeybee orientation: a backup system for cloudy days', *Science* (USA), **214**, 1041–2.

Edrich, W. (1969). 'Azimuthal calculations of the equatorial sun's course during the night by *Apis mellifera adansonii*', **XXII**. International Beekeeping Congress Summary, 118.

Edrich, W. (1981). 'Night-time sun compass behaviour of honeybees at the equator', *Physiological Entomology*, **6**, 7–13.

Eid, M.A.A., Ewies, M.A. and Nasr, M.S. (1980). 'Biological significance of the weight of newly emerged honeybee queens and weight changes during the pre-oviposition period', *Bulletin of Faculty of Apiculture, University of Cairo*, **29**, 137–69.

Engels, W. and Fahrenhorst, H. (1974). 'Age and caste dependent changes in the haemolymph protein patterns of *Apis mellifera*', *Wilhelm Roux Archiv fur Entwick-lungsmechanik der Organismen*, **174**, 285–96.

Fell, R.D. (1979). *The Production, Recognition and Treatment of Queen Cells in the Honeybee (Apis mellifera L.) Colony*. PhD Thesis, Cornell University, Ithaca.

Fell, R.D. and Morse, R.A. (1984). 'Emergency queen cell production in the honeybee colony', *Insectes Sociaux*, **31**, 221–37.

Ferguson, A.W. and Free, J.B. (1980). 'Queen pheromone transfer within honey-bee colonies', *Physiological Entomology*, **5**, 359–66.

Ferguson, A.W., Free, J.B., Pickett, J.A. and Winder, M. (1979). 'Techniques for studying honeybee pheromones involved in clustering and experiments on the effect of Nasanov and queen pheromones', *Physiological Entomology*, **4**, 339–44.

Fletcher, D.J.C. (1978). 'Vibration of queen cells by worker honeybees and its relation to the issue of swarms with virgin queens', *Journal of Apicultural Research*, **17**, 14–26.

Fournier, B., Darchen, R. and Delage-Darchen, B. (1983). 'Amounts of juvenile hormones in the salivary glands of worker honeybees: a new approach to the study of caste determination', *Comptes Rendus Hebdomadaires des Seances de l'Academie des Sciences*, **III**, 297, 343–6.

Free, J.B. (1967). 'The production of drone comb by honeybee colonies', *Journal of Apicultural Research*, **6**, 29–36.

Free, J.B. (1969). 'Influence of the odour of a honeybee colony's food stores on the behaviour of its foragers', *Nature*, **222**, 778.

Free, J.B. (1977). *The Social Organisation of Honeybees* (Institute of Biology's Studies in Biology, 81). London: Edward Arnold.

Free, J.B. and Williams, I.H. (1975). 'Factors determining the rearing and rejection of drones by the honeybee colony', *Animal Behaviour*, **23**, 650–75.

Von Frisch, K. (1967). *The Dance Language and Orientation of Bees.* Oxford University Press.

Getz, W.M. and Smith, K.B. (1983). 'Genetic kin recognition: honeybees discriminate between full and half sisters', *Nature*, **302**, 147–8.

Green, M.B. (1978). *Eating oil: Energy Use in Food Production.* Westview Press.

Hagerstrand, T. (1967). *Innovation Diffusion as a Spatial Process.* Chicago University Press.

Halberstadt, K. (1980). 'Investigations on the secretory activity of the hypopharyngeal gland of the honeybee, using electrophoresis', *Insectes Sociaux*, **27**, 61–77.

Hamilton, W.D. (1964). 'The genetical theory of social behaviour, I and II', *Journal of Theoretical Biology*, **7**, 1–16; 17–32.

Jaycox, E.R. (1970). 'Honeybee queen pheromones and worker foraging behaviour', *Annals of the Entomological Society of America*, **63**, 222–8.

Johnston, N.C., Law, J.H. and Weaver, N. (1965). 'Metabolism of 9-ketodec-2-enoic acid by worker honeybees (*Apis mellifera*)', *Biochemistry, New York*, **8**, 1615–21.

Juska, A., Seeley, T.D. and Velthuis, H.H.W. (1981). 'How honeybee queen attendants become ordinary workers', *Journal of Insect Physiology*, **27**, 515–19.

Kefuss, J.A. (1974). '*The Influence of photoperiod on honeybee brood rearing (Apis mellifera)'*. In augural Dissertation zur Elangung des Doktogrades der Naturwissenschaften. J.W. Goeth-Universitat, GFR.

King, G.E. (1933). *The larger glands in the worker honeybee: a correlation of activity with age and with physiological functioning.* Abstract of PhD Thesis, Graduate School of the University of Illinois (1928).

Koeniger, N. (1970). 'Factors determining the laying of drone and worker eggs by the queen honeybee', *Bee World*, **51**, 166–9.

Law, J.H. and Regnier, F.E. (1971). 'Pheromones', *Annual Review of Biochemistry*, **40**, 533–48.

Lensky, Y., Darchen, R. and Levy, R. (1970). 'Aggressiveness between queens, and of workers to queens, in the formation of multiqueen honeybee colonies', *Revue Comptes d'animaux*, **4**, 50–62.

Lensky, Y. and Slabezki, Y. (1981). 'The inhibiting effect of the queen bee (*Apis mellifera*) foot-print pheromone on the construction of swarming queen cups', *Journal of Insect Physiology*, **27**, 313–23.

Lindauer, M. (1953). 'Division of labour in the honeybee colony', *Bee World*, **34**, 63–90.

Lindauer, M. (1955). 'Schwarmbienen auf Wohnungsuche', *Zeitschrift fur Vergleichende Physiologie*, **37**, 263–324.

Lindauer, M. (1971). *Communication Among Social Bees.* Boston: Harvard University Press.

Lindauer, M. and Medugorac, I. (1967). 'A social time setter in the honeybee colony. L'effet de groupe chez les animaux', *Colloques International*, **173**, 15–25.

Lopatina, N.G. (1979). 'A comparative genetic study of the thresholds of neuromuscular excitability with respect to the signalling behaviour of the honeybee', *Genetika*, **15**, 1979–88.

Louveaux, J., Mesquida, J. and Fresnaye, J. (1972). 'Observations on the variability of drone brood production by honeybee colonies', *Apidologie*, **3**, 291–307.

Lumsden, C.J. and Wilson, E.O. (1981). *Genes, Mind and Culture*, Harvard University Press.

MacArthur, R.H. and Wilson, E.O. (1967). *The Theory of Island Biogeography*, Princeton University Press.

Maturana, H. and Varela, F. (1975). *Autopoietic Systems*. Biological Computer Laboratory Report no. 9.4.

Meyer, K.D. (1982). *Studies on the Brood Rearing Activity and Foraging Flights of Honeybees under Flight Room Conditions*. Thesis, University of Tubingen, GFR.

Michener, C.D. (1974). *The Social Behaviour of the Bees*. Harvard University Press.

Mobbs, P.G. (1980). 'The neural regulation of bee behaviour', lecture given to Central Association of Beekeepers.

Nowogrodzki, R. (1984). 'Division of labour in the honeybee colony: a review', *Bee World*, **65**, 109–16.

Oster, G.F. and Wilson, E.O. (1978). *Caste and Ecology in the Social Insects*. Princeton University Press.

Otis, G.W. (1980). *The Swarming Biology and Population Dynamics of the Africanised Honeybee*. PhD Thesis, University of Kansas.

Page, R.E. and Metcalf, R.A. (1982). 'Multiple mating, sperm utilisation and social evolution', *American Naturalist*, **119**, 263–81.

Pain, J. and Barbier, M. (1981). 'The pheromone of the queen honeybee: evidence of a deactivating system for queen substance', *Naturwissenschaften*, **68**, 429–30.

Pain, J., Roger, B. and Theurkauff, J. (1974). 'Determination of a seasonal cycle for the content of 9-oxodec and 9-hydroxydec-2-enoic acids in the head of the virgin queen honeybee', *Apidologie*, **5**, 319–55.

Punnett, E.N. and Winston, M.L. (1983). 'Events following queen removal in colonies of European derived honeybee races (*Apis mellifera*)', *Insectes Sociaux*, **30**, 376–83.

Rembold, H. (1969). 'Biochemical aspects of caste differentiation in the honeybee', *Bericht Physikalischmedizinischen Gesellschaft Wurzburg*, **77**, 84–92.

Ribbands, C.R. (1952). 'Division of labour in the honeybee community', *Proceedings of the Royal Society*, B, **140**, 32–43.

Rosch, G.A. (1925). 'Untersuchungen uber die Arbeitsteilung im Bienenstaat. Pt. 1: Die Tatigkeiten im normalen Bienenstaat und ihre Beziehungen zum Alter der Arbeitsbienen', *Zeitschrift fur Vergleichende Physiologie*, **2**, 571–631.

Rosch, G.A. (1930). 'Untersuchungen uber die Arbeitsteilung im Bienenstaat. Pt. 2: Die Tatigkeiten der Arbeitsbienen unter experimentell veranderten Bedingungen', *Zeitschrift fur Vergleichende Physiologie*, **12**, 1–71.

Rothenbuhler, W.C. (1964). 'Behaviour genetics of nest cleaning in honeybees. Pt. IV: Responses of F.I. and backcross generations to disease-killed brood', *American Zoologist*, **4**, 111–23.

Sakagami, S.F. (1953). 'Division of labour in a colony of honeybees: studies in the biology of the honeybee *Apis mellifera* L', *Japanese Journal of Zoology*, **11**, 117–85.

Sands, W.A. (1981). 'The social life of termites', lecture given to Central Association of Beekeepers.

Sato, M. (1982). 'A study of 9-oxo-2-decenoic acid and 10-hydroxy-2-decenoic acid from honeybee mandibular glands', *Honeybee Science*, **3**, 1–8.

Seeley, T.D. (1979). 'Queen substance dispersal by messenger workers in honey-bee colonies', *Behavioural Ecology and Sociobiology*, **5**, 391–415.

Seeley, T.D. (1982). 'Adaptive significance of the age polyethism schedule in honeybee colonies', *Behavioural Ecology and Sociobiology*, **11**, 287–93.

Seeley, T.D. (1983). 'Division of labour between scouts and recruits in honeybee foraging', *Behavioural Ecology and Sociobiology*, **12**, 253–9.

Seeley, T.D. and Fell, R.D. (1981). 'Queen substance production in honeybee (*Apis mellifera*) colonies preparing to swarm', *Journal of the Kansas Entomolgical Society*, **54**, 192–6.

Seeley, T.D., Morse, R.A. and Visscher, P.K. (1979). 'The natural history of the flight of honeybee swarms', *Psyche*, **86**, 103–13.

Simpson, J. (1973). 'Influence of hive space restriction on the tendency of honey-bees to rear queens', *Journal of Apicultural Research*, **12**, 183–6.

Simpson, J. (1974). 'The reproductive behaviour of European honeybee colonies', lecture given to the Central Association of Beekeepers.

Simpson, J. and Cherry, S.M. (1969). 'Queen confinement, queen piping and swarming in *Apis mellifera* colonies', *Animal Behaviour*, **17**, 271–8.

Skirkyavichyus, A.V. and Skirkyavichene, Z. (1979). 'Effects of age on the electrophysiological response of an insect's antennal olfactory receptors', *Khemo-retseptsiya Nasekomykh*, **4**, 23–43.

Southwick, E.E. (1983). 'The honeybee cluster as a homeothermic super-organism', *Comparative Biochemistry and Physiology*, A, **75**, 641–5.

Southwick, E.E. and Mugaas, J.N. (1971). 'A hypothetical homeotherm: the honeybee hive', *Comparative Biochemistry and Physiology*, **40A**, 935–44.

Southwick, E.E. and Pimental, D. (1981). 'Energy efficiency of honey production by bees', *Bioscience*, **31** 730–2.

Taber, S. and Wendel, J. (1958). 'Concerning the number of times the queen bees mate', *Journal of Economic Entomology*, **51**, 786–9.

Trivers, R.L. and Hare, H. (1976). 'Haplodiploidy and the evolution of the social insects', *Science*, **191**, 249–63.

Tsibul'skii, P.P. (1975). 'Influence of queens and brood on intensity of foraging by honeybees', *Nauchro-Issledovatel'skii'*, Institut Pchelovodstva.

Vaitkyavichene, G.B. (1982). 'Some characteristics of neuron responses of the deutocerebrum of honeybees to stimulation by the pheromones of the queen honeybee', in *Khimicheskie Signaly Zhivotnykh* (ed. V.E. Sokolov). Moscow: Nauka, pp. 168–78.

Vaitkyavichene, G.B. and Skirkyavichyus, A.V. (1979). 'Characteristics of back-ground neural activity in the deutocerebrum of hive bees and field bees', *Khemoretseptsiya Nasekomykh*, **4**, 45–62.

Visscher, P.K. (1982). *Foraging Strategy of Honeybee Colonies in a Temperate Deciduous Forest*. MS Thesis, Cornell University, Ithaca.

Waddington, K.D. (1982). 'Honeybee foraging profitability and round dance correlates', *Journal of Comparative Physiology*, A, **148**, 297–301.

Wells, P.H. and Wells, H. (1985). 'Ethological isolation of plants. Pt. 2: Odour selection by honeybees', *Journal of Apicultural Research*, **24**, 86–92.

Wenner, A.M. (1974). 'Information transfer in honeybees: a population approach', in *Advances in the Study of Communications and Affect*, vol. 1, *Non-Verbal Communication* (eds. Kramer, L., Pliner, P. and Alloway, T.).

Wille, H. (1974). 'Numerical development of the honeybee colony', *Schweizerische Bienen-Zeitung*, **97**, 304–16, 369–74, 420–5.

Williams, I.H. (1981). 'Chemical communication in honeybees', lecture given to Central Association of Beekeepers.

Wilson, E.O. (1971). *The Insect Societies*. Harvard University Press.

Winston, M.L., Orley, R. and Gard, W.O. (1983). 'Some differences between temperate European and tropical African and South American honeybees', *Bee World*, **64**, 12–21.

Winston, M.L., Slessor, K.N., Smirle, M.J. and Kandil, A.A. (1982). 'The influence of a queen-produced substance, 9-HDA, on swarm clustering behaviour in the honeybee, *Apis mellifera L.*', *Journal of Chemical Ecology*, **8**, 1283–8.

Winston, M.L. and Taylor, O.R. (1980). 'Factors preceding queen rearing in the Africanised honeybee (*Apis mellifera*) in South America', *Insectes Sociaux*, **27**, 289–304.

Witthoft, W. (1967). 'Absolute number and distribution of cells in the honeybee brain', *Zeitschrift fur Morphologie und Okologie der Tiere*, **61**, 160–84.

Yushka, A.A. and Skirkyavichyus, A.V. (1976). 'Statistical investigation of the behaviour of the queen in a honeybee colony. Pt. 1: Movement of a queen in a small colony', *Lietuvos TSR Mokslu Akademijos Darbai*, C, **3**, 143–57.

The Viable System Model: Interpretations and Applications of Stafford Beer's VSM
Edited by R. Espejo and R. Harnden
Published 1989 by John Wiley & Sons Ltd

7

Application of the VSM to the trade training network in New Zealand*

G.A. Britton and H. McCallion

School of Engineering, University of Canterbury, New Zealand

The Viable System Model was used as a basis to formulate government policy on vocational training in New Zealand by diagnosing organizational deficiences in the trade training network. It is shown that none of the levels of recursion is effectively viable. The model enabled the authors to determine the factors preventing viability and to suggest appropriate solutions.

Introduction

The authors used Beer's (1966, 1972, 1979) cybernetic model of viable systems to diagnose organizational deficiencies in the vocational training network in New Zealand. The results greatly exceeded our expectations. We believe the use of Beer's model should be more widespread.

Firstly, we believe it is an appropriate organizational model to cope with turbulent environments. Emery and Trist (1975) have noted the increasing salience of turbulence in modern times. Beer (1970), Emery (1977) and Trist (1980) have pointed out that bureaucratic structures are failing to adapt to modern (turbulent) environments. Beer (1970) and Emery (1977) emphasize the destabilizing influence of bureaucratic structures. Trist (1980) argues that new organizational structures are needed to adapt to turbulent environments: holographic structures.

---

*A version of this chapter was published in *Cybernetics and Systems*, vol. 16, 1985. Acknowledgement is made to Hemisphere Publishing Corporation for permission to use the material here.

Given that turbulence is increasing, and assuming Trist is correct in suggesting holographic structures are adaptable to turbulent environments, we can expect sociologists and organizational theorists to focus attention on holographic structures. Though Beer's model is not explicitly holographic, it is compatible with the notion of redundancy (i.e. it does not preclude redundant functions). We believe it is an appropriate organizational model for turbulent environments and we expect it to assume a new significance in the light of recent developments in sociology.

Secondly, the model is an extremely powerful diagnostic and predictive tool for analysing deficiencies in organizations and inter-organizational networks, as we shall demonstrate later in this chapter.

Thirdly, we believe there is independent validation of the model. Thompson (1967) has developed a set of propositions about organizational design and behaviour from a theoretical base different from that used by Beer. These propositions support Beer's work and vice versa. Some of the propositions will be discussed later.

Unfortunately, potential users of Beer's model face a difficult task in applying the model to real situations, as Beer acknowledges:

> 'So much space is being devoted to the correct identification of operational elements at this stage, because it turns out in practice to offer the worst problems, and the most traps to intending users of these cybernetics. . . . We live in an hierarchical culture, and the notion of recursion in the form advanced is foreign to our ways of thinking.' (Beer, 1979, p. 120)

> 'It is perfectly clear to me that the model has no power, predictive or diagnostic or prescriptive, if its logic is perverted. . . . I have seen many an alleged use of the model that confused two, if not three, levels of recursion. But the most common mistake is to seize on the existing organization chart of the institution, and blithely to assume that every division or department shown as depending from the boss is a viable system in its own right.' (Ibid., p. 204)

To assist potential users in correctly applying the model, we propose that case studies, demonstrating valid uses of the model, be published. The studies should cover a range of different organizations in different contexts.

Furthermore, we suggest that these studies be primarily descriptive. Our own experience with the model indicates that significant organizational deficiencies can be 'discovered' using the model at the descriptive level. More importantly though, if the mathematics is to come out right, the analogy must first be appropriate. (For a discussion on the modelling process, refer to Chapter 6 of Beer, 1966.)

Our intentions in presenting this chapter are: (1) to promote greater use of Beer's model, (2) to demonstrate the model's diagnostic capability, and (3) to provide an illustrative case study for potential users.

The diagnosis of the organizational deficiencies in the vocational training network is incomplete. To appreciate the reasons for this and the difficulties faced by the authors In implementing proposed solutions, some background information is essential.

Background information

In 1981 members of the New Zealand electrical and electronics industries approached several cabinets ministers to complain about the structure of the training network. In particular, they were concerned about the lack of coordination of the various bodies involved in training. Their action resulted in a subcommittee of the NZ Vocational Training Council being established to look specifically at the training requirements for electrical, electronic and computer personnel. Professor McCallion was appointed chairman of the subcommittee (known as METAC).

Professor McCallion invited Dr Britton to submit a research proposal to the committee to investigate the structure of the vocational training network. The project was initiated in March 1982 and completed late in 1983.

The research approach followed was that recommended by Ackoff (1981). Information was obtained from published documents and a series of approximately 40 interviews with people directly involved in the network; each interview lasted approximately one hour. This approach highlighted significant deficiences in the network.

However, two statements made to Dr Britton during the course of the inquiry were not dealt with. The first of these was a statement made by members of the Industries Development Commission that there was no coordination at the highest level. They suggested establishing a cabinet committee to deal specifically with training.

The second was a statement from members of the electronics industry that there was a lack of coordination of the various bodies involved in training. Britton (1984) was able to show that this was true, but only after extensive research. There was no way members of the industry could have known about the specific difficulties facing the network, yet they perceived a lack of coordination. The question that remained unanswered was: what were they perceiving?

These unanswered questions and a special request from the chairman of the Vocational Training Council for a report on the roles of the Council and the Electrical and Electronics Industry Training Board prompted the authors to try a rigorous analysis using Beer's cybernetic model. As will be seen, the authors succeeded in elucidating the points mentioned above and in clarifying the roles of the Council and of the Training Board.

The research project did have severe limitations placed on it. There was no right of access to investigate individual government departments. Thus those departments which were responsible for coordinating other bodies—for example the Department of Labour, the Treasury and the Department of the State Services Commission—could not be investigated. This gap in information is apparent in the diagrams that follow.

Furthermore, the conclusions drawn by the authors were and are at variance with those held by bureaucrats in the network. Thus there were and still are major problems in getting our proposals accepted.

Some theoretical issues

There are four theoretical issues about the model we wish to clarify. This will enable the reader to comprehend the model, and our application of it, more fully. The issues deal with System 2, the three channels of communication between Systems 1 and 3, environmental monitoring, and diagramming conventions.

System 2

System 2 exists to dampen oscillations resulting from interaction between the operational elements. Thompson (1967) describes the different types of interactions (interdependencies) that can exist between organizational groups. They are pooled, sequential and reciprocal interdependencies. Different coordination procedures are necessary for the three types (Thompson, 1967). Thus there are basically three different types of System 2.

Pooled interdependence is the weakest interaction. It is the minimum that can exist for the set of operational elements to be considered a system as opposed to independent entitites. It occurs when the operational elements all have access to the same resources; for example, a common computer service provided by System 3 for System 1. The appropriate coordinating procedures are timetabling and the allocation of priorities.

The next level of interdependence is sequential. This includes pooled interdependence but has the additional feature that operational elements directly interact in a sequence (interactions are uni-directional). An example is the flow of materials in a mass-production manufacturing unit. The appropriate coordinating procedures are standardization and scheduling.

The highest level of interdependence is reciprocal interdependence, which includes the two previous types. In this case, though, the interactions are also bi-directional; there is direct feedback between the elements. An example is face-to-face work groups. In this case, coordination is achieved by mutual facilitation.

Beer (1979) states that System 2 coordinates according to a prescribed routine. This can be achieved for coordination of pooled and sequential interdependence. We do not believe it can be achieved for reciprocal interdependence. The uncertainty of the interactions means no pre-programming is possible. Coordination is by mutual adjustment and negotiation (Thompson, 1967; Kingdon, 1973).

System 3 communication channels

There are three communication channels by which System 3 communicates with System 1. These are the operational monitoring channel, the command channel, and the anti-oscillation channel (System 2). Consider, now, a common service provided by System 3 for the operational elements of System 1 (e.g. a computer service). How are the interactions of this computer department with the operational elements to be described in terms of the three channels?

When the department actually processes work on behalf of an operational element it is in fact acting as a part of the element: it is carrying out essential System 1 activities. When it 'tells' an operational element what to do it operates via the command channel. When it determines priorities and timetables requests for service from System 1, it operates via the anti-oscillation channel. When it carries out an investigation to improve its services for the benefit of System 1, it operates via the operational monitoring channel.

Environmental monitoring

One of the common services System 3 may provide for System 1 is environmental monitoring. That is, System 3 may set up a service (e.g. a sales department or advertising department) to manage the environmental loops of the operational elements. Correspondingly, a System 2 will need to be established to coordinate the behaviour of the operational elements with respect to that service.

The environmental monitoring carried out by this common service is different from that performed by System 4. System 4 monitors the environment to develop System 1—to raise its potentiality and capability. The common service, on the other hand, is concerned with monitoring the environmental aspects affecting the current behaviour of System 1 (i.e. its concern is to raise System 1's actuality towards its capability). Note that System 3 interacts with the environment via System 1 or System 4.

The System 3 provision of a common environmental monitoring service proved to be a crucial issue in our work. To highlight this aspect, and for ease of pedagogy, we have shown a link from System 3 to the environment in our diagrams. Note that this is not strictly true, and that the correct interpretation is the one given above.

BA

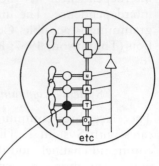

Recursion 0: THE WHOLE NETWORK
Elements: University training system (u)
          AAVA tech.training system (A)
          Trade training system   (T)
          Other training system   (O_i)

TERRITORIES

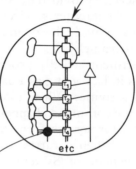

Recursion 1: TRADE TRAINING SYSTEM
Elements: Each individual trade
          training system (T_i)

ARKAS

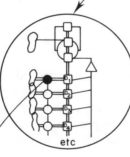

Recursion 2: EACH INDIVIDUAL TRADE
                    TRAINING SYSTEM
Elements: Districts (D_i)

ODS

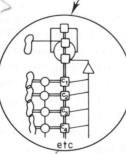

Recursion 3: EACH DISTRICT
Elements: Apprenticeship contracts (C_i)

Figure 1. Levels of recursion for trade training

DISTRICTS   SYSTEM IN FOCUS

Diagramming conventions

For ease of interpretation, we use a slightly different diagramming convention from that proposed by Beer. We show the command channel for each operational element *separately*, rather than drawing one channel for all elements.

Levels of recursion in the training network

We turn now to our application of Beer's model. The structure of the network can be divided into four separate parts:

(1) the training system for trades;
(2) the training system for 'AAVA technicians' (people trained to Authority for Advanced Vocational Awards standard; Technicians Certificate or NZ Certificate);
(3) the training system for university graduates;
(4) the training systems for other training programmes.

Figure 1 shows the four separate parts of the training network and the levels of recursion for the trade training system. Trade training only will be discussed here because it is the dominant training mode of the electrical/electronics industry (Britton, 1984). Most of the industry complaints about training relate to trade training.

It is important to stress that the diagrams which follow are incomplete. We do not claim to have included all publicly funded organizations that can affect training. The focus has been on the most conspicuous organizations and those most frequently criticized. Despite this limitation, significant inter-organizational deficiencies are discovered and highlighted by the analysis. For a description of all the organizations involved in training for electrical and electronic vocations, refer to Britton (1984).

One other aspect bears mentioning at this stage. It is not possible to consider training for the electrical/electronics industry in isolation from other vocational training.

Trade training

Our discussion is focused around training for electrical or electronic vocations. Most of the comments, however, are applicable to other trades. (The diagrams, naturally, will differ for each trade.)

The smallest operational unit which needs managing in trade training is the *apprenticeship contract*. Each contract consists of:

(1) one or more employers (depending on the type of contract: the Apprenticeship Act 1983 allows for joint contracts and group apprenticeships);
(2) one or more apprentices;
(3) a contractual obligation legally binding on all parties.

Each contract is managed (administered) by a District Commissioner of Apprenticeship with a Local Apprenticeship Committee, where one exists.

Our analysis starts at a level of recursion higher (Recursion 2); that is, management of a set of contracts. In each district there is a set of contracts which is managed (administered) by a District Management Unit. The management unit consists of the District Commissioner of Apprenticeship and support staff, and a Local Apprenticeship Committee if one exists.

Each set of contracts and its management unit is an operational element of the organizational model. There are as many operational elements as there are districts. The collection of all operational elements is the system (called System 1 in the model) which has to be managed by a higher-level management unit: the management unit concerned with managing *all* contracts in a particular trade.

Figure 2 shows the complete model of the training system for a particular trade. Figure 3 describes in more detail the specific features of each operational element shown in Figure 2. A brief description of the organizations in the model is given in an Appendix. An explication of the model follows.

Notes on Recursion level 2 (trade training)

The purpose of a trade training system is to provide sufficient numbers of people efficiently and properly trained in New Zealand in a specific trade, so as to meet the country's needs; yet ensure adequate opportunities for people trained in that trade.

System 1

This consists of different sets of apprenticeship contracts; each set corresponding to a number of contracts in a particular district. The objectives of each operational element of System 1 are:

(a) to ensure that sufficient numbers of apprentices are trained in the district to meet its needs, yet ensure adequate employment opportunities for apprentices when they complete their contracts;
(b) to ensure that each contract is successfully completed (i.e. runs its full period of apprenticeship);

(c) to ensure that each apprentice is trained in a satisfactory range of skills (deemed essential for that trade) to a satisfactory standard.

The District Commissioners are charged with the responsibility of promoting apprenticeships and assessing training capacity. Thus they have some influence in getting potential employers to take on apprentices, and actual employers of apprentices to take on additional people. The Vocational Guidance Division of the Department of Labour, through its field offices, informs potential apprentices about opportunities and may influence them to undertake apprenticeships.

A potential apprentice and a potential/actual employer interact (in the environment) and finally a joint decision may be made to proceed with an apprenticeship contract. This is indicated by the arrows converging on 'application'. The application is accepted or rejected by the District Commissioner/Local Apprenticeship Committee. If it is accepted the employer and new apprentice enter the training system; that is, the set of contracts in District A. Note that the set is 'notional' because its membership is continually changing.

The maximum number of apprentices who can enter training is determined by the training capacity of the employers in that trade. The capacity is determined by the maximum allowable proportion of apprentices of journeypersons—defined by agreement between employers and unions—and the actual number of journeypersons in industry. There is a further restriction in the trades requiring Electrical Registration. In his or her first few years the apprentice must be under the direct supervision of (in the physical presence of) a qualified tradesperson. This may limit the maximum number of apprentices an employer can employ rather than the maximum allowable proportion.

The District Commissioner and Local Apprenticeship Committee are expected to ensure that contracts are maintained and successfully completed. Maintenance involves ensuring the contractual obligations are met, that the range and quality of training given to the apprentices are satisfactory, and that apprentice behaviour is satisfactory.

During training employers and apprentices are simultaneously part of the training system and part of the environment. That is, they have training roles to play as part of the training system and other roles to play as part of the environment. Environmental influence on training is, therefore, effected through these roles interacting with the training role (i.e. role conflict within the employer and within the apprentice). Role conflict will affect the willingness of an employer and apprentice to remain in the contract, the willingness of an employer to provide the range and quality of training desirable, and the willingness of an apprentice to learn.

Interaction between an apprentice and employer in their training roles, between an employer and other aspects of the training system, and

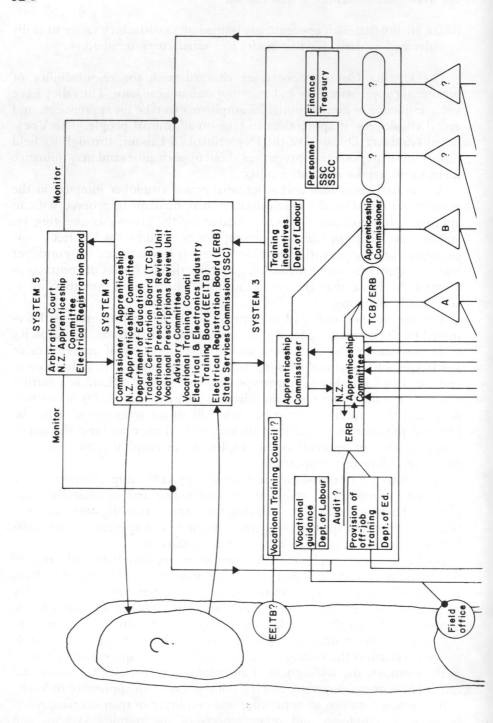

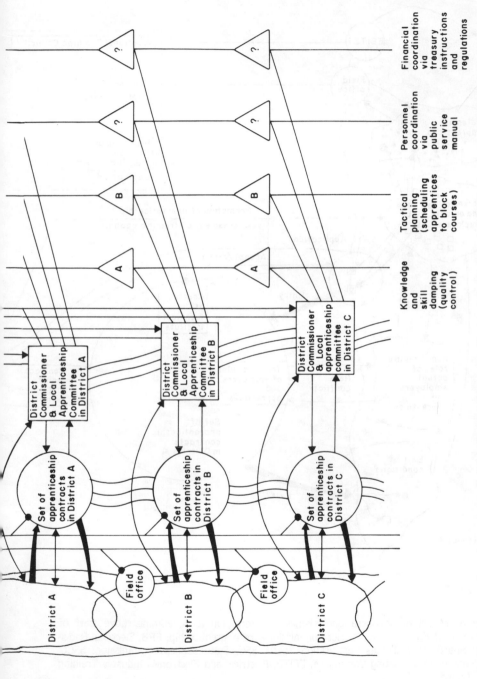

Figure 2. Model of a trade training system in New Zealand (Recursion 2). A, administrative staff of Department of Education; B, District Commissioner of Apprenticeship; SSCC, State Services Coordinating Committee

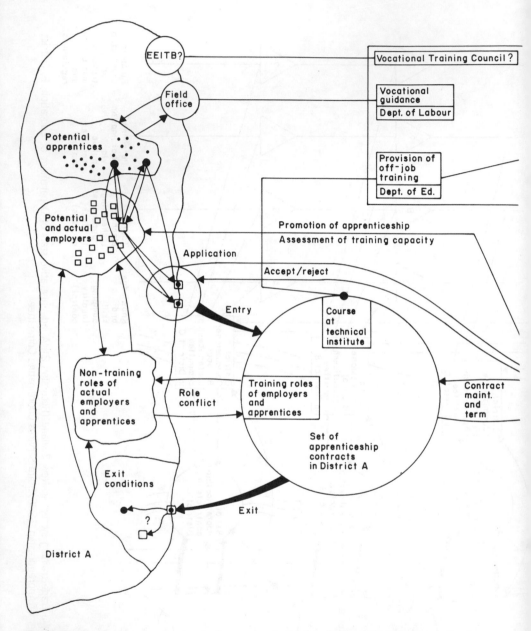

Figure 3. Model of an operational element at Recursion 2. A, administrative staff of Department of Education; B, District Commissioner of Apprenticeship; ERB, Electrical Registration Board; TCB, Trades Certification Board; SSC, State Services Commission; SSCC, State Services Coordinating Committee; EEITB, Electrical and Electronics Industry Training Board

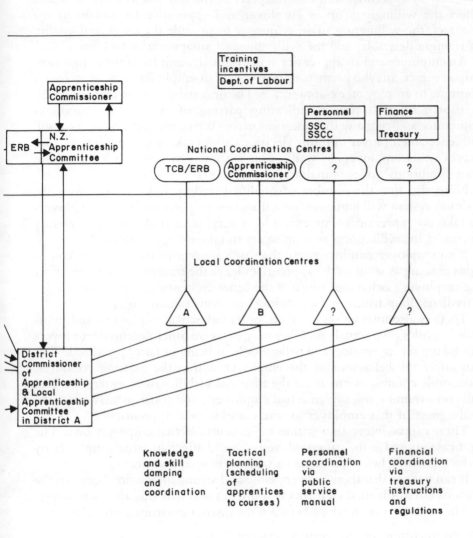

Knowledge and skill damping and coordination

Tactical planning (scheduling of apprentices to courses)

Personnel coordination via public service manual

Financial coordination via treasury instructions and regulations

between an apprentice and other aspects of the training system, will also affect the willingness of an employer and apprentice to remain in the contract, the willingness of an employer to provide the range and quality of training desirable, and the willingness of an apprentice to learn.

An employer and an apprentice will eventually exit from the apprenticeship contract, maybe prematurely. On exit an employer may or may not continue to employ an ex-apprentice. The uncertainty of the exit condition is shown by two arrows indicating parting of an employer and ex-apprentice associated with a question mark. Whatever happens on exit will influence an employer in future decisions about taking on apprentices (and maybe other employers as well). This is shown by a feedback loop from 'exit conditions' to 'potential/actual employers'.

Note also that the number of qualified tradespeople exiting from the training system will influence the willingness of potential/actual employers to take on apprentices. For example, a surplus of tradespeople probably decreases the willingness of employers to take on apprentices.

If an employer employs more than one apprentice then the exit conditions ensuing as some of the apprentices leave the training system can affect the employer's behaviour towards the other apprentices. This is shown by a feedback loop from 'exit conditions' to 'non-training roles'.

There is also interaction between 'potential/actual employers' and 'non-training roles'. An actual employer may perceive other potential employers not taking on apprentices and believe (s)he is being unfairly penalized. This can affect the behaviour of the employer within the existing contracts. Economic conditions can affect the financial viability of an enterprise (and thus non-training roles of an actual employer), and consequently affect the willingness of that employer to take on additional apprentices.

There can be interaction within the potential/actual employer box. The first case quoted in the paragraph above could affect an actual employer by reducing his or her willingness to take on new apprentices.

It can be seen that there is a very close and strong interaction between the various environmental elements in a district and that this affects training.

There are also interactions between the district environments. These are:

(a) competition for apprentices between districts;
(b) competition for employers between districts—a problem of regional development and a sensitive political issue;
(c) competition for and movement of tradespeople from one district to another.

The last is a problem concerning many in the electrical/electronics industry. In recent years the government has embarked on the construction of large-scale schemes, mainly in the petrochemical industry. The high rates of pay offered to qualified workers has made it difficult for other employers to recruit and maintain staff. Furthermore, the rapidly growing computer industry has been accused of 'poaching' qualified staff from the traditional industry.

The interactions between the operational elements are:

(a) movement of apprentices from one district to another;
(b) common use of national resources (e.g. block courses at Technical Institutes);
(c) the common mandatory electrical training requirements established under the Electrical Registration Act 1979;
(d) the common off-the-job training prescribed in and examined by the Trades Certification Board (TCB);
(e) use of public finance: the requirement to abide by the Treasury Regulations and Instructions;
(f) servicing by public servants: the requirement to abide by the State Services Commission Public Service Manual.

The description of System 1 indicates that the operational elements are environmentally controlled. The District Commissioners and Local Apprenticeship Committees are trying to match a high-variety environment with a low-variety operational system, without appropriate attenuators and amplifiers. It cannot and does not work.

This issue can be clearly seen when the response time of the training system is compared with the dynamics of the environment. The response time (term of apprenticeship contract) is 3–4 years depending on the trade. Significant and unpredictable employment opportunities (and hence training requirements) can and will occur during this time period. Thus it is not possible to achieve stability, either of employment or of apprentice intakes by training alone. There are three, and only three ways, stability can be achieved (the methods may be used singly or in combination):

(1) Decrease the response time of the training system (e.g. shorten the apprenticeships). This option is not favoured by the unions.
(2) Decrease the sensitivity of the training system to environmental changes. It might be possible to provide broad-based training so that a person can change employment with little retraining; or undertake training on an international basis, using international trainees to smooth the fluctuating demands on the training system. The first option is supported by the unions but is not favoured by many employers. Furthermore, there is considerable doubt as to its effectiveness. The latter option was not favoured by the national government.
(3) Stabilize or counteract the environmental changes. Government purchasing power could be used to offset free-market effects, or skilled labour could be imported on a temporary basis (with short-term

contracts of 1–4 years) to dampen out labour fluctuations. (The latter option was contrary to national government policy, which made the situation less stable.)

The last option was considered the best one for the electrical/electronics industry because of the dominance of the environment and the size structure of the industry: 88 per cent of the companies employ fewer than 50 people (DSIR, 1981). We wrote a paper outlining an integrated scheme for structuring the electrical/electronics industry (as defined by DSIR Report 1981) and its environment. This was submitted to four cabinet ministers. A slightly modified version of the paper was presented at a conference of the Institution of Professional Engineers of New Zealand (McCallion and Britton, 1985).

The major problems of System 1 are:

(1) The environment dominates the trade training system.
(2) No-one monitors the environment to determine how the environmental influences interact, how they affect entry into and exit from the training system, how they affect the behaviour of employers and apprentices during training, and how the training system affects employers and apprentices. Furthermore, auditing of the performance of the operational elements is inadequate.
(3) The range and quality of on-the-job training does not seem to be properly monitored and controlled. The range of training an employer can provide is restricted by his/her operations and will be taken into account during the apprenticeship application. However, the employer may not actually provide that range of experience to each apprentice.

The first problem has already been discussed. The second is really a System 3 function (of the trade training system) because the operational environments will have some aspects in common and also because they interact. The Vocational Training Council (VTC) and/or the Electrical and Electronics Industry Training Board (EEITB) are the appropriate bodies to undertake this work. The work involves continuous monitoring and some research. VTC and EEITB do not have staff capable of performing the research required. A temporary measure to remedy this situation is for VTC and/or EEITB to contract out work to expert consultants. In the long term, the staffing situation should evolve towards employing personnel with the required skills on a full-time basis.

The last problem is the responsibility of the District Commissioners and the Local Apprenticeship Committees. The latter meet infrequently and therefore cannot carry out the monitoring which is a full-time job. The District Commissioners are employed full-time and should do this, but the range of trades and number of employers and apprentices they have to cover may require assistance to be provided.

System 2

This exists to dampen oscillations due to interactions between district operations. In all cases, a System 2 exists for the interactions previously noted. With regard to interactions (c) and (d) (page 159), a problem mentioned during Britton's (1984) investigation is to get first-year apprentices registered in time for the First Qualifying Examination. TCB and the administration staff of the Department of Education do get apprentices registered but only after considerable effort by all parties. It should be possible to improve the procedure so as to reduce the effort and stress on all concerned.

No coordinating regulatory centre has been shown in Figures 2 and 3. Presumably this function is performed by the Apprenticeship Commissioner. (We have not been able to verify this, thus the centre has not been included in the diagram.)

System 3

System 3 is not satisfactory. The audit function is inadequate and monitoring of the environment (for synergistic purposes) is missing. This has already been discussed.

The interaction between Systems 3 and 4 is inadequate. The problem is mainly a consequence of inadequacies at the System 4 level, to be discussed next. (Note that Systems 3 and 4 interact primarily through the NZ Apprenticeship Committees and TCB Prescription Committees.)

System 4

The main problems of System 4 are as follows:

(1) How should the frequency of prescription revision required by the different trades be determined (Britton, 1984)?
(2) How can the prescription revision procedure be improved and the prescription structure be changed to allow frequent (2- or 3-yearly) revisions for the trades which require this (Britton, 1984)?
(3) How can it be ensured that the contents of the prescriptions are relevant to the requirements of industry and are matched to the psychological characteristics and needs of the apprentices (Britton, 1984)?
(4) No-one on a *full-time* basis appears to be monitoring the environment for changes that could affect the development of trade training (e.g. technological change). The current approach is inactive rather than anticipatory (Britton, 1984).

(5) Auditing of System 3 common services, System 3, System 4 and System 5 is inadequate. This is a problem of network development at the 3,4,5 levels.

(6) System 4 fails to provide a focus for integrating the various aspects of development. This is not surprising since no explicit model of the training system and its environment has been developed. A consequence of not having an explicit model is that interaction between Systems 3 and 4 occurs through personal persuasion (or authority) using personal models.

In addition, The NZ Apprenticeship Committees meet approximately four times a year. Therefore, large time lags can be expected in getting any new developments accepted and implemented.

With regard to (1) and (2), it should be noted that a newly established Vocational Prescriptions Review Unit is having a profound effect on prescription revision. However, there are a number of alternative prescription revision procedures available. These should be presented to the parties concerned so that they may choose the procedure which they believe will suit their needs. The procedures will need monitoring.

It can also be noted that during Dr Britton's investigation the Vocational Training Council initiated training needs analysis using the Dacum approach. A research unit is to be established to continue this work. The unit should be capable of dealing with problem (3).

Britton (1984) clearly showed the *ad hoc* development nature of the network (problems (4), (5) and (6)). Network development proposals should be integrated and the developmental process should be monitored. Furthermore, Systems 3, 4 and 5 need to be audited (monitored and investigated) to ensure they are performing well. This function and environmental monitoring are appropriate System 4 functions.

We considered the System 4 role to be appropriate for the VTC and/or EEITB. As it will be a full-time job, permanent staff should be employed who can undertake the required work. Britton (1984) recommended that a research unit be established under the auspices of the Vocational Training Council. Two of its essential proposed functions are to carry out management auditing of the training network and to have responsibility for monitoring of the environment. The latter suggestion is contrary to current practice. Development of trade training is primarily undertaken by the Department of Labour. We suggested that VTC be responsible because this kind of work has to be performed for other parts of the training network. It seemed sensible to have one body responsible for all parts. In practice, of course, it is envisaged that the Department of Labour will develop proposals for trade training and that VTC will then carry out a coordinating/integrating role to match trade training proposals with other network development proposals.

System 5

A major problem with System 5 is that it does not have an explicit model of the training system which it can use to monitor how well the system is currently operating and whether it is developing in a desirable manner.

The provision of such a model is a System 4 function, and again the VTC would be the appropriate body to undertake this work. An appropriate model has been suggested by Beer (1962). The main variables of the model are supply, demand, willingness and effectiveness. It is no coincidence that these variables are implicit in the description of System 1 given previously.

Notes on Recursion level 1 (trade)

The purpose of this system (level of recursion) is to provide sufficient numbers of people efficiently and properly trained in New Zealand in all trades, so as to meet the country's needs; yet ensure adequate employment opportunities for all tradespeople. The model at this level of recursion is shown in Figure 4. (Note that the diagram is indicative only; clearly not all the trade groups are represented.)

System 1

The operational elements of System 1 are the trade training systems previously described. There is, therefore, no need to describe further the operational elements.

In order to manage the collection of operational elements (collection of trade systems) it is necessary to understand both the environmental interactions between trade systems and the operational interactions (Beer, 1979). The latter are weak and can be summarized as follows:

(a) There is competition for financial resources to undertake trade training (mainly competition for resources from the Department of Education).
(b) Some trades have a common off-the-job training course (e.g. the Electrical Wiring Trade and Radio & Associated Electronics Industry Trade).
(c) With regard to the use of public finance, there is a requirement to abide by the Treasury Regulations and Instructions.
(d) With regard to servicing by public servants, there is a requirement to abide by the State Services Commission Public Service Manual.

The trade operational management units can be closely interlinked via the NZ Apprenticeship Committees. Common policy promulgation by

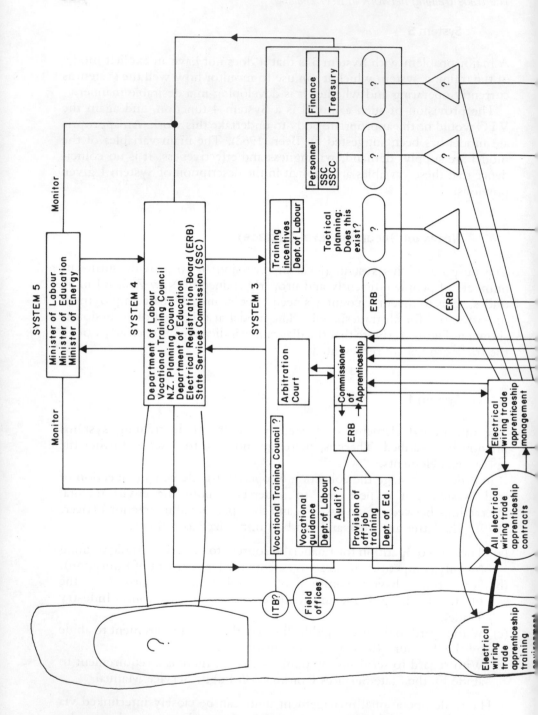

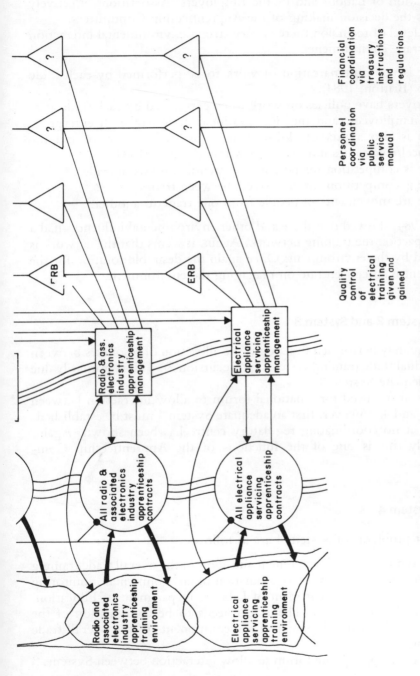

Figure 4. Model of the whole trade training system in New Zealand (Recursion 1). SSCC, State Services Coordinating Committee; ITB, Industry Training Board

the Federation of Labour and/or the Employers' Associations effectively constrains the decision-making of the Apprenticeship Committees.

At this level of recursion there is very strong environmental interaction between trade environments:

(a) There is union demarcation of work to be performed by each trade group (Britton, 1984).
(b) Employers have policies on work to be performed by each trade and/ or its employees, and the organizational design of each enterprise. There is also a mix of skills required by each enterprise and the electrical/electronics industry as a whole (Britton, 1984).
(c) There is competition for potential apprentices between trades.
(d) There is competition for employers between trades.
(e) There are movements of people from one trade to another.

Britton (1984) showed that these and other environmental influences had a major impact on the training network. Again, it seems that the network is dominated by the environment. Once again the desirable solution for the government is to concentrate on managing the environment.

System 2 and System 3

An inadequately performed function is the allocation of resources between the individual trade training systems (at Recursion 2). This is primarily due to an inadequate System 4.

There is also a need for a national forum to allow interaction between Systems 3 and 4; however, first an adequate System 4 must be established.

Note that no coordinating regulatory centre has been shown. Again, presumably this is one of the functions of the Apprenticeship Commissioner.

System 4

The major problems of System 4 are as follows:

(1) There is no *focus* for development and audit of the overall trade training system. There needs to be coordination of environmental monitoring performed by the Department of Labour, Department of Education, VTC and other bodies. There also needs to be coordination of the monitoring and development of the components of the overall trade training system.
(2) There is no appropriate forum to allow interaction between Systems 3 and 4.

The first problem could be overcome if VTC established a research unit (Britton, 1984). Developing an explicit model of the system and its environment would be a primary task of the unit; Beer's (1962) model would be appropriate.

The second problem could be resolved if VTC provided a national forum to allow discussion and debate on national trade training. Beer (1975, 1979) suggests an operations room for this purpose.

We believe it is necessary to have the forum controlled by VTC as it is the principal advisor to the government on training matters, and it is the only government agency authorized to coordinate all bodies involved in vocational training. Participants for the forum could be selected from the individual NZ Apprenticeship Committees and/or from the controlling committee of VTC.

Note that we are not suggesting another body with authority over the NZ Apprenticeship Committees. We are suggesting a 'conference' where members of the NZ Apprenticeship Committees (or their representatives) can come to an agreement on how trade training should develop in New Zealand.

System 5

The problem confronting the government ministers is how to monitor the operation and development of the trade training system so that they can intervene if it is necessary. This is another problem for System 4 and VTC.

Note on interfacing training with employers

Our analysis started at Recursion 2 because complaints from the electrical and electronics industry were about national organizations. Thus we were led to believe that the 'lack of coordination' occurred at the national level. However, it is clear that the diagnosis so far has not answered the question: What lack of coordination are the employers perceiving? The answer to the question lies at the lowest level of recursion—Recursion 3.

A major factor affecting trade training—perhaps the most important factor—is the willingness of employers to take on apprentices. Yet the trade training system as it currently operates presents *many faces* to an employer who trains, and *hinders* rather than assists. An employer of apprentices has to deal with a number of components of the network: at least one apprenticeship committee, the Electrical Registration Board, the Trades Certification Board, the Electrical and Electronics Industry Training Board, the Department of Education. Each component has its own rules, regulations and 'language' which the employer is expected to learn. This was the lack of coordination that was perceived by the employers and which apparently initiated their complaints.

In terms of Beer's model the interface problem occurs because there is no corporate regulatory centre for System 2 of Recursion 3. Consequently, the operational elements experience a variety overload 'due to the unnecessary diversity in presentation' (Beer, 1979, p. 473). There is clearly a need to design a corporate regulatory centre to coordinate the *approach* to employers.

Our recommendation was to establish a 'training consultancy service'. Such a service would interface the employer with the network (and vice versa) by: (i) handling employer enquiries relating to training, (ii) dealing with employers' training problems (thus removing this burden from employers), (iii) disseminating relevant information to employers, and (iv) developing and monitoring training files for smaller employers. (Note that the service would also interface employers with other training systems, e.g. AAVA.)

It was recommended that provision of such a service should be the primary task of EEITB with respect to the electrical/electronics industry. It was acknowledged that there would be problems in implementing the proposal:

(a) Lack of resources means the EEITB cannot effectively cover New Zealand on a national basis.
(b) There are already established agencies with statutory responsibilities in this area; for example, the Department of Labour (apprenticeship division). Conflict is certain to occur if the Board performs the role outlined here.

We stressed the need to confront and solve these problems and gave informal suggestions on how this might be achieved.

An overall view of training

Is the whole training network (Recursion 0 in Figure 1) a viable system? The answer is yes, *but* it is pathologically viable.

The concern of management at this level (Cabinet and Parliament) is the appropriation of finance for and allocation of finance between parts of the training network. In the financial appropriation and disbursement process, each part of the training network is treated separately. Universities obtain funding via the University Grants Committee (which is virtually treated as a separate Department of State for this purpose); AAVA funding is subsumed in Vote Education; trade funding is subsumed in Votes Education and Labour; and so on.

There is *no integrated training system* at this level of recursion. We believe that recent suggestions for a bureaucratic structure for all vocational training is a recognition by the proponents of the lack of integration and coordination at this level of recursion. However, we do not support the recommendation for a bureaucratic structure nor a recommendation for a special cabinet committee on training. Organizational redesign at this level is closely interrelated with the next higher level involving *all* state agencies. The creation of another quasi-independent structure without integrating it within the overall design of State agencies would simply add more complexity as far as Cabinet and Parliament is concerned. What is required is *reduction* of complexity, so that cabinet ministers and Parliament can understand what is going on.

It should be realised that System 5 of Recursion 0 is made up of Cabinet, Caucus and Parliament, and their associated committees. System 3 consists of individual ministers and cabinet committees and other 'bodies' that have delegated financial authority for training.

The need is for an appropriate System 4 and an appropriate System 2. Both these functions could be performed by VTC. VTC could be assisted in its System 2 role by the Industry Training Boards. In effect, this is a specification for a partial System 4 and a partial System 2 at the next higher level of recursion, because it only deals with training; it does not consider the other functions government has to fulfil.

As its System 4 role, VTC should gather information from the various parts of the training network and integrate this information into a model of (description and explanation of) the training network. This could then be presented to the Ministers of Labour and Education (the two ministers with major responsibility in this area) to aid them and Cabinet to perceive the training consequences of their financial allocations to training. Again, Beer's (1962) model could be used for this purpose.

It seems fair to say that the current mode of operation is to allocate finance independently requested by various parts of the training network, from which the numbers that can be trained are determined. That is, personnel resource planning follows from financial resource provision.

This is the reverse of what should be happening. Training targets should be set and then financial provision sought to meet these targets. Some of these training targets—particularly those relating to industry—could be negotiated within the VTC as all affected parties are represented on Council.

Finally, VTC should be monitoring and auditing the whole network so as to improve its performance. That is, it should be conducting management audits and devising and integrating network development proposals.

Prescription: Summary of functions to be performed by VTC and EEITB

Trade Training Recursion 3

(1) Systems 2 and 3:
Interface training with employers and provide assistance in training matters (EEITB).

Trade Training Recursion 2

(2) System 3:
Monitor the operational environments (VTC and/or EEITB).
(3) System 4:
(a) Assist prescription development (EEITB and/or VTC).
(b) Network audit and development and responsibility for monitoring the environment (VTC).

Trade Training Recursion 1

(4) System 3:
Monitor the operational environments, audit operations and investigate common skills/knowledge between trades (VTC).
(5) System 4:
(a) Provide a *focus* for network development and monitoring of the environment (VTC).
(b) Provide a forum to allow Systems 3 and 4 to interact (VTC).

The Whole Training Network (Recursion 0)

(6) System 4:
(a) Coordinate and integrate development proposals for the ministers (VTC).
(b) Undertake management audit of network and prepare network development proposals (VTC).

Conclusion

Application of Beer's model to the trade training system in New Zealand has shown that none of the levels of recursion is effectively viable; the highest level being the least viable. More importantly, though, the model enabled the authors to determine the factors preventing viability and to suggest appropriate solutions. Considering that the total training network for all trades consists of approximately 300 committees, quangos and government departments, the diagnostic power of Beer's model can be readily appreciated.

We shall conclude by briefly describing two performance indices which proved valuable in our work. The first is the relative support given to training programmes (the relative proportions of trainees defined according to an organizational classification): a public sector versus private sector classification was used. The results are shown in Figure 5. Using these results we were able to clarify several issues relating to trade and technician training.

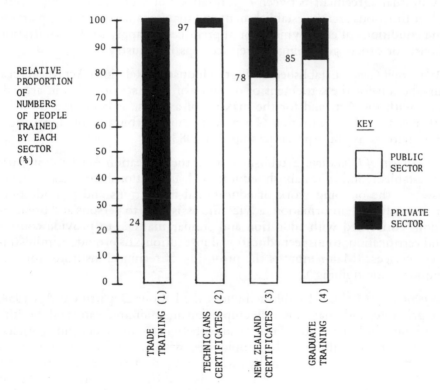

Figure 5. Comparison of public and private sector support of national training programmes (electrical/electronics only)

(1) Relative proportion based on apprenticeship contracts in force for year ended 31 March 1981.
(2) Relative proportion based on employment figures of winners of Technicians Certificates in 1980.
(3) Relative proportion based on employment figures of winners of NZ Certificates in 1980.
(4) Relative proportion based on employment figures of electrical engineering graduates from Canterbury University for the three years 1978–81.

The other important indicator is the drop-out rate from a training programme. High rates are a cause for concern. In our study, two high rates were obtained: 50 per cent of students sitting Engineering Intermediate fail to enter the Engineering Faculties; and 66 per cent of the students who start studying for a New Zealand Certificate of Engineering fail to complete the course.

APPENDIX: Statutory training functions of organizations listed in Figures 2, 3 and 4

Apprenticeship Committees: Established under the Apprenticeship Act 1983 by mutual agreement between organizations of workers and employers. Their functions are: the establishment and maintenance of training patterns and conditions of employment for apprentices; to apply to the Arbitration Court for orders governing apprenticeships in industry.

Arbitration Court: Established under the Industrial Relations Act to exercise jurisdiction for the settlement of disputes of interests or rights in accordance with the Act, and for the making of awards, as conferred on it by other Acts. (The Arbitration Court is appointed arbitrator on apprenticeship matters by the Apprenticeship Act 1983.)

Department of Education: Established under the Education Act 1964: in part, to establish and disestablish educational institutions, to make grants towards the ongoing work of educational institutions and provide bursaries; to furnish information, advice and assistance to persons and organizations concerned with education and similar matters; to provide courses and certification; to inspect educational institutions; to provide administrative services. (Major input is the provision of technical institutes for off-the-job trade training.)

Department of Labour: Established under the Labour Department Act 1954: to promote and maintain full employment, training, safe and healthy working conditions, good relationships between employers and workers, and the proper fulfilment by employers, workers and other persons of obligations placed upon them by awards and industrial agreements and by the Acts, Regulations and Orders administered by the Department. (Major input is via the Apprenticeship Division which services the Apprenticeship Committees.)

Electrical Registration Board: Established under the Electrical Registration Act 1979: to conduct or provide for the conduct of examinations for electrical registration; to authorize registration; to discipline or bring prosecutions against people who break the Act. (Prescribed electrical work in New Zealand can only be undertaken by people who are suitably registered.)

Electrical and Electronics Industry Training Board: Industry Training Boards are established under Section 14 of the Vocational Training Council Act 1982: to promote the benefits of vocational training; to undertake research and to make recommendations on the development and implementation of vocational training programmes; to cooperate with the Vocational Training Council in implementing vocational training programmes; to undertake vocational programmes if these otherwise cannot be met.

New Zealand Planning Council: Established under the New Zealand Planning Act 1977: to advise the government on planning for social, economic and cultural development; to comment on programmes and priorities; to advise on coordination; to act as a focal point for consultative planning; to foster discussion among agencies concerned with planning on a medium-term time horizon. (The Council has published reports on trade training.)

State Services Coordinating Committee: Established under the State Services Conditions of Employment Act 1977. The Act does not specify training as one of the functions of the Committee. However, the State Services Commission has responsibility for training within the Public Service and for other State Agencies. With the assent of the Committee, the Commission uses the Committee as a forum to discuss and resolve training matters common to the Public Service and other state Agencies.

State Services Commission: Established under the State Services Act 1962 in respect of the State Services and any other body whose expenditure is met wholly or partially by money appropriated by Parliament: to act to provide management consultative services; to act to review and approve establishments of staff; to act as the central personnel authority for the Public Service; to prescribe basic training programmes, furnishing advice on and assisting with the training of staff and making recommendations to the Minister of State Services on the facilities necessary for the proper training of staff.

Trades Certification Board: Established under the Trades Certification Act 1966: to set standards and establish prescriptions suitable for the various trades; to make provision for trade examinations and to issue certificates or diplomas to those who have passed such examinations.

Vocational Prescriptions Review Unit: Established under the Education Act 1964: to provide the authority for Advanced Vocational Awards and the Trades Certification Board with professional expertise in the development, evaluation and updating of prescriptions and syllabuses.

Vocational Prescriptions Review Unit Advisory Committee: Established to advise the Director-General of Education on policy matters affecting the Vocational Prescriptions Review Unit and, in addition, to advise the Council of the Controlling authority (Petone Technical Institute Council) on the progress of tasks to be performed by the Unit and other matters concerning the functions of the Unit.

Vocational Training Council: Established under the Vocational Training Council Act 1982: to make recommendations, carry out research, and collect and disseminate information on vocational training; to coordinate the activities of Industry Training Boards.

References

Ackoff, R.L. (1981). *Creating the Corporate Future.* New York: John Wiley.

Apprenticeship Act, 1983.

Beer, S. (1962). 'An operational research project on technical education', *Operational Research Quarterly*, **13**, 179–99.

Beer, S. (1966). *Decision and Control.* London: John Wiley.

Beer, S. (1970). 'Managing modern complexity', *Futures*, June and September, 114–22, 245–57.

Beer, S. (1972). *Brain of the Firm.* New York: Herder and Herder.

Beer, S. (1975). *Platform for Change.* London: John Wiley.

Beer, S. (1979). *The Heart of Enterprise.* Chichester: John Wiley.

Britton, G.A. (1984). *Training for Electrical and Electronic Vocations in New Zealand.* New Zealand Vocational Training Council.

DSIR (1981). *Electronics, Pt.2*, Discussion Paper 5, New Zealand Department of Scientific and Industrial Research.

Electrical Registration Act, 1979.

Emery, F.E. (1977). *Futures We Are In.* Leiden: Martinus Nijhoff.

Emery, F.E. and Trist, E.L. (1975). *Towards a Social Ecology.* New York: Plenum-Rosetta.

Kingdom, D.R. (1973). *Matrix Organization: Managing Information Technologies.* London: Tavistock.

McCallion, H. and Britton, G.A. (1985). 'An integrated scheme for the development of the electrical/electronics industry in New Zealand', paper presented at the 1985 conference of the Institution of Professional Engineers of New Zealand.

Thompson, J.D. (1967). *Organisations in Action.* New York: McGraw-Hill.

Trist, E.L. (1980). 'The environment and system-response capability', *Futures*, April, 113–27.

The Viable System Model: Interpretations and Applications of Stafford Beer's VSM
Edited by R. Espejo and R. Harnden
© 1989 John Wiley & Sons Ltd

8

Application of the VSM to commercial broadcasting in the United States

Allenna Leonard
Viable Systems International

The Viable System Model is used to examine the place of the television station in commercial broadcasting in the United States. The station, although it is the only unit licensed to broadcast, has a great deal less autonomy in practice than might be surmised. These recursive relationships include ownership, network affiliation, government regulation and membership in a community's information resource base. Each has its own criteria for satisfaction which are more or less compatible with those of the others. Using the VSM, these relationships are described and a diagnosis is offered to suggest where feedback loops are missing or channels of communication too constrained. Recommendations are made to enhance the activities of entities in these recursive structures to improve their viability.

Introduction

The structure of commercial television broadcasting in the United States is complex: the result of historical accident, political expediency and marketing technique, on the one hand, and the physical and geographical limits of channel capacity and signal transmission on the other. Moreover, the situation changes by the week with new technologies, unresolved legal questions and Congressional and Federal Communications Commission (FCC) attempts to deregulate or to reregulate various aspects of broadcasting. Neither law nor social policy regarding television have been compre-

175

hensive or coherent, and it has been difficult to keep pace with developments in technology or new business arrangements.

The principal legislation governing broadcasting remains the Communications Act of 1934 (As Amended). Although most of its provisions were written when our present system of communications was science-fiction, it has not been possible, despite several tries, to build a consensus for any particular rewrite. As the legal and social debate continues, attempts at a rigorous description of the situation should be a constructive contribution to the public discourse.

Cybernetics, particularly the management cybernetics of Stafford Beer, is well suited to a rigorous description and a diagnosis of broadcasting. In particular, his Viable System Model (Beer, 1979, 1981, 1985) provides a means to distinguish the contexts in which the issues arise and to look at the several structures in which the active instrument called a television station is embedded. The model specifically deals with communication channels and their capacities, transduction of messages from one system to another, and the management of variety.

The situation is complicated by the differing purposes ascribed to television by its stakeholders. The 'primary purpose' of television mentioned by a sample of broadcasters, government staff members and spokespeople for media-reform organizations was variously nominated as 'entertainment', 'information' and 'to make money for the shareholders' (Leonard, 1987). These answers reflected a differing set of identities and perceptions about the scope of responsibilities which accompany the broadcast license. Other stakeholders might well offer additional purposes

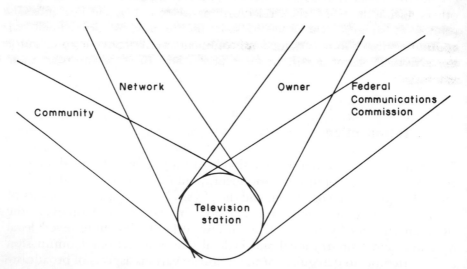

Figure 1. The circle and conic section picture

such as 'to sell our products', 'to contribute to the US balance of payments' (through foreign sales), 'to pass the time', 'to record history in-the-making' to 'to mould social attitudes'. A full treatment of the variety of these positions would be a book in itself, but it is well to keep them in mind.

In this chapter I examine the television station as the system in focus. I consider its commercial relationships with its owner; its affiliated network; its relationship with the regulatory agency (the Federal Communications Commission); and its relationship with its community as part of the society's information base (Figure 1). These non-commercial roles of the television station will be seen to be less specifically defined, with a corresponding lack of clarity about what would constitute desirable change and evolution. Because this clarity does not exist, the recursions describing these two roles will contrast my opinions of what would constitute desirable futures with the present situations.

Background

There are over 1300 television stations in the United States. (*Broadcasting*, 1988*a*). About a thousand are commercial stations, supported by the audiences, as measured by the rating services, to advertisers. Slightly more than 600 are affiliated, in roughly equal proportions, with one of the three major networks. Approximately 300 television stations, the remainder of the total, are public or educational stations: these are supported by a combination of viewer memberships, business and foundation donations and government grants. In 1987, cable system coverage passed 50 per cent and satellite signals were received by about 1.7 million homes (*Broadcasting*, 1988*a*). Cable systems offer local channels (often with improved reception), distant broadcast signals (such as the superstations), and channels dedicated to children or to movies which are available only on cable.

The station is the basic unit of broadcasting. It is independently licensed to transmit a broadcast signal to a community or communities in a specific geographical area. It is part of the community's commercial life, like the grocery store, and part of its resources of information and entertainment, like the library, the newspaper and the movies. Unlike the grocer or the newspaper, television (and radio) are subject to channel capacity constraints in addition to the market constraints which apply to the others. If the newspaper is not serving the community—or some substantial portion of it—another paper will soon be on the news-stands; if the lines are too long at the movies, there will soon be more theaters sharing the business. Not so the broadcasters. Once the spectrum space is allocated, there is no more. In contrast, cable systems are limited only by the capacity of their wiring. Although laying new cable may be expensive, it can be done if there is sufficient market demand.

Like any other market to which entry historically has been restricted, oligopoly conditions occur (Weiss and Strickland, 1982). Stations do not try to match the variety of audience interest and ability with programming variety but compete for the largest and most profitable audience for each time period; for example, adult women in the afternoon (Gitlin, 1983). When the number of stations increases, programming variety increases slightly. A fourth or fifth station may show childrens' programs or a talk show instead of another soap opera. Public television tries to fill some of this gap by providing programming for children and the well-educated, but it too is essentially a mass medium. Where cable is available, subscribers have a wider choice, from a dozen channels in the older systems to five dozen or more in the newer ones. Cable has made feasible sports, news, music, movies, soft porn, foreign language and children's channels, although some are available only for an additional fee. Because many cable channels originate as broadcast channels, spectrum scarcity continues to have a constraining effect on cable program variety. This could change rapidly with the implementation of new technologies. For example, the enhanced capacity of fiber-optic cable could make it possible for a subscriber to select from among hundreds of channels devoted to everything from agricultural research to chess matches.

According to the Communications Act, when the station receives the exclusive right to broadcast over a portion of the scarce resource of the spectrum, it is licensed to serve 'the public interest, convenience and necessity'. Because of the public interest provision, members of the community may negotiate with the station or may complain to the FCC if they are not satisfied with the station's service. The 'public interest' criterion, on which the resource bargain between the broadcaster and the community of license is based, has never been clearly defined. The Reagan Administration favored market forces to ensure appropriate services, although segments of the population—especially the young and the old— are not proportionately represented by the market. Nor are all segments of a license area. Until recently, the state of New Jersey had no VHF (the strong signal) stations and received its television programs from stations located in New York City or Philadelphia. Its cities pressed the case that they were underserved by local news programming and were finally rewarded when WWOR transferred its license to New Jersey.

Broadcasting in the United States began without restrictions on market entry. Government licensing came about in the 1920s because more radio stations were trying to broadcast in the same locations than could be accommodated by the radio spectrum. The result was so much interference that enthusiasm for radio, and the consequent purchase of sets, was rapidly declining. In addition to limiting the number and strength of signals, the government, through the regulations, tried to provide some procedures whereby the available licenses (a handful in each area) would go to the stations which would provide the best programming to the community. This was to be accomplished through competition by allowing other groups to approach the FCC and compete for the license at renewal time. To this day, the relationship between the station's programming service and its comunity remains the basis for the license, although the Communications Act specifically forbids censorship. The possibility for another applicant to compete for that license remains in the law although it has seldom been used. It is rare for a station to lose its license and even rarer for the loss to be associated with program deficiences. Most of the roughly one hundred lost licenses were attributed to commercial violations (e.g. misrepresentation, fraud, unauthorized sale) or technical violations.

The Viable System Model for television will be helpful in teasing out some of the complexity in the industry today. The scope of this chapter is by no means sufficient to depict it in full. Nevertheless, I hope that the skeleton will highlight some important features and indicate directions for exploration or improvement.

The station itself sits in the middle of a recursive structure. It is an organizational unit in several systems extending to higher levels of recursion and it contains subsystems within it. First, let us look at the station itself, with three subsystems.

The station: Recursion one (Figure 2)

The television station's most visible organizational unit is the one responsible for broadcasting itself: the transmission of programs within a geographical area (say, a circle with a 30-mile radius). The station promotes its programs both on air (promotional announcements or 'promos') and in other media (e.g. the daily paper). Feedback about its programs is received from the environment. It finds out how many viewers it has from the (independent) Neilson and Arbitron ratings. The ratings are a low-variety means of feedback: they report, within a certain margin of error, the number of viewers. They do not reveal anything about how attentive they were or how much they liked the program. Some additional, if less consistent, feedback comes from letters and telephone calls, market surveys and critical reviews.

This unit is concerned with the process of broadcasting. It transfers the programs and advertisements from the form in which they arrive in the station (over telephone lines, on video tape or live through studio production or remote hook-ups) to broadcast mode. Producers and technical

AXIOM NRT → Mention it

No Vertical Variety

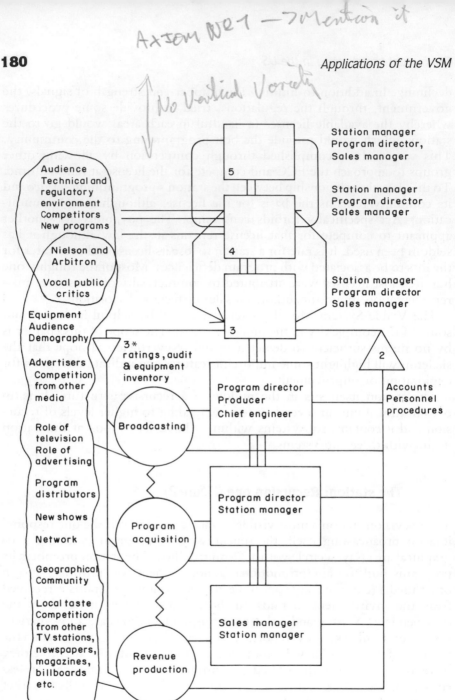

Figure 2. The TV station: Recursion 1

supervisors make up the management team, assisted by the traffic manager and librarian.

Acquiring programs to broadcast is the job of the next organizational unit. The program director, in consultation with other staff, matches the budget with the desired balance of programs and fills the available time with programming from different sources. The processes associated with obtaining programming through network feed, local production and syndicated program distribution are elements of Recursion 2.

The station's third organizational unit raises revenue to pay for its operations and to make profits for its owners. By far the largest source of revenue is the sale of advertising spots to local and national advertisers. A typical station will have more than 400 half-minute spots available each broadcast day. Most of them will be for sale to advertisers, with the price of each spot based on the number, age and sex of the viewers watching at that time.

If the station is affiliated with one of the major networks, it will sell about ten hours of its broadcast day to the network. In return, it receives programming which would be too expensive to produce itself or purchase from distributors, a fee based on the number of hours cleared for the network (called network compensation), a portion of the commercial spots to sell on its own, and the benefit of network shows to draw viewers to programs the station itself is running adjacent to them.

An additional source of revenue is the selling of time to political candidates prior to elections. At these times, messages from political candidates or parties replace some of the messages from advertisers. Promotions for forthcoming programs and messages from non-profit organizations are aired in the remaining time slots. The activities associated with each of these sources of revenue can also be considered to be elements at the next lower level of recursion. The revenue-producing unit will normally be the responsibility of a sales director, although some stations may farm out their sales functions to an agency.

The managers of the three organizational units often communicate among themselves in their System 1 roles as well as appearing in the three/four/five boxes. This can be confusing because the station manager, program manager and sales manager, in practice, may be involved in everything from a meeting with an important client, to the production of a particular show two levels of recursion down, to policy recommendations to owners one or two levels up.

System 2's function is to dampen oscillations between the activities of the component systems. In this role, it maintains consistency in routine accounting practices and personnel procedures and ensures that station identity messages and other materials are consistent with the style set by management.

System 3 manages the 'inside and now' of station activities. It coordinates the interaction between sales and programming and makes short-term financial and technical decisions. Station, program and sales managers may evaluate the impact of the ratings and decide to make schedule changes. In making the day-to-day decisions, they agree to budgets, ratify major decisions made at lower levels, and decide questions such as whether or not to cancel a show that have implications beyond the unit where the show originates. System 3 monitors the various subsystems to ensure that they are in compliance with government regulations. Management may institute a Three Star inquiry by hiring consultants to look at equipment inventories or personnel qualifications or to perform financial audits.

System 4 includes recruitment, training, public relations, market research and audience research as ongoing activities. It provides lookout points for programming trends, new technical developments and changes in competitive and regulatory climates. The station manager and program manager both participate in System 4.

System 5 makes overall policy and establishes the station's identity: 'We will not show R-rated movies before midnight' or 'We will maintain our role as sports promoters in this community' are examples of the choices made at this level. Despite the public-interest provisions in the Communications Act, there is no official community representation in System 5. There was a FCC regulation requiring stations to survey local leadership about the programming needs of the community on a periodic basis, but it did not carry any further obligations (e.g. to provide the nominated programs) and was dropped. The station manager typically functions as CEO.

The station's local environment includes its audience, advertisers, geographical community and competitors. There is a great deal of communication among the elements of this environment as well as between the station and the environment. At the national level, the environment includes suppliers of programs, equipment and personnel, government regulators, a wider range of competition, and the expectations of society. Criticism of the effects of television and the impact of debates about its role in society are also felt by the station.

Program acquisition: Recursion two (Figure 3)

Dropping down a level of recursion, we see the largest center of activity is in program acquisition. It components represent the sources of programs.

Programs can be received from the network (if the station is affiliated) in return for the option to sell most of the advertising spots in ten of the most popular viewing hours. Affiliated stations 'clear' most of the programs the network offers, but they retain legal responsibility for the programs aired

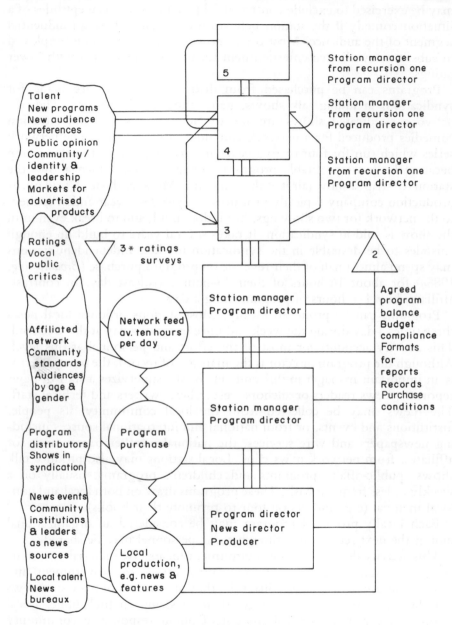

Figure 3. Program acquisition: Recursion 2

regardless of their source. The option to pre-empt network programming may be exercised to exclude controversial programs, or even episodes of a situation comedy if the station believes they might offend an influential segment of the audience. Most often, however, pre-emption is employed to substitute syndicated entertainment for network programs with lower ratings.

Programs can be purchased from distributors. Some are made for syndication, including talk shows, game shows and cartoons. The off-network syndicated shows are mostly adventure series and situation comedies produced for a network and aired in prior years. The network series which run for four or five years are over-represented in this group because the most profitable program schedule is one which allows the station to 'strip' a show (air it at the same time Monday through Friday). A production company typically produces 24 episodes a year and sells them to the network for two showings, but it does not begin to break even until the show is sold in syndication. It takes several years to build up enough episodes to be desirable in the syndication market. Independent stations may spend almost half of their revenue on program purchase (*Broadcasting*, 1988*a*) for about 16 hours of their 18-hour broadcast day. In contrast, affiliates buy 4–6 hours of programming a day.

Programs can be produced locally. Most stations produce local news three times a day during the week and once or twice a day at the weekend. This activity accounts for most of the effort and personnel at this level. Although the program director is the manager of record, the news director is an important manager in this unit. He or she supervizes a staff of news reporters, news readers or anchors, researchers, writers and techical staff. The 'news' may be collected from the local community, its people, institutions and events, or from national and international sources including newspapers and wire services, the national weather bureau and, for affiliates, from network news clips. Local stations may also produce talk shows, public-affairs programs and children's programs, usually on a weekly or less frequent basis. These programs draw on both local and non-local material (e.g. authors on tour to promote their books).

Each locally produced program may be considered an organizational unit at the next recursion, although some personnel may be shared.

This activity depends on the environment for its raw material as well its audiences. Some segments of the community (local government and institutions especially) have a dual role: they are sources of television news but they also use television to get their messages to the public. Local stations may provide a small amount of on-air response to community issues, such as allowing a few minutes of soapbox time to spokespeople for community organizations or for reading and answering letters.

Sixty or seventy hours of programming may be brought in from the major networks, CBS, ABC and NBC. The networks arrange for station management to preview and select the shows for each season. In practice, most of the offerings are accepted. Ongoing programs have a developed audience and new ones are tried out on sample audiences in pilot form before they are offered. Once a station has a network affiliation contract, decisions about whether or not to clear time for particular programs are made at the start of the season and reconsidered only in special cases.

System 2 for program acquisition is concerned with maintaining the agreed balance of programming and ensuring that records needed for financial and government reports are kept in harmony.

System 3 includes the station manager, the program director and the news director. The Three Star channel monitors the ratings of each program, comparing them with expectations and weighing alternatives to fine-tune the schedule.

System 4 includes the same managers as does System 3. It examines the market for available programs, talent and possibilities for competitive advantage. It may from time to time mount a special public relations effort (e.g. provide the school system with material about a children's show) or commission surveys to probe audience reaction to particular shows. It also develops the promo's announcing programs and participates in community-based sports and charity events.

We see the same managers again in System 5, hammering out the station's programming identity, monitoring the resource bargain of 'audiences delivered for dollars spent', and attempting to position the station's programming to change with its audiences. It should be noted that when Systems 3, 4 and 5 share the same management, recursions and responsibilities may not be sufficiently separated. In particular, the presence of the program and sales managers in multiple roles suggests possible blurring of the resource bargaining and command channels.

The environment for program acquisition includes the station's sources within and outside its community for news, talk and education, the program distributors for syndicated shows and the network if the station is an affiliate. It also inclues that portion of public and advertiser opinion about what is acceptable and popular. Because of prohibitions against censorship in the Communications Act, there are few government regulations dealing with content. Restrictions on obscentity, indecency and libel are their primary focus. At the time of writing, the Fairness Doctrine (which says that a station which airs one position on a controversial issue is obligated to provide some opportunity for other positions to be heard) has been dropped by the FCC and attempts by Congress to reinstate it have not been successful.

Revenue production: Recursion two (Figure 4)

There are three potential activities which are sources of revenue for a station. The dominant one is the sale of advertising spots to producers and retailers of goods and services, which is managed by the sales manager. Its staff maintains contact with clients and advertising agencies. The prices they charge are based on the ratings which indicate, on the basis of sampling, estimates of the number, age and sex of the audience for the time period in question. Although it is an infrequent occurrence, stations may refuse to accept advertisements for certain products or decline to air particular messages that are believed to create a false impression or to be in poor taste.

This system has a feedback loop which is atypical of market relations. In most commercial transactions, the user of the product pays for it directly. Even in situations where advertising subsidizes the product, such as newspapers and magazines, some direct payment is the rule. In the case of broadcasting, the payment is made from the advertiser to the station for the audience. The loop is closed in the environment when the audience buys the advertiser's product with, of course, the cost of advertising included in its price. The cost for 30 seconds of commercial air-time ranges from 600,000 dollars for national network exposure in Superbowl '88 to around 15 dollars for a single station to air a message during the late movie in a small market.

The second source of revenue for a station is political advertising; this may be handled by the sales staff or may be under the direct supervision of the station manager. This source of revenue is significant only in presidential campaign years. Although Congressional elections are held every two years and some states elect governors on alternate years, these races do less television advertising. Federal law on 'Equal Opportunity' requires a station selling time to one candidate for a national office to sell time to any other candidates for the same office.

The potential third element is network affiliation, which applies to slightly more than 600 commercial stations. The affiliation contract provides around 8–10 per cent of the station's revenue based on the amount of time cleared for the network's programs and advertisements. This revenue stream is typically under the supervision of the station manager.

System 2 for this recursion assures that agreed procedures for invoicing, credit and collection are followed.

System 3 is managed by the sales manager with some participation by the station manager. The sales staff members are seldom involved with programming, but they must know the schedule to ensure that spots reach the right audiences. They may check to see that particular commercials are not shown in contexts where they would be jarring. These instances are

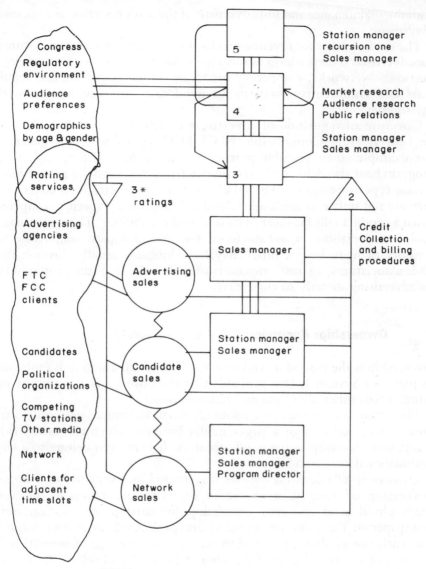

Figure 4. Revenue production: Recursion 2

rare, since most programs are designed to attract audiences for commercials. The Three Star channel is available to compare ratings and revenues.

System 4 concentrates on promoting the station as a vehicle for advertisers to reach their audiences. It also keeps watch on trends in audience characteristics, marketing and regulations.

System 5 contains the station manager and sales manager. It is concerned with the maintenance and improvement of the station's value as a means of advertising.

The environment for revenue production is dominated by the closure of the advertiser/audience loop as reflected in sales figures, market surveys or elections. Networks, competing broadcast outlets, other media, regulations and community standards of good taste are also factors in the environment.

Government regulation of advertising is shared between the FCC and the Federal Trade Commission (FTC). FCC regulations apply generally. For example, they prohibit practices such as 'host selling' (when the program host also delivers the commercial message), and the promotion of certain types of contests. The FTC regulates commercials to ensure that they are not unfair or misleading, and will investigate specific complaints when a product falls far short of its televised capabilities. These restrictions apply to advertising in any medium. For the most part, stations do not consider them to be onerous. They have lobbied, usually through their trade associations, against extensions of these rules (e.g. the proposed ban on advertising directly to children).

Ownership: Recursion −1 (Figures 5 and 6)

Ownership is the first of several structures in which a station is embedded as part of a System 1. It is probable that the business is a 'group owner' which owns other television and radio stations.

The group owner may be a relatively small enterprise with only a few broadcasting stations, or a larger media business which also owns cable companies, newspapers, or movie theaters. It may even belong to a large multinational corporation.

Ownership of television stations is restricted in several ways. Duopoly (ownership of more than one radio or television station in the same geographical area) and cross-ownership (ownership of a station and a newspaper in the same community) are prohibited; as is ownership of (currently) more than 12 television stations or coverage of more than 25 per cent of the population of the United States. In addition, the licensee must be of 'good character' (usually interpreted to mean free of serious criminal convictions) and must be or become a US citizen.

Each broadcasting station has a separate and unique geographical environment for its audience and its local advertisers which minimizes requirements for program coordination. Some group owners also maintain production companies. These may distribute programs to their own and other stations or facilitate the sharing of locally produced documentaries. The owner may interact with its stations on an almost totally financial basis

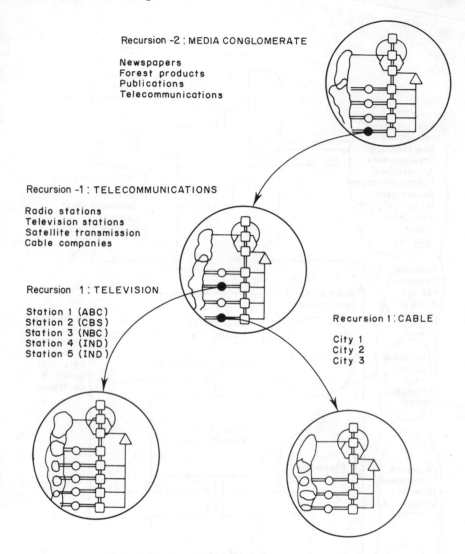

Recursion -2 : MEDIA CONGLOMERATE

Newspapers
Forest products
Publications
Telecommunications

Recursion -1 : TELECOMMUNICATIONS

Radio stations
Television stations
Satellite transmission
Cable companies

Recursion 1 : TELEVISION

Station 1 (ABC)
Station 2 (CBS)
Station 3 (NBC)
Station 4 (IND)
Station 5 (IND)

Recursion 1 : CABLE

City 1
City 2
City 3

Figure 5. Ownership: the unfoldment of complexity

or it may play an active role. Such owners may shift station managers and other personnel among their stations or suggest common formats for news or sports coverage. Owners vary in the extent to which they feature their own names prominently in promotions for their stations. A station manager is in charge of each.

System Two at this level of recursion oversees adherence to common standards for financial reporting and perhaps personnel policy and con-

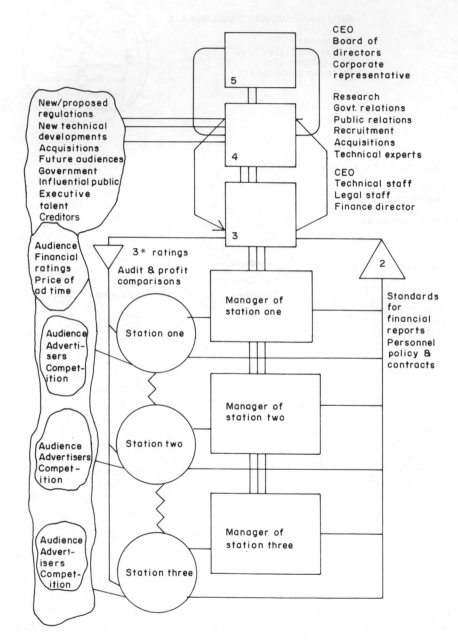

Figure 6. Ownership: Recursion −1

tracts. In some stations it may coordinate the use of a common 'house style' for on- or off-air presentations.

System 3 concentrates on financial management and may or may not become involved in shifting personnel or implementing program changes. More specialized legal staff dealing with federal regulations, copyright and libel laws are employed in this role. They also make policy decisions about what course to follow when there is confusion about federal regulatory policy. This occurs when a regulation is voided by the courts or when the criteria for enforcing a standard, such as the one dealing with indecency, change. In general, the resource bargaining channel is dominated by financial considerations and the command channel by legal ones. The Three Star channel performs financial audits and monitors the station's standing in the ratings.

System 4 of a group owner is likely to be watching the market for new broadcasting properties, technical innovations and opportunities to influence regulation. It may also be probing audience trends and preferences and watching for competitors. The public relations and organizational development staff will be active in System 4, as will financial and acquisitions experts, planners and engineers. If the relationship between owner and station is not just financial, the System 4 will be active in recruitment and training and will work with the stations' System 4s.

System 5 includes the CEO and board of directors or representatives from a corporate parent. If a company's enterprises are limited to television stations, System 5 will monitor the 3–4 homeostat that balances promotion of existing programming with investment in new programming, or it may sell one station and buy another. If the owner holds radio stations, cable companies, newspapers or other media properties, it will balance their several requests and opportunities on the resource bargaining channel. System 5 also decides if and when to mount or join lobbying efforts or to pioneer new technology, as proposed to it by System 4.

The recursive structure of ownership ranges from the case where the station and owner are one to the case where the owner is a large multinational. For example, Group W, which currently owns five television stations and is awaiting approval for a sixth, is ultimately contained in a structure called the Westinghouse Electric Corporation with more than 10 billion dollars in revenues and over 100,000 employees. Finances increasingly dominate these recursive relationships as the corporate management gets further away from that of the station.

Network ownership is a special case. Each major network owns and operates stations (called O and O's) in the larger markets. For example, NBC has a System 1 consisting of seven radio and five television stations. In the case of the O and O's, their identities as stations supersede the interests of the network.

They can and do pre-empt network programs when they believe it to be to their advantage. In addition, the stations compete with the other organizational units for resources, although the competition between programming units (news, sports and entertainment) is more direct: they compete for both resources and air time.

Network affiliation: Recursion − 1

More than 600 stations have an affiliation relationship with one of the three major networks: the independently owned Columbia Broadcasting System (CBS), the National Broadcasting Company (NBC) which is owned by its founder corporation, the Radio Corporation of America, now part of General Electric, and the American Broadcasting Company which was recently merged with Capital Cities, a group owner. Network affiliation is a contractual arrangement between the station and the network which trades access to audiences for more attractive programming and a network compensation fee.

The network may not force the affiliate to carry any particular program, but if the station pre-empts network programming too frequently the network may not renew the contract. As well, stations may decline to renew their affiliation agreements and affiliate with another network if a slot is open in their market.

Although competition from independent stations and cable has increased, programming provided by the three major networks still accounts for nearly 60 per cent of the audience viewing share (*Broadcasting*, 1987). Although when people speak of 'the networks', they are presumed to mean NBC, CBS and ABC, it should be noted that others distribute common programming to different outlets. Fox Broadcasting, for example, distributes a shorter program week to more than 100 stations, and there is a Christian Broadcasting Network for religious programming.

In the Viable System Model of a network, the affiliate relations division is an organizational unit, along with news, sports, entertainment and the O and O's. Within the affiliations division, there are some 200 affiliated stations. They differ in value and profitability primarily depending on the size of their market. The ABC affiliate in Boston, Massachusetts, will have many more viewers than the affiliate in Yuma, Arizona, regardless of the station personnel's skill or talent. There is little competition among affiliates as they may not overlap geographically and the programming available to one is available to all. The resource bargaining channel dominates the exchange between the affiliate and the network. Affiliates want popular programs; they are often less warm to prestige programming or to pre-emption of normal programing to cover major news stories. Not all of the entertainment programs offered to affiliates are warmly received, either. The network must sell its programming ideas to affiliates; if they do not like the pilot for a new series they will not buy it. On the other hand, it is in the interests of both the network and the station to attract the largest audiences.

System 2 in the network/affiliate model concentrates on monitoring the resource bargain with respect to program scheduling and clearance, advertising and promo's. Whenever possible, it sends copies of its recorded programs to the affiliates for advance screening. On the technical side, the engineering staff in System 2 sees that the 'network feed' runs smoothly and that the programming arrives on schedule.

Systems 3–4 and 5 manage the affiliate relationships, especially the ratings and clearance levels that relate to its resource bargains. With respect to the future, an eye is kept on the environment, watching for new audience trends, technical developments and regulatory issues which could alter the audience share of affiliates. System 4 also does periodic surveys and plans the program for the affiliate meetings. System 5 maintains the network affiliate identity and consolidates the affiliate position within the network in which it is a System 1.

The network: Recursion −2 (Figure 7)

At the next level of recursion, affiliate relations, revenue production (selling advertising), news, sports, entertainment and O and O's are network operating units along with (perhaps) cable companies, co-productions, technological joint ventures and publications.

The network System 2 facilitates the routine accounting and personnel procedures. It also coordinates technical standards for production and distribution.

System 3 oversees the programming schedule, making mid-season changes if ratings are not high enough. It uses the Three Star channel to examine the rating reports on programs and the effects of changes (like Neilson's switch to the 'peoplemeter' method of determining who's watching what).

System 4 works closely to monitor the environment for new programs offered by the production companies, and for new audience trends. It also does research on new technologies, the probable effects of different regulatory environments and the needs of specific segments of the audience (e.g. children or the hearing impaired). Finally it investigates the various ample of formal programming to cover major news stores is found in the ...

entertainment programs offered to affiliates, were amply represented ...
The network must sell any programming ideas to affiliates if they are not ...
like the affiliates it is easier that it will not become part of the different ...
the patterns through the network and the school to fill out the largest ...
optimum.

System 2 in the network affiliate model offers a variety of monitoring the ...
main channel with respect to entertainment, scheduling and services, director ...
scope and prices ... These corresponds in terms of policies of the recorded ...
programmes ... by the for regular broadcasting. On the technical side, an ...

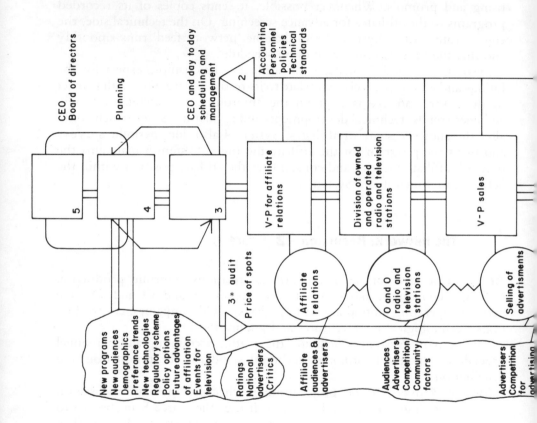

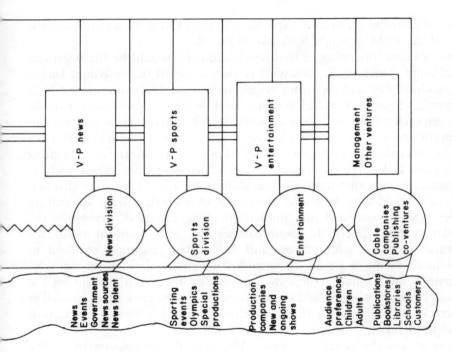

Figure 7. The network: Recursion – 2

roles of television and makes recommendations to System 5 on policy options.

System 5 brings the corporate whole together, building a corporate identity and making corporate policy.

The wavy lines of communication among affiliates are not very strong. They get together periodically for affiliate meetings and discuss common problems and reactions to next season's proposed programming, but there is little real coordination.

The Federal Communications Commission: Recursion −1
(Figure 8)

If the relationship between the stations and the FCC were set up along the lines of the VSM, it might look like Figure 8.

The stations, operating in their environments, would be the organizational units. Their oscillations with respect to signal transmission, length of broadcast day and reporting requirements would be monitored by a System 2. System 3 would monitor and enforce the resource bargain between each station and its community described in its license application. From time to time, it would undertake studies to determine whether or not regulations issued on an 'experimental' basis were having the desired effect or request task-force reports on issues which had been raised such as obstacles to minority station ownership along the Three Star channel. System 4 would have the resources to conduct research into the application of new technologies and run simulations of the effects of changes in the regulatory atmosphere. It could balance the assessment of needs for broadcast services with current and anticipated resources required to implement them, and plan the introduction of regulations accordingly. System 5, the Commissioners themselves and their staff, would set the tone of the agency and would be able to make policy proactively as well as reactively. In particular, with the benefit of a strong System 4 function, it could implement plans in a coherent manner.

This is not the actual situation. Although the establishment of the Federal Communications Commission represented an unusual circumstance where all the relevant parties wanted the government to intervene, there has never been full consensus on any but the regulations on signal transmission. Much of this difficulty may be attributed to variety problems. The regulations governing each licensee are the same whether it serves a metropolitan area with millions of people and many broadcasting stations or a sparsely populated region with few such outlets. This is just one example of the fact that the regulator does not incorporate a good model of what it is regulating, and therefore cannot do a very good job of it (Conant and

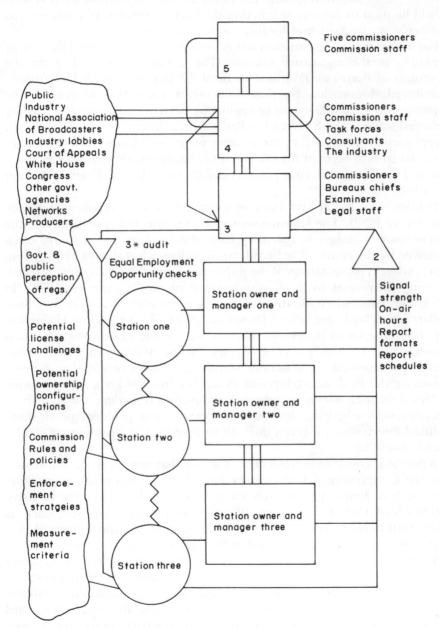

Five commissioners
Commission staff

5

Public
Industry
National Association
of Broadcasters
Industry lobbies
Court of Appeals
White House
Congress
Other govt.
agencies
Networks
Producers

Commissioners
Commission staff
Task forces
Consultants
The industry

4

Commissioners
Bureaux chiefs
Examiners
Legal staff

3

Govt. &
public
perception
of regs.

3 * audit

Equal Employment
Opportunity checks

2

Signal
strength
On-air
hours
Report
formats
Report
schedules

Station owner and
manager one

Potential
license
challenges

Station one

Potential
ownership
configur-
ations

Station owner and
manager two

Commission
Rules and
policies

Station two

Enforce-
ment
stratgeies

Station owner and
manager three

Measure-
ment
criteria

Station three

Figure 8. Federal Communications Commission: Recursion −1

Ashby, 1966). Moreover, there has never been consensus on what criteria would be used to determine whether the various regulatory goals such as competition, diversity and localism were being met.

For example the Commission has never adequately addressed the role of networks in the regulatory scheme. The Commission's policy for the allocation of licenses established in the 1952 Report and Order virtually guaranteed domination by the three major networks. Moreover, the Commission's oblique means of regulating network activities (such as the Financial Interest and Syndication Rules which limit network rights to sell independently produced programming in the syndication market) have never taken into account the effects of the economic efficiency of production and distribution to large numbers of affiliates (Besen, Krattenmaker et al., 1984).

Politics, too, has always been an important factor (Krasnow, Longley and Terry, 1982). The Commissioners are presidential appointments, the Commission's budget is voted by Congress, and its decisions are often reviewed by the courts. The broadcasting industry has always had a strong voice, while representatives of the public have found it much more difficult to bring their cases to the Commission (Cole and Oettinger, 1978). In 1988, the National Association of Broadcasters budgeted over a million dollars for its lobbying efforts (*Broadcasting*, 1988b). The White House has also been known to intervene. Threats were made against the television licenses held by the parent company of the *Washington Post*, whose newspaper reported the Watergate scandal and, more recently, an investigation by the FCC into television evangelist Jim Bakker's use of money collected through television appeals was diverted (*Frontline*, 1988). If these circumstances where the law is very clear became problematic through political interference, it is not difficult to imagine the fate of more murky problems.

Additional constraints result from the fact that neither the Commission nor the Congressional Committees have had the resources to conduct large-scale or long-term research and have had to depend on the industry for much of their information. When problems are addressed, such as when Staff Reports are commissioned, the mandate for the investigation is drawn so narrowly that the approach is at best piecemeal and at worst lacking in credibility. Providing the government with sufficient resources to do comprehensive long-range research would allow additional perspectives to be explored and decrease the comparative information advantage now held by large commercial enterprises. It would also be a step toward expanding the variety of the agency to match the variety of its constituents. The attempt to violate Ashby's Law of Requisite Variety is evident, and—as always—it fails.

The political environment for regulation in the United States varies from administration to administration but has never been warm to restrictions on the operation of private business unless a clear threat to public safety was involved. This is where the consumer movement has been most effective. Where the media are concerned, the dominant restrictions apply to government. For example, the First Amendment has been effective in constraining government censorship, but there are no comparable constraints on private censorship. Consumers have been less able to influence government when the products are intangible; still less when the desired products do not exist. Nor have government regulators been anxious to open new areas of inquiry unless a powerful and articulate constituency requests one. There are few outposts listening for 'what if?' questions on communications plicy and no channel to amplify them to public consciousness.

It is noteworthy that the recent explosion of video technology may offer an opportunity to draft a better set of regulations. The expansion of the number of outlets should allow the relaxation of some regulations and the possible introduction of others before happenstance creates new perceptions of property rights.

Community information resources: Recursion – 1

At this point, we move to less solid ground. Although the three to seven television stations in a community broadcast similar and highly standardized material, this does not result from formal connections at the local level but from national influences. The television stations in a community are not linked by the internal lines of communication mapped on the VSM. They do communicate extensively, but most of this communication takes place in the environment. Managers and staff see each other socially, work with the same local organizations and charities and share a common grapevine. On the business side, personnel at the station always seem to know the going rate for staff, especially on-air talent, and how each is doing with respect to selling commercial time. Also, new tips, successful innovations and even gimmicks tend to spread from one to another.

Coordination and self-regulation among stations is severely constrained by anti-trust law. For example, children's television advocates were dismayed that none of the network affiliates was offering any regularly scheduled programs for children from Monday to Friday. They suggested that the stations each offer after-school programs for children one day each week, but were told that it would be a violation of anti-trust law to do so.

What coordination does exist among stations happens at another level of recursion. News or production crews may share or alternate facilities for remote coverage of events. And, of course, in the event of a disaster, they cooperate in broadcasting emergency instructions and information.

Whether coordination or competition with respect to programming is more beneficial to the community has been a debated issue. The notion of sustaining programming (where each station broadcasts some worthy but unprofitable material) was dropped from the FCC's vocabulary long ago in favor of competition. However, competition among stations has been so concentrated on the commercial aspects of television that the benefits from competition on broader fronts have been less than was hoped.

There is no official coordination among all the television stations. They do work together through their professional associations, the National Association of Broadcasters (which includes both radio and television) and state broadcasting associations (Kraus, 1988). The NAB has fulfilled a number of the functions assigned to Systems 4 and 5, but it does not dictate station behavior nor issue sanctions.

Until the 1970s the NAB did perform some System 2 and 3 functions through the Television Code of the National Association of Broadcasters. Among other provisions, the code set limits on the number of commercial minutes per hour in children's programming. When the US Department of Justice disallowed this provision and others under anti-trust law, counsel advised the NAB that the code be discontinued.

Most of the NAB's coordination has been performed by providing services and information to member stations. It sponsors a large annual convention where station personnel can get together to discuss common problems, establish common positions and review trends in technology and regulation. Like other conferences, it also serves as a marketplace for recruiting staff and seeking out new business opportunities. Unlike many other conferences, the main themes of its public sessions have an impact on many in addition to the actual attenders because they are reported in some detail in the industry's trade magazine, *Broadcasting*. Smaller annual meetings are also held for subgroups of the association, such as personnel from small market stations or engineers.

Some System 4 activities, such as scheduling presentations by representatives from the technical, regulatory and market-research environments occur at meetings. In other cases, the Association may seek information from its members relating to specific issues.

System 5 for the Association is provided by its board of directors, which maintains and evolves the identity of the organization and, to some extent, that of the industry it represents. It meets twice a year, with a monthly meeting of the Executive Committees for radio and television. This group determines positions on legislation and regulation and sets policy for the staff to implement.

Most of the real regulation of television at the national level is market driven. Market research, advertiser preferences, networks, production and distribution companies, the ratings and training protocols all contribute to a common 'look' of American television. The viewer looking for something to watch at a given time may have a choice of only two or three program genres. If our viewer uses the zapper to check out other channels, he or she will find even the timing of the commercials is the same. This situation is changing but the desire for predictable programming products to sell may be expected to constrain viewer choice for some time yet.

Community information resources: Recursion −2 (Figure 9)

Although communities vary enormously, most will have seven to nine different kinds of information resources available to the public under different conditions. They include television stations, (perhaps) one or more cable companies, radio stations, newspapers, magazines, bookstores, computer bulletin boards, libraries and adult education facilities. These outlets, with the exception of public radio and television (if available), libraries and adult education, are usually private profit-making organizations. They bring a mixture of information and entertainment into the community. Often it is information about the wider world rather than the local area. Access to these information resources may be limited by distance (libraries, bookstores, adult education), by cost (all except radio and television broadcasts and the library), by sophistication and by time.

System 2 for a community's information outlets is practically non-existent at this level. There are 'community standards' which allow that what is appropriate in Manhattan may be unacceptable in Des Moines, Iowa. There may be zoning regulations or conventions about hours of operation, but there is no channel to Systems 5–4–3 to determine what constitutes an oscillation, and still less, no process by which a System 2 can say: 'We don't have enough information about (say) toxic wastes.' If anything of this sort is done, it occurs at a lower level of recursion.

Nor is there much activity at System 3. There may be some relevant bylaws, but most of the laws and regulations are made at the state or national level. In some cases, a local voluntary agency may conduct a Three Star inquiry into whether or not the minority community is being sufficiently and/or accurately reported, or if children's access to information needs improvement; but such efforts have limited scope.

If the community is wired for cable, there will be a franchise agreement with local government with its resource bargains. At this time, cable franchises commonly include channels set aside for local educational, government and public use. Last year, the legality of community requirements for these channels and for extending cable service to the whole

Marketing research in advertising, broadcast, and telecommunication and distribution companies is often similar to and complements research in consumer products companies as well. A consumer looks at American television the way he does cars; he watches as a given time may have a video effect on his television programming gear; if our conversation so the rapport reflects on just what he or she will do, even if it's timing on the community, but is the same. His situation is changing, but this date for predictable programming products — as will just the established contents have versatile for and one picture.

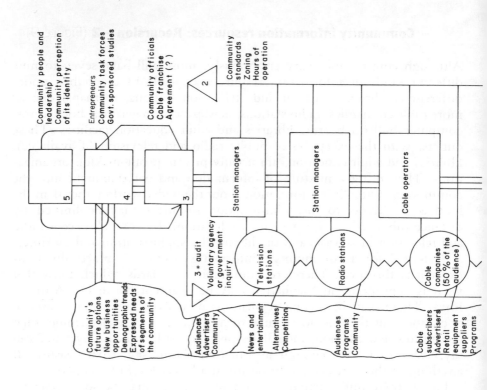

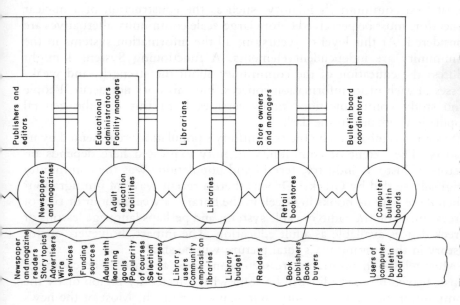

Figure 9. Community information resources: Recursion –2

community was challenged and overturned in a California court. Local governments responded by questioning the current exclusivity of the franchise agreements. This would leave cable operators vulnerable to overbuilding and competition within communities for the same subscribers. So, for the time being, it looks like cable will retain the community channels. Also, although many have not been programming very long, the established community channels have earned themselves a constituency.

System 4 is not in place as a solid presence. Some four functions are performed by entrepreneurs looking for a market niche. Others are performed when local government sponsors demographic or planning studies or convenes a commission to address a specific issue, such as perceptions of the availability of vocational information. It is probable that a substantial threat to a community's identity, such as the construction of a nuclear generator, must be perceived before large-scale community alternatives are considered. At this level of recursion, all the information systems in the community are independent elements. A functioning System 4 might address the education of the community about the strengths and weaknesses of each of its information sources, but this does not occur. People who study communications in universities learn this but the general population remains in the dark.

System 5 is inhabited by the community's population, represented by its leaders. The extent of community identity expressed here depends on factors of history, industry, terrain and population. Was the community originally settled by Greeks? Or was there a recent wave of immigration? Then there may be books in Greek in the library, a weekly hour or two of Greek music on the radio or cable system, and perhaps language lessons or Greek dances available as adult education. The impact, however, of this history in the community's total picture will be small.

Most television programs are seen from coast to coast. The choice of music played on the radio will be somewhat affected by region or by the composition of the community, but again, not much. Most of the news reported in the papers will come from wire services and most of the books offered by bookstores and libraries will be nationally distributed.

This is not to say that no coordination exists, although most of it is informal. The public and non-profit portions of the community's information resources do communicate among themselves along the wavy lines. There is a great deal of political will behind public sector coordination, at least, because cooperation allows for more efficient use of resources. For example, the library may set up its catalogs and computerized databases to be accessed by schools, government offices and anyone else with a modem on a 24-hour basis. This allows for a substantial expansion of services for very little cost. Many libraries already maintain archives of local and national newspapers and receive texts used in adult education courses, videos of educational programs broadcast on public stations and of movies shown by local film societies, and documentation prepared by and about local organizations.

This type of coordination could be made more formal and commercial television could be included without infringing on the First Amendment rights of broadcasters. Currently, it is difficult for most people to gain access to previously broadcast programs, even the news and information features which are not often offered for rebroadcast. Increased coordination might allow stations to expand local programming and coverage by sharing the resources of other information providers. In cable communities, even the one-way-flow constraint could be overcome. The Chicago Access Corporation is using a touch-tone telephone as a keypad to request information from a community information channel and to participate in straw polls (Conhaim, 1988). The capacity of fiber-optics creates the possibility for community access and information channels at a national level. The possibility of hundreds of channels, devoted to numerous areas of interest, is technically feasible. The expansion of outlets also creates the possibility for 'workshop' television and a range of production styles comparable to that employed by cinema and theater. These and many other options exist, but may not appear unless enough political will is generated to make them happen. I, for one, would be pleased to see such a platform.

Conclusions and diagnostic

Television is a product. That fact is determined by so-called market forces. Its social role in linking people with their environment in continuous communication on a feedback loop has been abandoned to marketing. Thus, most of the interactive variety available is lost. The product called television reaches the public through a legally accountable entity called a station. Because the economies of scale do not favor ownership and operation of a single station, network affiliation and group ownership has become the rule. While legal accountability remains at the level of the station—our system in focus—it has little actual power to change or evolve. Thus, the recommendations advanced here primarily concern the several 'minus-one' levels of recursion embeddings where this power resides.

Television stations and networks have been profitable businesses for the past 40 years. Although some of the smaller independent stations have failed in the past year owing to overbuilding and high debt service and programming costs, the industry remains profitable on the whole. There are warning signs, however. The business links with the environment are dominated by ratings, which have a short time-span and, because they are essentially head-counts, a very low variety. It has been very difficult to come to a consensus on qualitative measures. Nor, with the exception of gross measures of age, sex and income, are there channels to give greater weight to the opinion of one head over another. The result is a deceptive sort of leveling. It is possible that a majority of the audience might find low satisfaction in television, but that the reasons for any one person's dissatisfaction would disappear into the averages. Survey research and focus groups may help, but they also lack variety and, because of the conditions under which they are commissioned, do not perform the function of the algedonic signal.

Nor does critical success without popular success carry any weight. In contrast, critical success in book publishing and film production retains some respect—if for no other reason than as a source of growth and development which might someday contribute to popular success. In television, new concepts and programs are tested on a short-term basis with average audiences. This procedure does not provide enough exposure to develop new tastes. Americans who spend time abroad often find characteristics to praise in foreign television, but there is no machinery in place to change on more than an incremental basis. Without qualitative measures and a greater element of risk in programming, it is unlikely that the industry will be able to adapt as conditions change. In the past, broadcasting has been dominant enough to attract audiences on the basis of 'you like what you get'. But with more video options, this condition might not hold for much longer. The lopsided focus of identity on the business aspects of television may threaten the long-term stability, if not the survival, of the industry.

Economic factors also contribute to a lack of the kind of risk-taking behavior that encourages adaptation. The frequent changes in station ownership (a process called flipping) increase the costs of station operation without contributing to its ability to earn profits. The likelihood of an ownership with either the interest in carving out a new station identity or the financial cushion to ride out the transition diminishes with the expected duration of ownership—which is getting shorter. At the other end of the process, program production has become so expensive that the situation comedies and adventure shows do not begin to break even until they are in syndication, and are not attractive for syndication unless they have run for several years. This further reduces risk-taking with the product.

A VSM redesign of the commercial broadcasting structure would lead to major changes. At the moment, the Five-Four-Three management is concentrated in System 3; and in the station and network recursions, it may also be engaged in the operation of one of the operating units. These functions need to be teased apart and given more formality, especially at the station level where the same individuals play multiple roles. The three Axioms underwriting the VSM need to be interpreted correctly to ensure that this is done to proper effect.

System 4 badly needs broadening and strengthening. Information about the environment does not come in from a large enough variety of sources. Because of its former success in molding the regulatory environment, the industry is not as sensitive to changes in government and regulatory agency perceptions as it needs to be. Currently, for example, there seems to be agreement among both Democrats and Republicans in Congress that broadcasting either retains its status as an industry with some protection from competition (and accepts the regulations that go with that status) or it gives up most of the protection along with the regulations. Industry efforts appear to be geared toward holding the line with government rather than to seeking stable alternatives in a changed regulatory environment. Barriers are also raised against some sources, such as public-interest groups and others critical of the industry. The industry would be well served by amplifying these critical channels, thereby increasing its own information, rather than insulating itself from them.

Within System 5, a narrow definition of corporate interests and a reluctance to admit that the structure of television creates adverse effects adds to the reduction of variety admitted to System 4. If it were accepted that the system called television entailed adverse effects through both its technical and its marketing activities, then compensatory efforts within or outside the medium might be implemented to everyone's benefit.

As a regulatee of the FCC, television performs its technical functions well but has been criticized by scholars from several perspectives and from both sides of the political spectrum.

The Commission's regulatory goals such as 'choice', 'competition', 'diversity' and 'localism' are vague and are differently understood by different participants in the policy process (Leonard, 1987). This means that appropriate feedback loops to indicate whether the desired results have been achieved are nearly impossible to design. Moreover, unambiguous standards, such as ownership restrictions, have not been proven to have any substantial effect on what appears on the screen.

The amount of time it takes to catch up with and incorporate new technology into regulation has been a problem from the beginning. When the FCC reallocated the spectrum space for FM radio and made it incompatible with the FM sets then in use, there were serious economic consequences for both FM manufacturers and broadcasters. In addition, public acceptance of FM was delayed. When new television license applications were under review after the Second World War, the FCC was prevailed upon not to reallocate spectrum space and disrupt the existing broadcasting system, although the result led to a three-network domination and to a skewed distribution of television services which is still not fully resolved. In turn, cable was delayed and restricted, and now satellite distribution is under review.

Related to both these difficulties is the fact that, although it is an 'independent regulatory agency', the FCC does not have either the resources or the political independence to be a good watchdog, still less to anticipate and prepare for change. One reason is that it has no independent source of revenue, such as might be acquired from spectrum fees or television-set licenses. Another is that jurisdication is divided with other agencies, including the Federal Trade Commission, on commercial practices and the National Institutes of Health on problems associated with, for example, televised violence. Thus the first aspect of a court case dealing with television is the sometimes lengthy process of establishing who has jurisdiction and who has standing. This, of course, takes time.

As an element of the community's information resources, the role of television becomes more important as the years pass. It has already become the main source of news for the majority of Americans. There are serious public policy questions raised by such dependence for news and information—no matter who is in charge or what the regulations are. Television is simply not good at communicating some kinds of information. It falls short when context or logical argument are to be conveyed. Above all, it does not deploy Requisite Variety toward its pretended social purpose.

Surely we face the fact that television often distorts information in its very presentation. As Mavor Moor says:

> 'television news, everywhere, is designed as entertainment and structured as drama—giving fact the sheen of fiction, and fiction the weight of fact. We can no longer tell one from the other.' (Moore, 1988)

This problem has not been faced squarely by communities nor, in the United States, by the nation itself. The problem does not rest with television alone; it rests with the lack of perception of information as a public as well as a private resource. From its beginnings as a nation, the United States has espoused the necessity of a well-informed citizenry as a condition of maintaining a democratic form of government. If the regulation of access to information is not addressed soon, according to valid cybernetic canon, this assumption, tested by events, could fail.

A VSM design to improve the situation would have only limited effect at the community level. Too much of the information content is generated by and about events in other communities or at the national or international level. We need a systemic, holistic view of societary cohesion to handle this, that goes far beyond television in its scope. It requires the application of Beer's Law of Cohesion to all of society's communication channels. This recommendation may be rejected owing to concerns that its implementation would restrict freedom of speech or promote censorship. This is unfortunate, because the result is that debate focusses on questions of content, and issues concerning context, structure and access are excluded from consideration. It has been clear since McLuhan that not only the content of information, but its selection, presentation and distribution counts. We have not begun to appreciate the extent to which the transition to an information society is recasting the realities of privilege in terms of access to, and facility with, information and its supporting technologies. Until this occurs, it will not be possible to address the role of television in a serious way.

References

Beer, S. (1981). *Brain of the Firm*. Chichester: John Wiley.

Beer, S. (1985). *Diagnosing the System for Organizations*. Chichester: John Wiley.

Beer, S. (1979). *Heart of Enterprise*. Chichester: John Wiley.

Besen, S. M., Krattenmaker, T.G. et al. (1984). *Misregulating Television: Network Dominance and the FCC*. Chicago and London: University of Chicago, pp. 115 and 135.

Broadcasting (1987). 'What 1987 wrought; 1988 prospects', 28 December, 33–8.

Broadcasting (1988a). Table, 2 May, 14.

Broadcasting (1988b). 'NAB Board New Year's Resolutions in Hawaii', 18 January, 48–9.

Cole, B. and Oettinger, M. (1978). *The Reluctant Regulators: the FCC and the Broadcast Audience*. Reading, Mass.: Addison Wesley, p. 277.

Conant, R. and Ashby, W.R. (1970). 'Every good regulator of a system must be a model of that system', *International Journal of Systems Science*, **1**, 89–97.

Conhaim, W.W. (1988). 'Automated cable channels: new media for Chicago's nonprofits', *Information Today*, 11, 38 and 39.

Frontline (1988). 'Praise the Lord', 26 January.

Gitlin, T. (1983). *Inside Prime Time*. New York: Pantheon.

Krasnow, E., Longley, L. and Terry, H. (1982). *The Politics of Broadcast Regulation*, 3rd edn. New York; St Martin's Press.

Kraus, S. (1988), speaking for the National Association of Broadcasters (May 1988, telephone communication).

Leonard, A. (1987). *Broadcast Regulation: A Comparison of Perspectives*. PhD Thesis, University of Maryland, College Park.

Moore, M. (1988). 'Is all-news TV good news, bad news or news at all?', *Globe and Mail*, 6 March, C1.

Weiss, L. and Strickland, A. (1982). *Regulation: a Case Approach*. New York: McGraw-Hill.

The Viable System Model: Interpretations and Applications of Stafford Beer's VSM
Edited by R. Espejo and R. Harnden
© 1989 John Wiley & Sons Ltd

9

The evolution of a management cybernetics process

Stafford Beer

*Chairman, Syncho Ltd.**

Note 5 of *The Heart of Enterprise* describes the developing relationship between the author and a large mutual life insurance company. This chapter completes the story.

Introduction: On Success†

No-one in his right mind would expect a management suddenly to submit to the revelation that management cybernetics provides a miraculous solution to the problems that beset any complex organization. Nor do we expect that management to assert that 'as of next Thursday' the institution has been handed over to cyberneticians.

But if the arguments that there are laws governing the structure and dynamics of *any* viable-system are valid, then *all* successful enterprises will be found to respond to those laws. They may nonetheless respond too slowly, too hesitantly, too uneconomically; too formally or too informally; too aggressively or too anarchically.

What is the meaning of all those statements of 'too'? Too whatever for *what*? Why, too whatever for maximum benefit—whether of profitability, satisfaction, or general ease: in a word, of *eudemony* (or well-being). Everyone has had the personal experience of achieving something that works: with satisfaction, but with the realization in hindsight that it could all have been done with much less stress and strain.

---

*Visiting Professor of Cybernetics, Manchester University Business School.
†The first eight paragraphs of this Introduction are drawn from Note One in *The Heart of Enterprise*.

In the continuing process of organizational reform, this is where theoretical cybernetics and practical cybernetic experience each has something to say. It is both silly and agonizing to reach successful organizational outcomes by trial and error, if the rules of the game are already known. And, in the limit, it is perfectly possible that the trial-and-error approach (which is itself good cybernetics, given a suitable learning framework) will blow the institution apart (in the absence of a learning framework). After all, one of the trials may result in an error so great that viability is altogether lost. In private organizations, this is called bankruptcy. In government organizations, it is called PSBR: the Public Sector Borrowing Requirement.

For these reasons, managerial cybernetics ought in the first place to be seen as diagnostic. By mapping both the organization and the development process in which it is engaged on to the Viable System Model, it is possible to understand strengths and weaknesses in terms of the axioms of viability. In the second place, it is almost possible to prescribe for whatever turns out to be pathological.

The use of the word *almost* is of exceptional interest to all managers concerned with the proper use of these tools. In talking about the pathology of organizations, about diagnosis, about prescriptions, we are using a medical metaphor. It is noteworthy that, in our culture, doctors take it upon themselves to determine treatment. It is a case of 'mother knows best'. Many commentators on the medical scene now think that physicians have over-reached themselves: only the patient (or the comatose patient's representative) has the ethical right to say what is to happen to him or her. Of course, this presupposes that the physician is able to explain the issue to the patient, while medical autocracy is based on the benign assumption that the patient is a fool. Be that as it may, the manager who consults a cybernetician is certainly not a fool. It should be taken as read, then, that the institutional 'patient' is going to prescribe for himself.

The cybernetician *almost* prescribes in this sense. The collaboration between the patient and his physician, that we would like to see in medical practice, is well modelled by the collaboration that often exists—and certainly ought to exist—between the manager and the management scientists. Only the manager is entitled to take the decisions. It is the duty of the cybernetician to press his expert views; but he must not bully or cajole beyond the threshold of the manager's personal accountability.

All of this, so far, ought to be self-evident to anyone engaged in managerial work, from whatever standpoint. Thus I find it surprising to be asked so often: 'Where can I see "all this" in action?' The questioners appear to be using what I have called the Joshua model of success. Jericho, it seems, really was a closed system, which is something very hard to find. We may read in *Joshua* 6.1 that 'none went out, and none came in'. If a system can be envisaged as truly closed, it is possible to believe that there exists a unique algorithm that would lead to 'success'. At any rate, Joshua had such an algorithm. His people had to process around Jericho once a day for six days, and make seven such processions on the seventh day; seven rams horns had to be blown by seven priests, and the people had to give a great shout. The walls of Jericho then fell down flat. It would be highly satisfactory if this model of success were generally applicable. But we do not confront closed systems, and we do not have the algorithm; nor is it easy to rally the great shout, unhappily, since that might be enough on its own. But the people are going about their business, or watching television; and when the people do show any strong signs of working up to the great shout there are all too many agencies keenly interested in stifling it—as we have repeatedly seen happen in our own lifetimes in country after country.

The real-life story presented here covers my own involvement with a major mutual life assurance company self-consciously engaged in what I just called 'the continuing process of organizational reform'. Our close association lasted for nine years. But the company was there reforming itself for decades before that. So why call in a consultant in October 1973 precisely? The answer is that the reforming process had gathered such speed that the senior management felt that they were not fully in control of developments, and that they did not know why. Then why tell the consultant 'That's enough of that' in January 1983 precisely? Using the same mode or style of explanation, it was because things were in a more stable condition—or at least the senior management felt that they were. But doubtless, after nine years, both sides were getting rather tired of each other. That's not something you write in a letter. Anyway, they are still reforming themselves today.

It is for this reason that I have always been suspicious of the traditional case study beloved of certain business schools. If one goes into the story deeply, an *episode* in the life of an organization has been dramatized to demonstrate a point. Now if a company has been turned around from bankruptcy, or a rebellion overthrows a government, that point may be well made: even so there will be an elaborate web of factors that the case study does not acknowledge nor probably understand. When it comes down to the small 'vacation project' of a student, there is often no trace of the 'solution' a few years later. 'Yes, I suppose we may have got something out of it' is probably as good as you will get.

The argument is that this good-as-you-will-get is good enough— because it is central to the viable system that it should be in a constant state of evolutionary flux. There are no problems and solutions, except as abstractions of reality. For instance, the annual accounts are an attempt to 'stop the world'; no wonder that the stopped world's inertia centrifuges funny money around its financial capitals. There are, however, confusions and terrors—and these may get better or worse, on an elastic time-scale. This being so, I hoped (30 years ago) to create a metric that would at least make it possible to detect whether consulting assignments had led to a general improvement, or not. But the ambition is delusory: such a metric *could not have requisite variety*. There is in principle no way to determine what might or might not have happened, if what happened had not happened. Witness: attempts to do so invariably divide observers into two camps—the protagonists and antagonists of whatever action stands proxy for the cause of the happenings concerned. There are actually no causes either; *consequences* there are.

Then expectations, whether of client or consultant, ought not to be too high. Looking back, my own have been invariably the higher of the two. But if the consultant is doomed to disappointment, at least he has a better chance of his client's satisfaction than he thinks! As to the enquirer whom we recently met asking where 'all this' can be seen in action, let him brood upon the Joshua model of success. We do not operate in a closed system, as Jericho was. We have no revealed algorithm, as Joshua had. And our enterprise does not have the class of support enjoyed by the Children of Israel at that time.

Unfortunately, our enquirer may well be the kind of entrepreneur who prides himself on his willingness 'to take a calculated risk' (which means a risk he cannot calculate), after he has had assurance that there is no risk at all (because it has all been done before). So much for innovative management in our Western culture: it says much for its decline.

The story here is addressed to more enquiring minds, who wish to understand *evolutionary process* in an organization, and how management cybernetics may interact with the process that is occurring anyway. In this example, I was not on the staff, and my base was on the other side of the Atlantic; but I distrust purely episodic or sporadic activity. Thus I eventually developed the term 'Intervention' to refer to a visit, from which consequences of some kind would flow, which would then be monitored by considerable communication and interaction on audiotape between Wales and Canada, until entropy took the latent energy out of the conversation. Then there would be another Intervention. This system began to work well after the Third Intervention, and truly blossomed (as the history will show) in the Fifth. The relationship became continuous, but it 'peaked' by agreement on visits whose timing evolved from the unfolding situation. As a result, I was not under anyone's feet, and the visits were usually fun—a most underrated desideratum for managers, who will not always admit to it.

Everything began with a great many interviews with the senior officers of the company, jointly and severally, excepting the President. He and I met later, and the meeting was not a success: we had to struggle for our relationship, which I think in the end was based not on shared ideas and plans, but on mutual respect for integrity. Otherwise the meetings went well. To those who originally looked askance on such unspecific, top-level, and theoretically loaded outside consulting, I told the story that when Sir Nigel Foulkes (then heading Rank Xerox in Britain) used me in the same role, he wrote the letters GPF after my name on copy documents. It stood for 'Guide, Philosopher and Friend'—which is better than MBA.

Thus I 'got up to speed' with the corporate experience and current situation. To convey some idea of that, we go back to *before* 1973.

BC: Before Cybernetics[1]

This tale has been told in a company report. It explained that a task force had been set up to develop an approach to reorganization, which was perceived to be necessary, and that this task force identified 'three major areas of weakness'. In the words of the report:

> 'The old company had been organized essentially along functional lines. The style of management was what might be called 'paternally autocratic', a style quite common in our industry at the time.'

> 'Not only was our functional approach constraining us in the marketplace, but the orientation towards specialists was not producing the generalist kind of management talent necessary for running a complex organization.'

> 'And finally, we lacked a capability for systematically sensing and reacting to the changes that were building up in the environment.'

Following these conclusions, the report continues, 'a series of major changes was set in motion'. Three phases were identified by the same company report:

> 'In Phase 1, marketing divisions were created, integrating some of the major functions on a territorial basis, rather than the total company level, in order to place responsibility closer to the scene of action.
>
> At the same time a version of accountability management, or management by objectives, was introduced and a conscientious attempt was made to move towards a more participative style of management.
>
> Phase II quickly followed. The decentralization of authority unleashed a remarkable enthusiasm. Activity blossomed and expenses rocketed, highlighting the fact that we were operating with an inadequate control system. Plans had been laid to develop a broad new information system; but these had to be accelerated, with initial emphasis on expense budgeting and monitoring.

---

[1]What follows is taken more-or-less verbatim from its first publication in Note Five in *The Heart of Enterprise*, up to and including the Tenth Intervention.

Phase III. During the whole period of reorganization, there had been a marked increase in human stress and strain. The change in management style placed heavy demands on both the new management and on personnel. There was a shift of emphasis from specialists to managers; there were changes in career patterns; there was a new emphasis on productivity, and pay was geared to performance.

Again, developments that had been planned in the human resource field had to be accelerated. Management made unusual efforts to communicate with the staff and each other. Special meetings were held. The house organ was changed from a glossy monthly magazine to a bi-weekly news sheet providing a running commentary on the train of events; letters to the editor provided an avenue through which personnel could draw attention to problems; a confidential channel was created in which people in distress had access to professional help; major efforts were made to help the managers learn their new jobs, and to help people learn to work in groups.

In a somewhat different aspect of Phase III, the internal audit was rejuvenated, with new authority and new vigour, to provide another means of ensuring that the new delegation of responsibility did not turn into abdication.

And finally, senior management created a semi-formal management committee to help coordinate the operations of the company.'

It was at this point that a company officer was introduced to the cybernetics of the viable system. He made an exhaustive study of *Brain of the Firm*. He mapped the experience recounted above on to the model, and came up with the following identifications (in his words again):

'The attempt to create responsible marketing divisions matched the postulation of viable components in System One.'

'The introduction of accountability management granted at least a degree of autonomy to these quasi-viable units.'

'Phase II emphasized the need for a mechanism beyond the command and accountability channel which could foster self-regulation through feedback.'

'Phase III dramatized the importance of rich intercommunication and the absolute necessity of providing parasympathetic channels to relieve stress and strain.'

'Taken together, these points support the proposition that internal stability is to be achieved by a clever managerial use of these three channels.'

'The formation of the management committee was an attempt to use the collegiate form of authority as a means of obtaining reliable results from fallible human components.'

'The search for a satisfactory approach to corporate planning acknowledged a concern that we were close to being a 'decerebrate cat'—that is, that we were virtually cut off at the neck in lacking a coherent mechanism for creating a future in a rapidly changing world.'

All this being so, the cybernetic process was already clearly in action. But now it was recognized for what it was. It was not until this point that I myself was called in.

First Cybernetic Intervention

It was October 1973. And it was exciting to find that so much of the cybernetic reality of this viable system was already understood. I also found it commendable and humanly satisfying that a top management group should be so open in their first interaction. For, when all the cybernetic criteria had been listed and examined, the question posed was this:

Can you tell us how it is that this team,
which is working with all honesty and frankness,
and which devotes unlimited hours to free-ranging discussion,
finds it almost impossible
to reach consensus about anything at all?

By the time this question became explicit, I had already concluded that System One of the first level of recursion was in reality divided into two viable systems on the horizontal axis. The management team was not working on this basis. They thought of territorial markets as one possible model of System One, which would therefore have many components—as many as their marketing areas across the world. They also thought of functional activity (such as marketing itself, accounting, actuarial acitivity) as possible components of System One, without realizing that no *functional* activity can itself be a viable system. It short: you cannot hive-off a function to an independent existence, because it cannot carry its substantive content with it. The content belongs to the corporation as a whole.

The two components of System One that were immediately recognizable at the first level of recursion were quite straightforwardly: insurance and investment.

Now those men who had arrived in the top management group via the *insurance* component of System One gave this kind of account of the company (I caricature for brevity):

> 'We are an aggressive, market-oriented firm. It is our job to get out into the field, and to *sell*.
>
> PS: This produces a lot of income. That had better be competently invested, because policies will eventually mature.'

On the other hand, those top management men who had arrived via the investment and functional component of System One gave a rather different account of the company (still caricaturing);

> 'We are an extremely knowledgeable actuarial and investment firm. It is our job to use our expertise in the markets, in order to maximize return for our policyholders.

PS: We need a lot of money to provide our stock-in trade. Will someone sell a lot of insurance, please, to provide the cash.'

Well, I did say 'caricature'. Actuarial specialists most certainly worked across the board; and both groups were wholly aware of each other. The point was, however, that there was *no formal mechanism* directed to their linkage (the squiggly vertical lines between the circles of System One). The linkage was achieved in practice by powerful effort at the level of System Five. Thereby was the company successful. But in cybernetic terms, there was a diagnosis here. It was also the answer to the question as posed.

The top management team were in difficulties about their concensus, not because of issues relating to market areas nor yet to functional responsibilities, but because their perceptions of the total system were at total variance. A sketch entitled 'Descriptive and diagnostic modes of the existing organization' was prepared and demonstrated with slides. This was year one of the Intervention (November 1973).

A cybernetic assault of this sort is hard to accommodate. And naturally the usual processes of an interpersonal kind within the top management team were continuing. Soon one member was to leave the company; soon, a differentation of status between remaining members would become apparent. Another member would leave and be replaced. But the cybernetic evaluation of affairs was now part of the whole picture. And I felt very much party to the whole affair.

Second Cybernetic Intervention

A year passed. During this time much happened. The company tried to apply the model of the viable system, in some detail, to the whole corporation. As a result, questions were formally addressed to me—all relating to the measurement of variety and the handling of information. But these questions were related to the model that they had created. I was very uneasy about this. It did not seem to obey the axioms of viability; therefore the questions could not be directly answered.

A memorandum was submitted criticizing this particular application of the model, because of what I considered to be confusions about the five subsystems and the viable systems involved in System One. I also felt that different levels of recursion were being confused. But it certainly seemed that the questions could not be answered until the issue of the nature of the corporation had been resolved. The memorandum called this the Joint Normative Decision. It would be joint, because 'both sides' (investment and insurance) would be in agreement. It would be normative, because it would exercise a *governing intention* for the firm. It would be a decision, because the top group would have made a firm commitment as to the nature of the corporation.

These contentions were of course derived from the original model, which diagnosed the fundamental weakness of the insurance-investment components of System One: namely that they did not have any machinery to implement the four principles of organization (see Chapter 1) as between themselves—requisite variety, channel capacity, transduction, and dynamism.

The memorandum went on to talk over the issues raised about managerial information in the light of this criticism; and it drew up a Plan of Actions intended to get to grips with a proper use of the viable system model that would then (but only then) be available to answer questions about information. A week-long meeting was planned to discuss the Plan of Actions. It aborted; too much else was happening. A major shift of power in the top echelon was occurring. And it is worth noting that many good efforts to improve matters in any organization are often withdrawn for completely adventitious reasons. It also seemed at this time that much activity was being promoted without a secure foundation. Therefore the memorandum included this statement:

> 'A great appearance of competence and expected victory in a forthcoming battle might be created by marching soldiers around the battlefield in complicated manoeuvres, digging trenches, shouting commands, and blowing bugles. But if the generals have not yet agreed upon the reason for the war, nor identified the enemy, nor formulated their campaign, all of this activity is nugatory.'

The plea was addressed to facing up to the Joint Normative Decision, to building the model properly in all its recursions, and to educing the consequences. It was another assault, again hard for the enterprise to accommodate. This abstract is included here to demonstrate something about consultant–client relationships. It is necessary to speak thus directly; it is necessary to be heard to be speaking thus directly. People do not want to waste their time and money on soft-soap.

On the face of things, nothing much happened. But consultations continued; and so we came to the next instalment.

Third Cybernetic Intervention

It was now year three (July 1975). A company officer and I were enabled to spend two weeks together. We determined to analyse in full detail the insurance component of the organization—since that was his background. We made recordings of many of our conversations, which were subsequently edited. These had a considerable impact on the President. In fact we 'covered the waterfront'. But the detailed model of insurance proved to be a turning point with an insurance-side man, who was soon to become Executive Vice President of the company.

The work resulted in a hugely elaborate insurance model. But the *corporate* model was still not in evidence. Therefore the fundamental issues pinpointed in the first year could not be addressed; and (as was seen later with hindsight) the detailed model of insurance activity managed to confuse three different levels of recursion. This experience underlined, in cybernetic terms, the absolute need to determine the total recursive system before engaging in such detailed work. Well: never mind hindsight. The fact was known to me as a cybernetician in advance as a methodological principle. But the cybernetician qua *consultant* knew very well that he could work only on those aspects of the total system that were transparently available. The black boxes always have to wait.

Was this exercise, hardly achieved, therefore a waste of time? Far from it. The global cybernetics were a mess. But the sheer force of the analytic capability of variety engineering made its point. In particular it made its point with that member of the top management team who was, as has been mentioned, soon to emerge as Executive Vice President. And he was not even supposed to be present at the meeting called to discuss this somewhat recondite work. Why did he come? The answer to that question has everything to do with the effective management of any institution. We perforce rely on the human genius to know how, precisely, to apply itself. This is in reality, all cybernetics apart, *the heart of enterprise.*

Even so: by working in such detail on a major component of the business, the seriousness of the approach was established. The cybernetic process was continuing. And so was the interpersonal process of command. . . .

As a result, there was a slight hiatus. Although I was in sporadic contact with the corporation, many threads of enquiry were now floating free. And still no-one would underwrite the basic necessity—to get agreement on the global model of the company, considered as a multiple recursion of the viable system.

Fourth Cybernetic Intervention

Year four arrived. It was early in 1976. The hope that a multi-recursion model could be made was still thwarted. There were doubts about its value, and disagreements about the proposed arrangement of the recursions involved. It is difficult for people without training in epistemology to understand the nature of such a model; they tend to think that it has to be either 'right' or 'wrong'. But these words are inappropriate, because the model is not an organization chart, but an *account* of the firm's activity in terms of the criteria of viable systems—and this account is either useful or not. The top management group eventually took this point on trust, but were not yet ready to do so.

Meanwhile, we had set out to make a variety analysis of the insurance activity. We had the structural model (including information flows) that we had together constructed the previous year; we had access to all the relevant data. We proceeded, by many iterations, towards a quantification. The discoveries we made led to a new initiative.

It has frequently been emphasized that the measurement of variety is not an exact science. We are seeking an understanding of the variety amplifiers and attenuators whereby the firm meets the requirements of Ashby's Law. I set out to measure the variety of the system whereby the company's insurance model of itself was conducted. This self-model, reflected in the total information system (and in its computer classification sytem, and therefore in the cost of its computers) was contained in a set of heavy paper manuals—the 'bibles' of the business. It took a whole day's work to understand the system as thereby exemplified.

The company identified itself as operating in nine lines of business, for which eight different funding arrangements were available; it operated in four distinct areas: life insurance, deferred annuities, equities, and vested annuities. In each of these areas, there was identified a variety of *type of plan*, and within these *sub-types*, together with a demarcation of both paying periods and of benefit perods.

Then the ostensible variety of the insurance operation, as revealed by the 'bibles', and reflected in the information coding structure, appeared as in the table. This ostensible variety is multiplicative, and works out to something more than thirty million million possible states of the system. Thus in order to identify a particular item—a business transaction with one client—we need to select *one* out of that total variety. I call it the 'ostensible' variety because the informational structure of the codes shows this generating capacity.

Technical note: This measure determines the *selection entropy* of the system. This is given by the equation

$$H = -\Sigma\, p_i \log_2 p_i\,,$$

where the probabilities within any one selection (for instance each one of the 56 plan sub-types) are disbalanced. In this context, however, these probabilities are not heavily biased, and the equation collapses to

$$H = -\Sigma \log_2 V\,,$$

where V is the variety of each component in the selection.

Table 1 Ostensible variety of the insurance operation

	Field	Number of possible states
	Fund	8
	Line of business	9
Insurance	Plan type	6
	Sub-type	71
	Pay period	49
	Benefit period	49
Deferred annuities	Plan type	2
	Sub-type	56
	Pay period	49
	Benefit period	49
Equities	Plan type	1
	Sub-type	8
	Pay period	49
	Benefit period	49
Vested annuities	Plan type	9
	Sub-type	51
	Pay period	49
	Benefit period	49

Thus all that is necessary is to determine the number of *bits* (that is, $\log_2$) represented by each selection. These are then additive instead of multiplicative, which makes 'ball-park' computation very easy. It is also a direct measure of the number of yes–no decisions required to focus on a final selection, which is helpful in the managerial as well as the computer context.

Well, the system as so far investigated (in its self-model of the 'bibles') has requisite variety to lay roughly eight thousand insurance contracts on every person alive in the world: it postulates some 2^{80} variety. This seems an excessive provision, even given the notorious population explosion. But it is not surprising, nor at all unusual, that reductionist categorization leads to implicit proliferations of variety which large computers are required to handle—ranging largely, of course, over empty space. My driving licence, for instance, registers me under nineteen alphanumeric characters, and in another square under twelve others (this second trivially encoding my date of birth, which I am permitted to chop off the licence where it appears 'in clear': so much for privacy).

Such proliferation of variety is clearly nonsensical, and no management could handle it. The 'real' system, however inefficient, must be far less grandiose. Then why is it not systematically identified, instead of being presented as a lexicon? It is because people do not think in terms of *regulatory* systems but of *classification* systems. This confusion leads to vast and escalating expense in libraries, hospitals, universities . . . and every other social sytem, including insurance companies.

The new version of the insurance system depicted in Figure 1 replaces the classificatory account with the beginnings of a regulatory one. The arrows in this diagram are meant to show determining routes. It seems that to select a line of business determines the fund to be used: thus the variety of $8 + 9 = 17$ is a purely additive function, and this variety of seventeen feeds the entire system—without combinatorial implications.

Next, it seems clear that the sub-plan determines the plan. This notion is surely counter-intuitive to the reductionist mind, which says: first determine the plan, and then discriminate its sub-type. But the realities are otherwise. The insurance agent is attempting to match his available products to the client's needs. It seemed to me that Ashby's Law demands that he must be doing this at the sub-plan level, whereupon the plan concerned would simply be a name in large type for the name in small type of the sub-plan—since it is *that* which matters to the client.

Assuming that the insurance agent is dealing across-the-board with the sub-plans, he embraces a variety of $71 + 56 + 8 + 51 = 186$. That is less than eight bits: a professional can easily handle that much variety. The 'plan' is the heading for the sub-plan, and the funding is no concern of his. The paying period and the benefit period interact, according to the nature of the plan; but we certainly do not need to invoke multiplicatory varieties between them, because each helps to determine the other.

At the end of the line, therefore, the variety with which the managerial regulatory system needs to cope is 646 at the maximum—less than ten bits. Whereas the classificatory-cum-reductionist approach implies eighty bits. My submission to the company at this point said: 'the variety we are handling is *very much less* than many people have obfuscated themselves to believe', for I had heard the story. It went on to mention the 'vast redundancy of variety in the classification system, and to say: 'this is irrelevant to the control system; but the control system is bemused into thinking that it *is* relevant'. We need redundancy, but not nearly so much.

What the new systemic diagram (Figure 1) really shows is that attention must be directed to the contract: the sub-plan that relates to the individual. But obviously, everyone in insurance knows this already! Then, says the cybernetician with curiosity, why is this not mentioned in the 'bibles', nor reflected in the company's regulatory system (except at local levels), nor matched within the huge computing arrangements? Somehow, that key relationship, which the variety analysis so far shows as *determining* everything else, is spread all over an information system in which all variety sources appear to be combinatorial. But if the 'real' system of attenuation

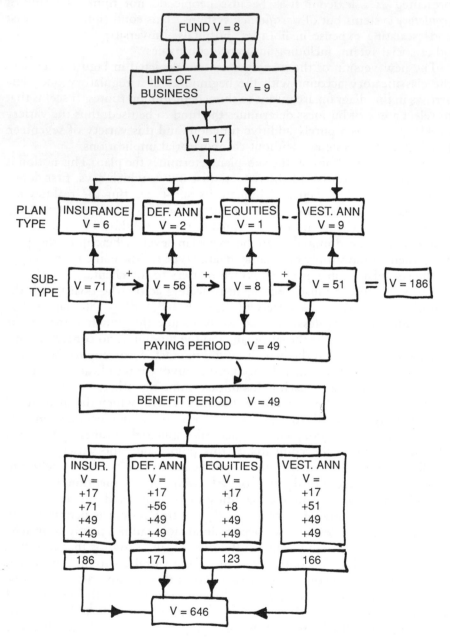

Figure 1. The system of attenuation which actually procures requisite variety (the arrows here **specify**)

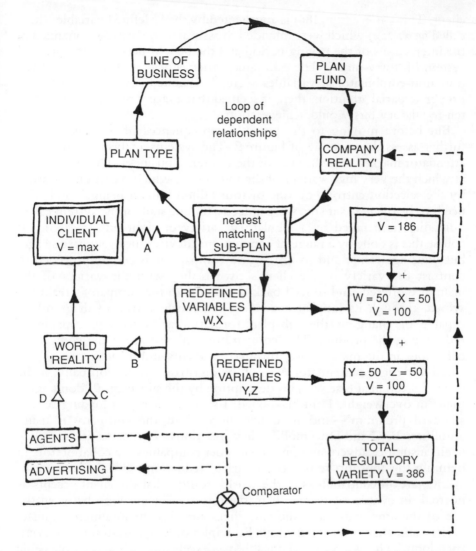

Figure 2. A redesign for Figure 1, offering a cybernetic model of the regulatory process that minimizes control variety

(80 down to 10 bits) is how regulation is actually made possible, then Figure 1 ought to be reformulated according to the best variety engineering that can be accorded to the basic facts. This was attempted in a diagram of which Figure 2 is an improved version, which cut the requisite regulatory variety once again—this time by nearly half (646 to 386).

The argument was that the crucial item in controlling selection entropy is the sub-plan. This is what actually absorbs the variety of the individual client. This variety ($v = 186$) is augmented by the redefined variables (here called w, x, y, z) which were intended to assimilate the variety contained in the interactions of the paying period and the benefit period in the original system of 'forty-nines'. The estimated varieties of $v = 50$ for each of the four non-combinatorial variables were, I reckoned, overestimates; with proper actuarial attention, then, it seemed that a target of eight rather than ten regulatory bits would achieve control. . . .

But before moving to the possible consequences of that, we should further assess the meaning of Figure 2. The systems analysis points to the attenuator marked A as the crux of the insurance business. This is the point at which the very high variety of the individual, which we might measure by the selection entropy of 'one of four billion human beings', and who needs to be treated as if s/he were indeed a unique state of the system called humanity, is handled by a regulatory variety of eight bits ($= 256$). Of course this is done by a range of conventional devices (not *your* precise age, but your age *group*, and so on). But a comparator is certainly needed to contrast the variety of the client's own reality with the variety of the insurance offered, and to feed back the results to the company's reality— which, through its loop of dependent relationships (top of diagram), is capable of changing the sub-plan format: that is to say, capable of redesigning the product. This comparator will also affect the company's behaviour in the marketplace, through its agents and its advertising.

All ill-designed attenuator A could be disastrous to the company, if the client felt too ill at ease with his description by the sub-plan. ('I don't agree that I'm overweight, I am big-boned; I shall not accept a loading on the standard premium'—and so much more.) But the company is using amplifiers B, C and D to mollify these reactions. The terms of the policy itself, insurance advertising (in which other companies are collaborators in defining the appropriate degree of variety reduction), and the training of agents, are all contributors to the world 'reality' that the client accepts as normal. In cybernetic terms, then, there is a balancing loop between the use of the attentuator and the amplifiers aimed at maintaining requisite variety and answerable to the four Principles of Organization that govern such loops. This is a System Four job, par excellence. It is at least plausible that the first company to redesign this loop (which includes all three feedbacks from the comparator) would rapidly gain a reputation for meeting clients' needs more sensitively—since the variety matching of attenuator A would now be a learning device.

The aim of designing a regulatory system based on eight bits of information relates to the existing range of products, since these currently *define* what sort of person the client is allowed to be. Now this is not a static definition, at least in the company concerned. New products are constantly being designed and marketed; and the system sees itself as adaptive to its market in terms of social as well as actuarial trends. The fact remains, however, that when an agent interviews a prospective client, he must encourage him to climb on to one of the Procrustean beds manufactured as the company's 'current range'. The question poses itself: whether it would be possible to use the learning loop analysed above to make the company responsive to the individual client in real time, without reducing this variety by categorization in advance. For, however adaptive the existing system tries to be, it is certainly committed to this principle at present.

In probing this question, I set out to model the process whereby the existing sytem (common to all life insurance companies) arose in the first place, in order to understand how it might be cybernetically redesigned. Here is the argument:

At time t_1

Start an insurance company. The variety of insurable states of insurable people tends to infinity. It is not possible to offer an infinite number of plans.

Then divide 'risk' arbitrarily into x specific risks.

Divide 'people' into y arbitrary categories.

Work out the actuarial consequences of xy plans.

This is a first statement of attenuator A: it has $V = xy$ variety, conceived as a *constraint* on $V \rightarrow \infty$, with which we started.

It would be possible to fund this insurance in a very large number of ways. Constrain this number to z ways.

So the business plan has variety $V\star = xyz$.

We have done much attenuating, but we have invented a variety generator *within the business*.

At time t_2

Observe that the xyz mix is losing market opportunities. Enrich the range of risks covered from x to X. Make more discriminations among people: y goes to Y. Make more ingenious funding arrangements to exploit the investment market, forcing z up to Z.

$V\star t_2 = XYZ \ggg xyz$.

We are now amplifying, by relaxing constraints. The business is becoming more complicated by virtue of its *internal* variety generation.

At time t_3

The authorities (government) observe the variety proliferation $V^\star t_1 \ll V^\star t_3$, **and predict** $\ll V^\star t_{n+1}$.

They decide to 'constrain' $V^\star t_{n+1}$ **in advance**, in order to protect people from possible exploitation.

Thus regulations are introduced at time t_3 which would *attenuate* $V^\star t_{n+1}$, but effectively *amplify* $V^\star t_3$ when they are introduced, because they have to be met at that point.

Then at any time t_n—which is **now**—we are dealing with an existing system that has

(a) grown by accretion in response to commercial opportunities, whereby it has equipped itself with huge variety generators (enshrined in clanking computers and creaking software) which are capable of proliferating ludicrous amounts of variety; and
(b) has had its variety hugely increased by government actions that amplify variety **now** in response to a *feed-forward attenuator*.

I had been well-schooled by the company to acknowledge that it necessarily lived in two different realities which much increased the complexity of its managerial process. One is the business reality in which the company must remain a viable system in the sense of this book; the other is the legal reality in which it acknowledges the law and obeys it. Usually in business, the second is merely a constraint on the first. But in insurance, the two realities are complete and different accounts of the business—so that the enterprise can survive only in the intersect between them. The above analysis, which is reflected in Figure 3, stumbled (I thought) on the reason why; and this had consequences later, as shall be seen.

Now if the company has the capacity to proliferate variety adequate to the expression of two different activities, which—in their sum—raise a necessary eight bits of information to eighty bits, then somewhere in all of this is embedded an eight-bit variety generator. This ought to be isolated, and Figure 4 is its picture.

In making this isolation, I tried to take account of the question that led to this analysis: 'whether it would be possible to use the learning loop (of Figure 2) to make the company responsive to the individual prospective client in real time . . .'. Thus I distinguished between

(a) *description rules*, which relate the personal profile elements of the individual's variety to the elements of sub-plans; and
(b) *transformation rules*, which embody the actuarial expertise that turns attenuator A into a contract-producing machine.

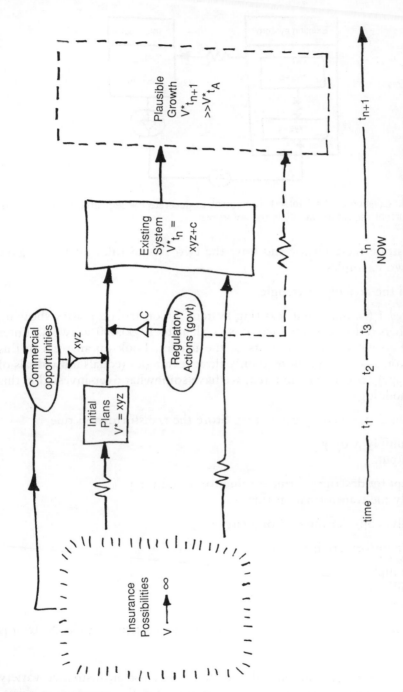

Figure 3. Accretion of variety, much of it illusory, in existing system. Note the effect of amplifier C (see text)

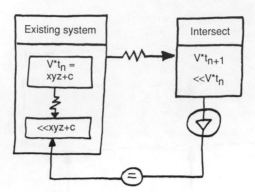

Figure 4. Isolation of the Intersect, competent to generate the **legitimate** component of current regulatory variety in the enterprise

To illustrate these terms, and how the two sets of rules interact, I gave these two examples:

To find the area of a rectangle:

Method 1 (compare our 'existing system'): Construct a matrix showing numbers 1 to 1000 on each axis. Calculate 1,000,000 answers. Enter these answers in the elements of the matrix. Look up answers in this Lexicon as required. Note: if only three rectangles (equals individuals of *this* type) are ever encountered, we have somewhat over-invested in this methodology.

Method 2 (the variety generator): Store the transformation rule

multiply x by y
output.

Accept by description rule a value for x and for y.
Apply the transformation rule.

Secondly, consider the case of a circle:

The transformation rule is

multiply x by x
multiply by π
output.

The description rule measures the radius of a particular circle. Note: π is invariant.

These simple expedients introduced the design for an insurance variety generator which would use descriptive rules for the prospective client, transformation rules for the actuarial expertise to be applied to the description, pick out the statutory invariants that apply in the 'second reality', and cement all of this into a box of microprocessing chips. This box could then be handed to the agent, at very low cost (much lower than his use of a terminal in the branch office to speak to the mighty but somnolent dinosaur computer at head office). It points the way to highly sensitive selling (because it handles Ashby's Law by design rather than by historical inadvertance), and to a vast reduction in cost. I do not give any further details here; partly because they were preliminary, and partly because this company is still my client. . . .

It has to be acknowledged that the ideas of this Fourth Intervention appeared and remain recondite to the company's officials. The company's account of the investigation says: 'the investigation may have within it, when the time is ripe, the seeds of an improved method of generating control variety for the company'. With affection I can say to my friends that I have heard this talk before; and maybe I shall have carved on my tombstone: 'his time was finally ripe'. But see 'Introduction: On Success', and await the Tenth Intervention.

In any case, the acquisition of these insights had two important consequences for me. Firstly, the variety analysis suggested a completely new way of designing the attenuation system. Secondly, the variety analysis raised very grave doubts in my own mind about the effective handling of variety reduction *between recursions*, in terms of the Law of Cohesion. But the recursive model through which to examine the issue did not yet exist. Both these results will be referred to again in this case study. At any rate, and at that time, it was clear to us all that the company could not handle my cybernetic initiatives in this regard. And this fact gave special prominence to the long-standing criticism that there was something radically wrong with System Four activity at the corporate level of recursion. It is not that the ideas were rejected (that would have been legitimate) but that there was no machinery for investigating them.

Besides, while all this was going on, further significant changes were taking place in the top management structure of the company: there were many preoccupations. The cybernetician was, for a time, lost in the consultant. But it was clear to me that the next intervention had to result in the multi-recursive model—or no more specifically cybernetic work could be accomplished.

Fifth Cybernetic Intervention

In year five (1977) that very breakthrough occurred. If the 'GPF' after five years of GPF-ing cannot (as was said earlier) have some advice accepted on trust, it would be a pity. The advice was, of course, to invest at last in the creation of the total model of the company as a viable system. It was finally accepted; despite all the reservations that had been made in the past, and which were not withdrawn. Such a moment ought to be regarded as a real triumph for the manager–scientist relationship. It well demonstrates how that collaboration really works.

This is the moment to remark that the collaboration spanned the physical distance of the Atlantic Ocean. Meetings were frequently held, on both sides of 'the pond', but much work had been conducted through the medium of audio-tape. All concerned were practised in its use. Even so, it was overwhelming (in both senses of the word) to find such warm-hearted effort in helping to make the model. Tapes arrived in Britain from all eight members of the senior management group. All were expansive and candid. All were long, ranging from three-quarters of an hour to two hours.

What follows is an account of the modelling and diagnostic processes, which the company officer and I jointly undertook in Wales. Of course, I accepted full responsibility for the cybernetics and the conclusions; but it cannot be over-emphasized that work of this kind requires the full-scale involvement of the client organization, and that this normally has to be channelled (as it has been in this case) through a committed officer of the company who enjoys the total confidence of his colleagues at home.

It is also of interest to record that the following account was first communicated to the senior management by means of the diagrams and several hours of audio-tape. This medium offers something approaching requisite variety between human beings who already know each other, and the nuances of each other's voices, that printed reports often lack. The words used here are an amended version of the written report made at the Eighth Intervention; of course, a great deal had been done about the criticisms made in between, and what follows should therefore be taken as an 'historical document'.

The recursive mappings

It was said earlier that a recursion of viable systems is not properly based if it merely forces the organization chart on to the Procrustean bed of the viable system model. We have instead to look at the criteria of viability, and detect within the actual organization a recursive sequence that meets those criteria at each level.

It is usually best to begin with the basic operations of the enterprise, calling these the System One of the lowest level of recursion. Higher levels of organization certainly exist; at first sight it may look as if these *must* be viable systems themselves. But higher levels of organization are in fact abstractions: they are arbitrary groupings of the basic operations, working with aggregated data. For example, the territorial divisions of the company and the functional divisons of the company both look as if they might be viable systems; but neither set responds to the cybernetic criteria. In fact, the first set turns out to be a discrimination of environments; while the second turns out to be the professional embodiment of activities that occur on the connections between subsystems of viable systems, and indeed between whole recursions of viable systems (in accordance with the Law of Cohesion).

This is in itself an important conclusion. For the supposition that these two sets both constitute viable systems, whereas in practice they manifestly *interlace* (rather than *nest*), causes confusion. The standard resolution of the confusion is the so-called 'matrix organization' approach. If two aspects of the enterprise are running orthogonally to each other across the managerial space, then why not create a matrix to represent that space . . .? Each element of the matrix then represents the interaction between one territory and one functional specialism. Well, this may be a useful conceptual model of a practical problem; but if it leads to a committee for each matrix element, the practical problem will certainly be exacerbated. The confusion is better resolved by understanding the recursion of viable systems that underlies the organizational conventions.

Now the basic operations of the company consist of insuring and investing: these are the operations that create wealth. Therefore they were chosen as representing the fundamental level of recursion—as distinct viable sytems. For (referring here to the First Intervention) each activity could in principle exist independently. Insurance could place the proceeds of its activity with an investment group; investment could obtain its operating funds elsewhere.

Looking first at insurance, we find that the basic operation is the management (in every sense) of a product. This surely constitutes a viable system: *any* product can be designed, made, sold, and so forth, if it is a good product, on its own strengths. But there are a great many products within the company. Then we may look for a next level of recursion, wherein the viable system comprises a *set* of products, jointly managed. Immediately, we find the obvious example: group insurance. It follows that the next level of recursion will constitute insurance as a whole, and as distinct from investment (although, since investment has *not* been hived off, there will be insurance–investment interactions at every level of recursion). Then we have three insurance recursions, embedded in the enterprise itself. It follows that the fundamental product recursion is the fourth recursion.

Turning to investment, the picture is totally different. The effort of modelling the investment activity as a viable system made the obvious distinctions between types of investment; and this produced the discrimination of the elements of System One. But variety analysis of this basic level of recursion, conceived as a viable system, indicated a mode of management that permits variety to be absorbed to the full at this fundamental level. This was a remarkable discovery for me: I had not met a similar situation before. That is to say, the company needs to discriminate four levels of recursion in order to conceive of itself as an insurance company, and to discriminate two levels of recursion to conceive of itself as an investment company. There seem to be three reasons for this asymmetry.

In the first place, the environments of investment (although covering very similar geographic zones to those of insurance marketing) are quite differently divided. There are few countries within which funds will be moved; and the exchanges within these countries are closely integrated between themselves—because of international fiscal policy. In the second place, *and therefore*, the management of investment (which might appear to be divisible over several levels of recursion) is in fact impacted. That is to say, there exists a nest of portfolio managements at the fundamental level of recursion which necessarily reflects the integration of the exchanges in the environment. Thirdly, there is so powerful an interaction between the operations (circles) that the 'squiggly lines' concept of their connectivity has to be altogether replaced by the potent 'looping' concepts to which the Four Principles of Organization directly apply.

Variety analysis underwrites this compression of the viable system called investment. Remarkably, the necessary variety generation can be, and is, handled—because variety is continuously absorbed in all three domains of the vertical plane (namely, environments, operations, central command) that are given in objective reality. How is it suddenly possible to invoke such a notion as 'objective reality' in the context, whereas the basic cybernetic teaching calls all systems subjective? The answer is clear: such a process as investment-under-constraint (where the constraints are both professional caveats and governmental edicts) creates a special universe of discourse. The 'objective reality' is, as a consequence, a function of the conventions which are accepted by the financial confraternity: it is virtually the invention of a shared subjectivity from which no-one entitled to do this business can escape. Whether this is fair comment or not, there can be no doubt of the variety attenuation involved.

The mapping process therefore yielded a recursive system which is cursorily represented in Figure 5. In fact, it would be very possible to model one element of *Recursion Two Investment* at the third level of recursion, namely the management of physical assets. The Real Estate operation is certainly a viable system. However this was not investigated as the managerial problem in its own right that it undoubtedly is, because its role in Recursion Two is so well understood. This is not true of the insurance recursions, as shall be seen shortly.

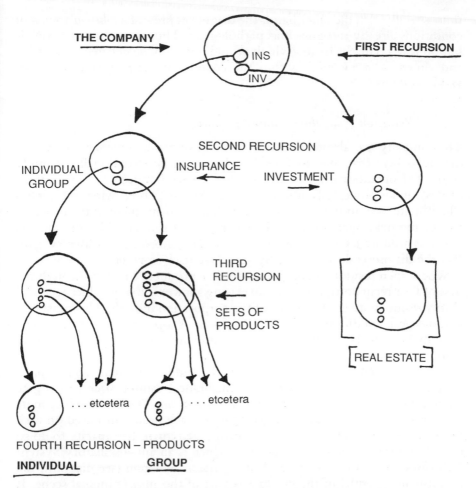

THE COMPANY

FIRST RECURSION

INS

INV

SECOND RECURSION

INDIVIDUAL
GROUP

INSURANCE

INVESTMENT

THIRD
RECURSION

SETS OF
PRODUCTS

REAL ESTATE

... etcetera

... etcetera

FOURTH RECURSION – PRODUCTS

INDIVIDUAL

GROUP

Figure 5. The general scheme of recursions of the viable system for this company

The diagnoses

A serious and thrustful company such as this is constantly gnawing at issues which are perceived as based somehow in unresolved problems. That is to say, symptoms of difficulty are noticed at various places and levels in the enterprise, and become a syndrome that presents itself to the corporate consciousness. When a thoroughgoing model is employed diagnostically, we would expect these issues to be pinpointed. It is unusual (though not unknown) for totally exotic issues to be raised: indeed, the familiarity of the issues themselves is part of the validation of the mapping of the model. So the diagnoses do not hit the enterprise altogether traumatically. On the other hand, the diagnoses present a *coherent account* of conditions already recognized as pathological. Thus the cybernetic contribution is to relate the issues (which would normally be debated separately, and almost in a vacuum) within the context of the recursive model of viable systems.

Recursion Four: the insurance product

The company was already debating the issue of product management. The diagnosis says that System One (see Figure 6) is exceptionally strongly connected operationally (see the strong arrows between the operational circles), and is powerfully administered by System Three. Moreover, it has a legitimate System Four, concerned with the enhancement of the product, and of the field operations relating to it. All of this provides the illusion that the product level of recursion is already a thoroughly viable system. But the diagnosis declares that System Five itself is void.

Where then resides accountability for the product? Tracing through the nest of four recursions, the quite surprising answer is that product accountability resides with the President of the corporation himself. This must be a pathological condition—that the President is operating as a surrogate System Five in the *fourth* level of recursion, whereas his post obviously belongs to the *first* level.

Can the model explain this vast anomaly; and in particular can it explain why System Five at Recursion Two (Insurance) cannot offer a surrogate System Five for the fourth recursion? Indeed it can. The product has, in its very inception, an investment *doppelganger*. Each premium raised on each policy is earning—must earn—an appropriate return. Its capacity to do so is underwritten by actuarial expertise, which is an anti-oscillatory System Two function at this fourth level of recursion, wherein (see diagram) the investment potential of the product is part of the environmental scene. It cannot be other; because the investment recursion is not product-oriented. The investment recursion is oriented wholly towards financial markets. The crossover-point between insurance and investment is not reached until Recursion One, where actuarial activity is far from being a System Two damping function (as shall be seen) but is the major *regulator* of the company's whole activity. The Chief Actuary is directly responsible to the President, and *that* is why the President operates as a surrogate System Five at the fourth level of recursion.

Obviously it is a matter for management to decide what to do about this situation. There are many possible solutions The role of the cybernetician is to monitor the discussion of incipient organizational changes to see if they are themselves answerable to the criteria of viability. But the diagnosis itself points to a key issue (and perhaps this was *not* recognized). If there *were* a concept of product management, then its System Five would want to know all about the investment potential of the product. Hence, if the President were to relinquish the role of surrogate System Five in Recursion Four, there would have to be totally new linkages between the investment nest of recursions and the insurance nest. Probably, then, there would be four investment recursions as well; because the same issue arises

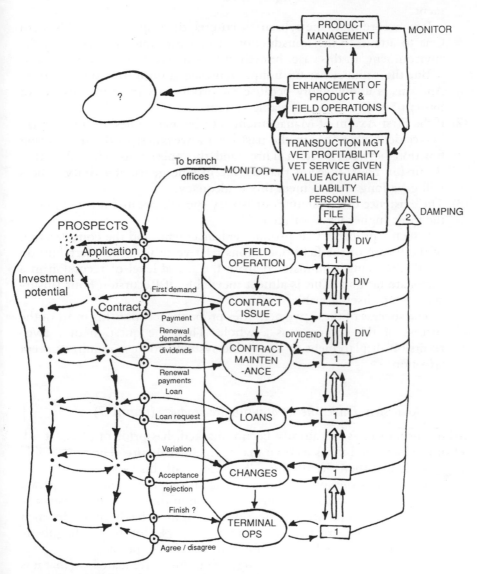

Figure 6. Insurance, Recursion Four—an insurance product

at each level. If so, the strangeness of the four–insurance/two–investment recursions would be seen to be not merely an aesthetic disbalance. . . .

There are three more diagnostic remarks to be made at the product level of recursion.

(1) There is a very unusual intervention, coming apparently from System Three, in the *transduction* capabilities of System One with its environment.

It is, according to the cybernetic criteria, the responsibility of System One to attend to the transduction of its operations (circles) with the environment. In this case, branch offices are the primary transducers.

But the product has no unique representation in this transduction. And this is undoubtedly because the product (witness the absence of System Five) is not yet a viable system.

(2) If the First Axiom of Management is to be met by design, rather than by accident, there has in the past been a weakness in the monitoring function between System Three and the operations (circles) within Recursion Four. It seems that the recent emergence of activity indices will go a long way to meet this inadequacy.

(3) The existence of a System Four is very special, because it is missing in the three higher levels of recursion.

It seems probable that product enhancement, and the consideration of the future of the field operations relative to it (which certainly do exist), are being administered from the *second* level of recursion in a surrogate fashion. This is almost inevitable, if Recursion Four (Insurance) is not itself a viable system.

The success of these two activities probably accounts for most of the success of the company as a whole. Then the question of a two-recursion displacement in the viable systems context requires urgent attention.

Recursion Three: the set of insurance products (Figure 7)

Group Insurance, as has already been remarked, has pride of place. It *is* a set of products, and it has its own management structure. The remarkable diagnosis is that there are no more 'businesses' of this type.

As with the issue of product management, the issue of separate 'businesses' has become a major preoccupation of the corporate consciousness. It is of no consequence whether this development has been conditioned by the continuing cybernetic analysis over the years. But, by the diagnostic stage, the cybernetic analysis has much to say on the point.

How is the variety of the whole range of products attenuated, so that it is possible to regulate the activity of insurance (Recursion Two) as a whole?

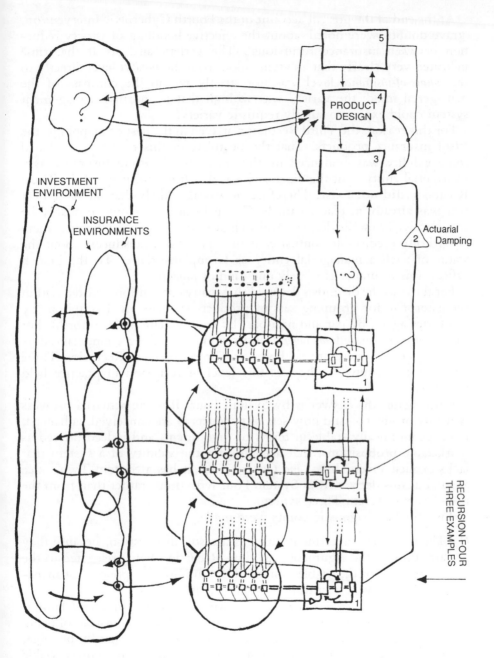

Figure 7. Product insurance, Recursion Three—a **set** of insurance products. Three examples of Recursion Four may be seen embedded, by turning the page through 90°

At the end of the present account of the Fourth Cybernetic Intervention, 'grave doubts' were noted 'about the effective handling of variety reduction between insurance recursions'. The variety analysis at that time indicated very clearly that an attenuation from the *product* level directly to the *whole-of-insurance* level was not strictly within the compass of the managerial mind: the variety was so high that no normal management system could expect to supply requisite variety.

For this reason, the Third Recursion arrived in the nest of mappings: the *set* of insurance products. That the postulate of this recursive level had some validity was evidenced by the existence of Group Insurance as a managerial entity. But the diagnosis was that (this entity apart) the Third Recursion **did not exist**. Therefore, it was also likely that the piece of it that was already in place, namely Group Insurance, would probably be misconceived to some degree. And so indeed it is contended: because there is a wholly predictable confusion at the marketing transducers about the status of such a managerial entity as Group Insurance. To the branch offices, this insurance is merely one kind of *product*.

For the rest, the diagnosis declared that many conventions existed within the enterprise for grouping sets of products in terms of description (by territories again, by par and non-par, by endowments, and so forth), but there were no agreed *managerial* units which could be countenanced as viable systems. There are no Fives or Threes, then, at Recursion Three (apart from Group). So how is this level of recursion in practice held together?

Once again, the answer is by System Four. It is the actuarial prowess whereby products—and now sets of products— are conceived, enhanced, made economically stable in terms of investment, and indeed offered as marketable propositions, that accounts for the validity of a system that lacks explicit Five, Three, *and One* activity. It has a System Two, once again; but this derives from System Four reflections, rather than from the non-existent System Three synergies.

Two further diagnostic points were made:

(1) The insurance-investment linkages are still in question, because they are still (see the argument regarding Recursion Four) relegated to the *environment* of the viable system. They are not components of managerial effectiveness until Recursion One is reached.

(2) Insofar as managerial effectiveness for Recursion Three and Four is pushed upward to Recursion Two (which is the case), instead of being delegated downwards *from* Recursion Two (which is hardly possible, in the absence of the metasystemic components of these quasi-viable systems), control of expenses must be very difficult to accomplish.

In effect, this kind of regulation—which is a System Two function—is working out of Recursion Two, and attempts to span two more recursions as well as its own. Then it would be very surprising indeed if symptoms of instability at the two levels of recursion were not in evidence.

Recursion Two: the whole of the insurance activity

We are confronted here with a viable system that is largely complete. It answers to the criteria of viability in general. The diagnoses now are more minute; but because of the higher level of recursion, they are by no means unimportant.

Firstly, as far as Figure 8 is concerned, the convention whereby the operational circles are drawn to be roughly proportional to their 'value' is surely helpful. And we notice a System One 'investment only' operation within Group, which is very small—but could in principle become very large. Diagnostically speaking this is of great interest. It refers to cover provided for other insurers; from the viewpoint of this enterprise, then, it is an insurance activity (because it belongs here) that does not happen to be insurable. . . . Diagnostically, then, perhaps it is a cancerous growth? Consideration should be given to moving this to the investment side of the house.

Secondly, a close enquiry into the articulation of System Two is called for. Four species of regulatory damping are indicated on the diagram. They refer to: housekeeping (e.g. computer manuals); accounting practice (i.e. the conventions whereby the books of the company are made consistent with each other, and are made to conform to legislation); actuarial damping (i.e. the *highly* professional rules whereby this business may be conducted at all); house style (e.g. the letterheads, the newsletters, and so on, that help to give the company an identity in the minds of all its publics).

These diverse activities are not perceived (this is a diagnosis) as mutually interactive, in the role of oscillatory regulation. If they were, and if the performance of System Two were deliberately studied as a component of the viable system at this level of recursion, there would (this is a prediction) be changes in all four modes. Moreover, other modes of System Two activity would be identified that have not so far been identified. What, for example, is the System Two role (*not* the System Four role) of Sales Conferences at Recursion Two?

Thirdly, we are *yet again* faced with the interaction of insurance and investment, which is *yet* again, relegated to an environmental interaction—except insofar as actuarial expertise brings them together. Certainly, this works. But such functional activity on the part of Recursion One is **not** a significant aspect of the *management* of insurance (Recursion Two). Therein lies the diagnostic judgment.

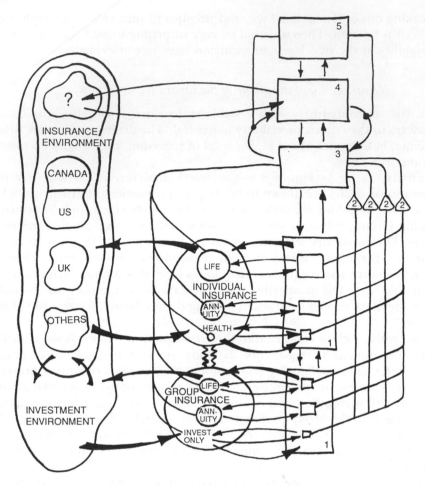

Figure 8. Insurance, Recursion Two—total insurance

Fourthly, we reach a very strange diagnosis indeed. There is no System Four at this level of recursion. The diagnosis is strange because it was *only* a System Four that was discovered to represent the Five–Four–Three meta-systems at Recursions Three and Four. Now, suddenly, we have a manifest System Five and System Three; but there is **no** System Four. The reason is obvious: the Four function for total insurance, Recursion Two, is already dissipated in the third and fourth levels of recursion. The System Four problems of Recursion Two as such are not addressed. That is because they are viewed as the sum of the problems that relate to the lower levels; but this is not true.

Many new things could be done with the concept of insurance. It may well be that a mutual life company has already and firmly decided not to enter the fields of motor-car, ship, aeroplane, or house insurance. If so, that is fair enough. But the System Four activity in this company, if there were one, would ask itself other questions than these. 'We insure Life, What is Life?' It surely turns out that Life is more than the guarantee of material worth to one's dependents. Life includes the capacity to earn, the opportunity to indulge in leisure pursuits, the security of intellectual capital, and so on. Which insurance companies anywhere are considering these insurance possibilities? The answer in terms of this account is probably none. And that is because none of them commands a System Four at Recursion Two. Only this company, of course, among insurance companies, knows what that last sentence even means. Herein lies the strength of the model as a diagnostic tool: it points especially to voids.

In short, System Four of Recursion Two is not 'all about marketing existing products'—or even extensions of these. Therefore this putative System Four, which people imagine to be in place, is all about Recursions Three and Four in their own eyes; but they are mistaken.

Fifthly, it is only in the light of such considerations that anyone can talk about new methods of marketing *at this level of recursion*. It may well be that new methods of marketing have nothing to do with Recursions Three and Four. They may have everything to do with insurances that have not yet been invented.

Sixthly, it has to be noted that when great attention is paid to such issues (if it is), such activity is in itself potentially destabilizing to the existing business. People will surely say, to be brief: 'What the hell is going on?' This, in its turn, is a diagnostic remark. For it ought to be accepted, but is not, that such things are indeed going on. If they are not going on, the enterprise in the long run is not viable. But of course any such going-on-ness is threatening to everyone on the payroll. Therefore there is, here as in any other enterprise, an onus on senior management to explain why it supports a System Four at all.

Recursion Two: the whole of the investment activity

We find here a very strange version of the VSM. Here is the massive 'looping' involvement of the operational circles between themselves. Here also is the 'impacted' set of System One management boxes which reflects that operational involvement. These features have already been explained, as has the extent to which they provide a variety-attenuating system of such potency that it soaks up the whole of the variety of the two further levels of recursion that might otherwise have been expected to exist (see Figure 9).

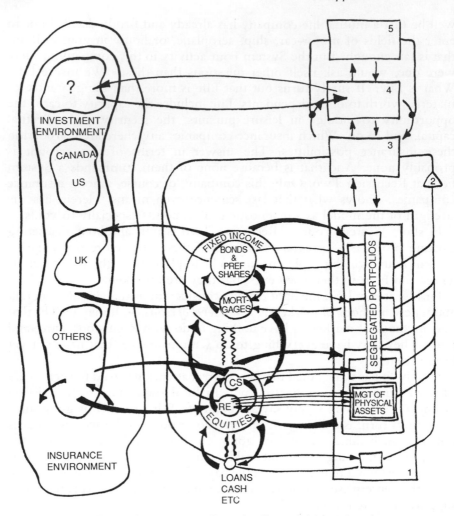

Figure 9. Investment, Recursion Two—total investment

These are diagnostic remarks, when they are taken in conjunction with the diagnoses already offered on the insurance recursion. But the diagnosis uniquely relevant to *this* diagram is that the impacting of System One management activity seems to *subsume* the roles of both System Three and Five. For if all synergistic functions are cared for at the System One level (by the impacting) there is nothing left for System Three to do. And if there is no System Four, then System Five cannot be discriminated, and must be part of the impacting as well.

Well: System Four does appear to be void. We remarked earlier on the conventions (professional and governmental) that are applied to the investment activity. These are so severe that it seems likely that any genuine System Four developments would lead to proposals that were either unethical or illegal, or both!

★★★The alarm that this situation generates in the cybernetician, because he is concerned with viability criteria, is matched by concern within the industry: although the industry does not understand cybernetics, nor use its language, it knows very well that it is hidebound by regulatory processes that could easily stifle evolution. This is a classic case, from the societary standpoint, of counterproductive regulation. In order to protect the citizen, professional and governmental rules are in force which restrict insurance companies in their marketing, in their investment, and above all in the financial cover that they must provide against the liabilities that policies yet to mature constitute, to the point of unrealism. What is an *enterprise* to do, if it is not allowed to be enterprising?

★★★The answer to this question, cybernetically, is that it is System Four's job to tackle the problem. Yet, at Recursion Two, there is no System Four in either insurance (since that kind of activity has been soaked up in recursion Three and Four) or investment (since it is impacted within System One). In the final diagnosis, we shall ask if there is a System Four in Recursion One, the corporation itself. But the answer is already entailed. If there were such an entity, the two Systems Four at Recursion Two could not conceivably be void. . . .

Recursion One: the entire corporation

The diagram at Figure 10 looks, but is not, simple. It looks simple, because System One (as was stated at the beginning of this story) is reduced to the two operational components of insurance and investment. It is complicated because the 'squiggly line' connection between the two operations is best described as a huge dynamo that in fact 'runs the business'. The model calls it 'the asset balancer'.

In all the years of applying this model, I had not seen so *intimate* a relationship within the operations of System One. There are often *commanding* relationships: as when iron ore is first mined, then reduced to pig iron, then converted to steel, then rolled or forged and so on. Recursion Four insurance exhibits this kind of relationship. Very often, at Recursion One, the interactions between the components of System One are restricted to the competition for capital—as happens in a conglomerate corporation. But in this insurance company, *intimacy* is the only word, and the story of the first cybernetic intervention explains why.

Textbooks say that in this situation management needs a whole collection of committees. Variety analysis says that their sporadic, variety-attenuating activities cannot possibly succour the dynamic interaction

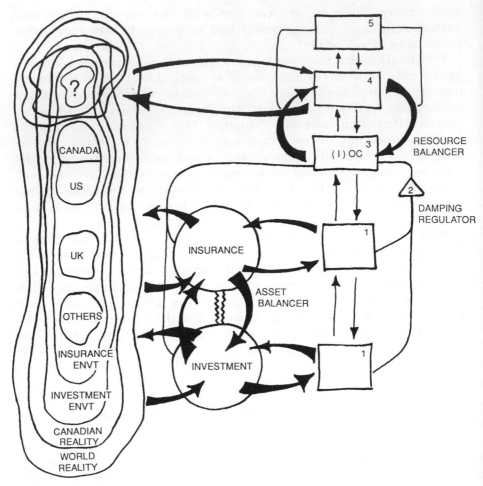

Figure 10. The company, Recursion One—total system

required. Now: if the corporation had not solved this problem, somehow, the firm would not exist. They had solved it in human terms. And humans, having roughly equivalent variety (ten-to-the-ten neurons, a need for sleep and so on) can usually cope—if the institution provides for their requisite interaction. The firm, it seemed, had achieved this, but at an astonishingly informal level.

In year one of this story, there had been an executive committee and an investment committee which between them seemed to be dealing with all managerial matters of major importance. Examining these processes, I came to the conclusion that they constituted a System Three masquerading as a System Five. This was a very early diagnosis. Over the years of the cybernetic interventions, the character of the key committee of internal management was radically changed. If it really was a Three Committee, it should be so constituted: for example, the President should not be a member. His role should be reserved for presiding over some kind of policy group in System Five.

Five years later, at the time of the diagnosis, the key managerial committee had become the *Insurance* Operations Committee, matching the *Investment* Committee, and therefore, in terms of this model, each had the appearance of a System Five at Recursion Two—which ostensibly made a great deal of sense. But by careful study of the activities of the two bodies via the structural model, a different picture emerged.

The Insurance Operations Committee was not fulfilling the function of System Five at Recursion Two Insurance, but the function of System Three in Recursion One. The Investment Committee, on the other hand, was fulfilling the function of System One in Recursion Two Investment— because of the impacted nature of the management—and also some part of the System Five function in Recursion One. That in turn was because the President, who had indeed left membership of the Operations Committee, was still very much in the chair of the Investment Committee.

The diagnosis of Recursion One therefore argued that the prefix 'Insurance' in front of 'Operations Committee' was no longer appropriate, and should be dropped. This was agreed (but the old name withstood the decision!). This was not a fussy terminological issue. Reference has been made to 'the asset balancer', and to the apparent informality of its operation. We would expect, via the model, to find a synergistic management activity in System Three that would reify the concept of dynamic interaction between the two System One operations of Recursion One. This activity had at last emerged in the Operations Committee; and the overt sign that System Three was properly constituted for this recursion level was the fact that the presence of the most senior *investment* officer of the company had at last its impact. This had been brought about in various ways, but it was real. Hence the advocacy for the change of name.

If this could be formally recognized, that would be a big advance in making clear what the metasystem (Three–Four–Five) was, and how it actually worked. Hitherto, the metasystem had simply been seen as that group of human beings who clearly wielded authority within the company. But now, since System Three was pinpointed, and if System Five (though not yet pinpointed) *fairly* clearly involved a small but influential group gathered round the President, this also served to point to the void that should be System Four. The diagnosis about this void had always been firm—that is, before the total model was constructed. No person, no committee, had any accountability for the Four function; no task force, no specialist group, existed to work in this area. It follows that the Four function must have been spread across the minds and discussions of the senior management. But this is never an adequate response to the challenge of the future in a rapidly changing society.

In fact, said the diagnosis, there had to be an investment (perhaps in money, but certainly in time, talent, care and attention) in System Four activity at Recursion One; and the sum of Four activity at *lower* levels of recursion in no sense took the place of this activity. A cultural myth existed that, somehow or another, new ideas 'bubble up' through the organization. No doubt they do. But in the nature of things such ideas will relate to the existing paradigm as to the nature of the business, and therefore they will stop bubbling up at the level of recursion to which they relate. And if, alternatively, such an idea were directed to fracturing the paradigm, how could it be received, properly heard, in the void of System Four? It would result only in an echo. Not only was this (diagnostically) predictable: it had already happened to the conclusions of the fourth cybernetic intervention. . . .

The diagram went on to emphasize the role of what the diagram calls 'the resource balancer'. This is supposed to monitor the investment (as above defined) between Three and Four activity, and is cybernetically speaking a crucial role in the maintenance of viability. But it is difficult to organize the reification of this concept in the absence of a System Four. Finally, the diagnosis discussed weakness in System Two, but declared that these could not be attended to until the logic of viability in the lower levels of recursion had been improved.

Sixth Cybernetic Intervention

This intervention was marked by a diversity of affairs.

In the first place, there had to be long and detailed discussions about the model and its diagnoses. As was said at the outset, it is not right to expect that drastic actions should instantly be taken in response to such inputs; for as was said later, it would be destabilizing to the enterprise if they were. I have often noticed, in the past and other contexts, that when consultancy is used in that way it is often merely an excuse for deliberate destabilization: that is, the management is in deadlock, and seeks to blame necessarily vicious action on an outsider. That is no way to evolve a viable system.

Instead, key implications of the cybernetic process to this date were carefully worked out in several fields. Many developments within the company, particularly its planning processes which were rapidly evolving, were examined through the insight provided by the cybernetic model. Surely this is the main value of such work: to provide a framework for viability, and to inject the language of viability into discussions which otherwise revolve mainly on the conflict of personalities and the apportioning of personal power.

In practice, there was no repudiation of the model, of the diagnoses, nor of consequential advice, by anyone in the senior management team. If there had been, there would have ensued great difficulties. But, as has been argued throughout, this is not a claim by the consultant to total acceptance, and not a commitment by the company to gross reform. It is all part of a cybernetic *process*; and the greek word κυβερνητησ simply means steersman.

It will be no surprise that, out of the whole diagnostic application of the model, two issues were in the limelight. One was the notion of a missing Recursion Three Insurance; and this was being paralleled institutionally by the notion that the company in fact consisted of a number of 'businesses'—a notion to which the cybernetic analysis leant much support. The other was the set of problems in Recursion One: the role of the Operations Committee as constituting System Three, the absence of System Four, the consequent failure to notice an institutional embodiment of 'the resource balancer', and the imprecise constitution of System Five in the corporate perspective.

At this stage, therefore, I wrote a two-page note entitled *Articulation of the Metasystem*. This was a very direct paper, involving personalities. Its main cybernetic points were to emphasize that System Three now existed (and should be recognized as such, dropping the prefix 'Insurance'); that System Four did not exist, and should be created in some form; that the resource balancer was finding an embodiment, even in the absence of System Four (which meant that Four was being created almost incidentally); and that the System Five policy group stood in need of definition. Obviously, from the cybernetic standpoint, this group consisted necessarily, but not necessarily exclusively, of the President himself, qua Five, of the Chairman of the now-accepted Three, of the missing Four, and of the heads of the two components of System One: investment and insurance. Equally obviously, the group could be formally constituted only as the result of the continuing negotiation between the personalities involved. This negotiation was impeded by the total obscurity surrounding the Four role.

Therefore, in an effort to clarify the realities of the metasystem, I now asked the protagonists to undertake a small experiment. Clearly, each senior official saw himself (correctly) as having a role in *each* of Systems Three, Four, and Five. If each would roughly allocate his time (of 100 per cent) between the three roles, and if each would estimate the time allocated to the three roles by his colleagues, then it would be possible to see how the metasystem saw *itself*. The hope was that this exercise would dramatically reveal inattention to System Four. It did not. It did give the impression that the System Four function at the corporate level was not yet understood.

Seventh Cybernetic Intervention

It was now the sixth year (1978), by which time I confess that I felt myself an integral part of the senior management team. I had no authority at all, but was very conscious of a responsibility. (That is obvious from the story; it is noteworthy that some glib management slogans maintain its impossibility.)

Every conceivable issue was alive and well, and mostly they were being handled in due time and with managerial aplomb. The void at System Four in Recursion One, however, matched by similar voids in both components of Recursion Two, had become a major focus of attention. What should be done?

The first requirement was to exemplify System Four issues at the corporate level—since the allegation had been that they were misconceived to be the kinds of issue extant at lower levels of recursion, but writ large. Three points were made.

(1) Just suppose that it is implicit in the state of society, its legislation, its economy, its technology, and so forth, that the corporation has no long-term future. It is a possibility. (For example, a government could virtually nationalize life insurance.) What sort of future does the company, consisting of all its personnel and all their knowledge and expertise, do then? Does it liquidate itself? Does it negotiate its assimilation by something else (such as a government scheme)? Does it use its powers to undertake new enterprise altogether? And what is the status of its policy-holders in such a scenario? This is extreme thinking, but it is a good place to start for just that reason. Original thinking is difficult to extract from the existing paradigm. Therefore it is a useful device to change the paradigm. One such notional change is (as above) to envisage the potential destruction of the enterprise as it is.

Another, which was devised and has been widely employed by Professor Russell L. Ackoff of the Wharton School at Pennsylvania University, is to envisage the enterprise as *idealized*. In short: what would we ideally like the enterprise to be and to do, if it were not shackled by all the constraints under which it currently operates? Ackoff has been known to contend that if this study is done, it often turns out that people say: 'But this is terrific! Let us drop the constraints.'

In either event, System Four is in action. It is loosening-up the enterprise thinking. It is asking the company to breathe.

(2) Working down to a more realistic version of point (1), System Four must be in a position to wrestle with the admitted constraints, rather than merely to accept them.

What exactly are the external regulatory systems that, as was mentioned earlier, inhibit the enterprise's activity? It is not the regulators that ought to be considered. They are sitting impassively, static; they defy abolition. But they are simply monuments to a dynamic regulatory *system*, which continuously behaves, and can therefore be influenced in one direction or another at all times—if the mode of behaviour is understood. It is a System Four job to acquire that understanding, even under a System Five umbrella that contents itself with deploring the regulators that the regulatory system imposes.

Once the understanding is there, strategies, influencing the external regulatory system can be cogently debated . . . .

(3) A totally different example of a realistic version of point (1) concerns technological development.

Enterprises for which *information* is the stock-in-trade are in a very special case where electronic technology is concerned. They have all been transformed, within twenty-five years, by the computer revolution. But, despite all warnings in advance, they have allowed the main-frame manufacturers to dictate the mode of automative developments. Therefore the company houses a museum of computer dinosaurs. And if a nice new dinosaur is carefully hatched from the carefully nurtured eggs of this so-recent past, it would be a dereliction of archaeological duty not to add the newcomer to the museum . . . .

Meanwhile: micro-processing. I had long been urging the opinion that within only a few years there will be a data-handling revolution that will make the original computer revolution of the fifties seem trivial.

There *could* be an entirely new method of selling insurance, whereby the high variety of the individual at risk could be matched by requisite actuarial variety carried in a small box of integrated circuits. It was to this possibility that the variety analysis of the Fourth Intervention referred. The idea may be vacuous: who can say? What can be said is that only System Four can handle it, and that System Four is void.

These three points were advanced as exemplars of Four activity. They were accepted as 'making sense'. And of course the management came to see what they were up against in embodying a System Four. The diagnosis had said:

(a) To what degree ought a response to such matters be institutionalized?
(b) It should not be institutionalized beyond the degree to which it can integrate with the dynamic whole of the viable system that is already in place.

(c) You can always raise capital, but you cannot easily raise managerial resource at the top level of the company.

This was indeed the rub. Everyone was very busy. The idea of bringing in a Director of Development from outside had no verisimilitude. And it was just for these reasons that System Four did not exist.

Then everything pointed to a *continuing negotiation* of the articulation of the metasystem, which had been going on for years and in all aspects of the viable system; but focused with deliberation and care, for a short while, on the whole 'Four' issue.

★★★★★★★★★★

The diagram at Figure 11 shows how the missing System Fours at Recursions One and Two are dynamically related, and how surrogates for them might be generated by the existing managerial subsystems (Fives and Threes).

Therefore the proposal: the Policy Group itself, however constituted, meets in a System Four capacity. It identifies a set of Four activities, based on the previous discussion and this diagram, and thereby creates a *normative plan* for future development. It decides how each activity should be staffed, organized, steered and monitored. It decides how this set of activities should be integrated. This would be tantamount to nominating a Director of Development from within. If no obvious *individual* 'emerges' from this process, maybe the policy group decides to continue in existence for some time—to fulfil that role collegiately.

The diagram points to the three void Systems Four, at Recursion One and Two, and connects them with a heavily marked dynamic circle—for of course they **must** be so strongly connected.

Here are the terms of reference for the first meeting of the Surrogate System Four:

In preparation

All nominated members, having reconsidered the cybernetic process in the company, and having considered especially the processes of the Sixth and Seventh cybernetic interventions in which they were all implicated, should prepare a list of issues which ought to be addressed under the heading of System Four: Each member should ask himself how best the necessary work could be initiated. In particular:

(a) Who are the people to join which activities, and who should be in charge of each?
(b) Is any external consultant required to each activity?
(c) To whom should each development team report?

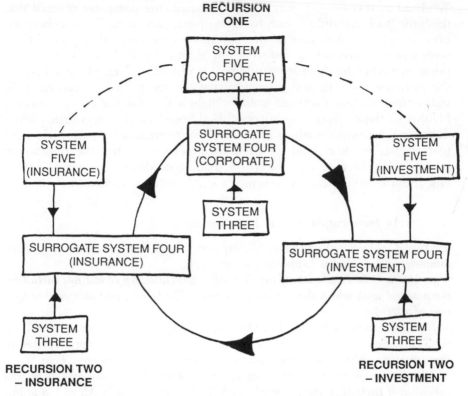

Figure 11. Proposed system for the creation of a surrogate System Four at the corporate level of recursion

(d) If more than one sponsor is nominated, how should the sponsors themselves be integrated?

(e) What is the time-scale of the initial reviews?

The agenda

(a) The agenda for this meeting ought to take care of itself. This Surrogate System Four must be, before anything else, a *self-organizing system* in the cybernetic sense. That is to say: if there is an 'order of business' indicating priorities, we can be sure that the order is pre-emptive. Nobody will know what the priorities are until *after* the meeting—if then.

(b) The first person to speak is the person who, because of his preparation (see above), demands to speak first.

(c) The meeting should grow organically out of such demands, which calls for an insightful chairmanship.

While all of this activity was being incubated, the company realized that the time had certainly come to talk openly about its own cybernetic process. Thus it was part of the Seventh Intervention that preparations were made to disseminate information 'down the line'. Work was commissioned whereby the story would be introduced to all employees through the in-house journal; and a preliminary conversation was recorded on video-tape, so that everyone could obtain a feel for the entire exercise. Neither of these efforts was successful, although neither was in any sense damaging. Communications *across* levels of recursion are extremely hard to facilitate, in short; and we need to understand better the transaction process whereby messages are encoded and recorded to these ends. (Most talk about communications assumes a single level of recursion.)

An Intervention in these Interventions

This meta-intervention will prove slightly complicated, but it is necessary to make completely clear what is happening in this story.

In March 1978 I was making the Seventh Intervention **up to but not including** *the proposal with which that intervention ends. The break-point was marked by a row of asterisks.*

Discussion with the President and his chief officers had centred almost entirely on the System Four issue. Nevertheless I was concerned that they had not yet perceived the 'cybernetic drift'. They might jump to conclusions about Systems Four.

Therefore, in order to establish that 'cybernetic drift', I wrote this case study, and submitted it **including** *the proposal (with its diagram) in early May. Up to this point, what has been written here (following the title BC: Before Cybernetics)* **was** *that submission, with minor changes, and the addition of the detailed variety analysis—which had been circulating separately. The present introduction was obviously unnecessary; the original introduction was the initial version of Note One. Attention is drawn to the 'three points' of Intervention Seven, and to the remarks about all Systems Four in the diagnoses. Two paragraphs there are now marked like this: ★★★. Although they reflected much discussion, they were not so marked in the original text.*

Eighth Cybernetic Intervention (aborted)

This occurred in July 1978. The intent had been to discuss this very report—which is to say, its account of the first seven Interventions, and in particular the proposal made in the last diagram.

After all, despite any difficulties there might be in the *staffing* of the corporate System Four, it seemed that it was at last becoming clear what the notion meant to the corporation. Moreover, we now had the viable system model in four recursions, and there are two strong points about this:

(1) Chapter 9 of *Heart* says 'the elaborated model provides exactly the "screen" on which to focus System Four activity'.
(2) It argues in detail the cybernetic requirement for System Four to contain a model of itself (see Figures 40 and 41 of *Heart*).

Finally, there was the very practical matter of generating some *action*, to which the proposal for discussion was addressed. We had been talking about System Four for long enough; and so my notes contained a third point. It was a remark that had been made a long time previously by a member of the senior management team; and it was to be (I thought) the text for a sermon at this point. He had said:

'It's time to stop reading the driving manual, and get out in the traffic.'

In the event, I found myself severely up-staged. The phrase 'no System Four, eh?' became tiresome. The firm was in the throes of absorbing a large block of business from another company. Thereby it would add significantly to the stability of its home base, and improve the service it could offer in the market.

The press and television as usual handled this naively. They over-simplified the issues. In the host country, this was to be 'a merger'. In the country of the taken-over company, it had 'formulated an intention to withdraw'. But it was by no means a simple matter of the taking-over company's 'moving in'. What I observed was an example of how a System Four mobilizes itself for action, despite massive legal, fiscal, and even political constraints, and despite the non-existence of an *articulated* System Four itself! Therein lies a problem. If System Four is disseminated, and has no *focus*, there is bound to be trouble later. . . .

But when we discuss (see Introduction: On Success) what counts as 'implementation' in management science, and what is to be called 'success', it is surely enough to perceive that the cybernetics played a certain role in such an event as this. It is a bonus (other managers of other management scientists who may well need this incentive: please note) to be told as much by the company president.

Ninth Cybernetic Intervention

This was made in writing, in the following terms:

(a) There were, in cybernetic terms, many loose threads floating about as the result of the foregoing diagnoses.

(b) One of them concerned the whole question of stochastic filtration (see Note Four of *Heart*) of managerial information.

(c) A second concerned the impact of microprocessing on the future of the insurance business (see this report).

(d) A third concerned the consolidation of a system Four that had so effectively sprung into being to handle the merger affair.

(e) All issues of this kind still needed to be advanced.

But—in straight managerial terms—*how were any such matters to be compared with the immediate task of creating the new company?* The management had no choice but go get on with that task, as from day one. All sorts of urgent decisions had to be taken day by day. Thus I wrote, in part:

> What will happen about the amalgamation of the two companies cannot be predicted.

However, it is predictable (in terms of the model of the viable system) that either:

(1) the other company will be assimilated, piece by piece—and dissolve; or

(2) unresolvable problems will be generated by the other company's will to retain its own identity.

There is no point in *predicting* which will happen.
There *is* point in asking which is *intended*.

Does anyone REALLY KNOW?

For the moment, it is unarguable that two large enterprises are trying to get together. Therefore, at least in the short run, we are considering 'Recursion Zero'.

Recursion Zero

This constitutes an amalgam—of whatever sort—of two companies who wish to be, and jointly remain, *a viable system*.

> Recursion Zero has a bipartite System One. It consists of two components, which are exactly the two companies.

Then all conceivable problems fall neatly into place.

As to System One:

Where are the 'squiggly-line' connections?

As to System Three:

This (and forgive me, only this) is what you are successfully embodying right now.

As to System Five:

JUST EXACTLY WHAT IS THIS INTENDED TO BE? *That* is a diagnostic remark. Surely (since this is an amalgamation of *Mutual* Life Companies), System Five includes the policyholders of both concerns. Hitherto, they have not known each other. This matters.

As to system Two, Recursion Zero:

Create it fast.

As to System Four, Recursion Zero:

There absolutely **must** be (even yet) a new resurgence of the notion of your System Four (which upstaged me), to become Recursion Zero's System Four. *It is yet another level of recursion to consider.*

Here, at this very moment, is all my concern.
Exactly what constitutes the new company?
Think of ambivalent intentions; think of unpredictable events. . . .

In conclusion:
Please do not forget—
the model of any viable system—
especially,
not now.

Well, that is it.

Tenth Cybernetic Intervention

There was uneasiness expressed in that report. Extreme and dedicated effort was being expended in attending to necessary issues as they arose: that was clear. But viable systems are highly resistant to change; and if they are to be changed then they have to be viewed in the context of viability at a higher level of recursion. . . .

The merger deal in fact fell through. Obviously, the next intervention began with a post-mortem about that. But the President and his senior officers all took a positive attitude to the affair: they wanted to learn from the experience, which was not necessarily over.

We had reached the seventh year of this work, and the Tenth Intervention: but the problems referred to in the paper *Articulation of the Metasystem* (Sixth Intervention) had yet to be resolved. The President himself was by now keenly aware that the matter was urgent.

Intense conversations were therefore held, and (as of this writing) appeared successful. There was agreement about the shape of the meta-system, and about the roles of the key people concerned. The problem of 'distancing' was much discussed: that is to say—how could the meta-system be prevented once again from collapsing into System Three? Obviously, came the agreed answer, by the creation of an appropriate System Four. It had always existed (it *has* to exist), but it should be 'pointed to'—made more aparent; *focused*. Moreover, there would have to be a definite and senior accountability this time.

Suddenly this idea became plausible in the context of the actual people, departments, and information systems involved. Action is now intended. But if the articulation of the metasystem, and its proper internal 'distancing', are to become effective through the pinpointing of System Three (completed) and of System Four (imminent), where stands System Five? It was the discussion of this problem with the President and the Executive Vice President which might turn out to be the most productive outcome of the Tenth Intervention.

The other major topic of this intervention concerned the missing third recursion. The notion that insurance products should be grouped was beginning to gain acceptance. There was much discussion of the requisite varieties involved in linking the fourth to the second level of recursion. And here was the practical snag: under such groupings, the 'new busi-nesses' thereby created might easily become dangerous competitors of each other in the company's own branch offices. Considerable time was spent in trying to elucidate which sorts of groupings would and would not be safe. It was my inhouse cybernetic confrère, the 'company officer' referred to so often in this note, who had provided the key to that. . . .

At this point . . .

At this point, the account already published in *The Heart of Enterprise* ends. The book was published in 1979; so the story had been updated to the very last moment. What was the dramatic intervention of my company con-frère, and why was it not recounted? It was because he had tossed a very hot potato into the air. Let us recapitulate.

The note I had written during the Ninth Intervention was by any standard very strong. It had to be. The company was faced with an incipient merger for which no foundation had been laid. It had presented itself as a 'target of opportunity'. The *ad hoc* System Four of experts that had been cobbled together to strike the deal had its work cut out to handle the pressing exigencies of the day-to-day negotiations, and was not think-ing about 'Recursion Zero' at all. My own view is that this is the underlying reason why the merger collapsed. Each party had a different conception of the deal—as to why it was mooted, as to how it would be consummated, as to the purpose of the joint enterprise, and so on indefinitely.

It must be said to the amazed onlooker that this is wholly usual. Mergers are normaly fiscal power struggles, and have no deeper rationale than does arm-wrestling. The story released is invented by the PR people 'on the day'. This is because System Four in acquisitive companies comprises only people on the lookout for killings, who are neither qualified nor interested in organic company development. Corresponding with a national CEO about another matter, in which such a rationale had been created *post facto* for a particular move to decentralize authority (and turned into an advertising hype), he wrote: 'It was at least seven years before we treated any of the subsidiaries as anything other than a small department of Chief Office!'

At any rate, the merger debacle let loose major debates—and a considerable jockeying for position among key players. If there really was to be a major shift of focus re System Four, and some kind of regrouping within operations, then everyone had their own ideas as to what should happen. Teams were set up, task forces mobilized, and so on. Everyone started sending tendentious memoranda to everyone else. My colleague moved into this mêlée with great determination. Recording his arguments, telegraphically:

> We had gone to great lengths in 1977 to construct a Viable System Model. Much use had been made of this diagnostically, and had led to various changes. Moreover, many said that it had provided fresh insights into the business. The fact remained that this model had not been used at all in current debates concerning possible major changes in company structure. Why not? What could be done about it?

Deliberations resulted in a very long tape of a deep discussion between him and the Chief Executive, reviewing the current situation. In this tape I was asked to provide a commentary on the current prospects for *cybernetic change* in the company. This phrase refers, both by definition and by context on the tape, to changes in effective organization (as distinct from 'political' manoeuvres), with all their implications in the *regulatory mode* (as distinct from who-reports-to-whom). My good friend the CEO had lobbed the hot potato straight over the Atlantic once again.

Eleventh Cybernetic Intervention

The commentary began with a detailed analysis of the self-image of the company. It was July 1980; apart from the tape, I had had long meetings in Canada with the senior officers during March. It was clear that the self-image was still primarily territorial.

'Group Insurance is seen as a "specialized" activity. Investment is seen as something internal (sic) to the company. Thus these two are rendered no longer anomalous to a "territorial" model.'

The description is as historical and immovable as it is cybernetically grotesque. Investment is a worldwide activity (hence the *sic*), which necessarily balances the worldwide insurance operation in the massive homeostasis so often described here. If that is 'internal' then 'external' must apply solely to the sales forces. But if sales are 'external', how can Group be operating from Chief Office? Group Insurance is of much the same size as Life Insurance: if that is a 'specialized' activity, operating from the centre, and cutting across all territories, what price territorial autonomy? The arrangement will obviously lead to the muddle and contradiction that I had observed, and which nobody denied.

The explanation of all this is stunningly simple, I believe. After populating the metasystem of the company with the President, the Senior Vice President (CEO), the Chief Actuary, and the Directors of Services (that means Computers), the management team further consists of a number of men who are also front-runners. Each of them runs either a territory, or Group, or Investment. All of these people constitute simply a list of The Elect. It is important that they enjoy mutual and equal esteem. Thereupon they are *interchangeable*. Apples and oranges mix very well under the label of 'fruit'.

Consider in the light of this the odd business, recounted in the Fifth Intervention, about the representation of investment on the Insurance Operations Committee. The prefix 'Insurance' ought to be dropped: 'This was agreed (but the old name withstood the decision!). This was not a fussy, terminological issue.' No, indeed. In fact, it took a great deal of manoeuvring to have the senior investment officer put on to that vital committee, as the original VSM made clear was required, and it had taken some years to effect. The above account says, tactfully: 'This had been brought about in various ways. . . .' The fact is that there was great resistance—surely because it entered this officer as interchangeable too. He joined the club—and to remarkable effect, as we shall see.

The years of work that went into describing the company according to cybernetic canons bore various fruits; but it did not change the paradigm which this section started by describing. Well, if the 'stunningly simple explanation' of interchangeability holds, where is the evidence?

This passage is inset because it does not wholly belong to the Eleventh Intervention. Writing with hindsight, I am able to compress the story of the then Elect right up to the press-date of this book. The names refer to the *holders of the top jobs* concerned at the relevant times, and the period covers six years.

Services moved to Group, while USA went to Services, and Number Two Services went to USA. Canada left: Group went to Canada. Services moved to Far East, USA went to Planning, and Number Two UK went to USA. The CEO went to UK, Investment became CEO *as well as* Investment, UK went to Canada, while Canada nearly became the missing System Four. UK (ex-CEO), widely supposed to be vying with the Chief Actuary for the Presidency, was as good as his word to me: he retired. The President retired, and Investment took the job—carrying CEO and Investment with him! The 'nearly System Four' (ex-Services, ex-Group, ex-Canada) retired, and was not replaced. The Chief Actuary left.

It is a remarkable tale. Not only did Investment succeed to the Presidency; he became CEO, and kept Investment too. Now this was exactly the situation with the outgoing President, when this saga began in 1973. Some social patterns are very tenacious indeed.

This game of musical chairs replaced the attempt to steer the organizational course by managerial cybernetics. The results were not very successful, as a matter of public record.

Maybe however, the hibernating VSM in the system will reawaken. It is currently being used in a major project dealing with valuation controls.

It seems that the model of all this is Cabinet Government. No ostensible qualifications are required to be one minister rather than another. When the reshuffle comes, the performances that are rewarded and punished are not evident—except to the Elect themselves. They become evident to the public when the media have evolved an explanation. Similarly, when the Prime Minister or the Pope is replaced, the succession is determined in ways that the media will rationalize for the public after the fact. The real factors are essentially internal to the Elect.

These reflections replace the elaborate analysis, detailed and incomprehensibly domestic, of the report insofar as the replacement of self-image is concerned. It is little wonder that the job of a management cybernetician often feels as if it falls halfway between psychiatrist and confessor. . . .

Inevitably, the commentary comes back to the problem of System Four. As always, I had been pushing for any acceptable design that would focus Four-like activity, and had indeed proposed one in Intervention Seven. A new Corporate Planning Committee had been set up, and the commentary says that it looked for a time as if that 'rather well fitted the design proposed. But, if I have not misunderstood, the *whole* of senior management soon became involved—and thus the focus was inevitably lost'.

Having re-examined the problem as it now seemed to exist, I proceeded with the 'tape' remit by constructing three scenarios through which to debate the possibilities for cybernetic change. It was necessary to start from the above demonstration that we could not shift the 'territorial' paradigm.

Readers know that I always favoured decentralization by territories. But that assumed that each territory would operate as an autonomous System One. The objection to the 'territorial' in-house paradigm was that the metasystem of each territory stayed in Chief Office! The centre wanted to retain control, and the adolescent units were timorous.

The Structural Scenario now put forward attempted to deal with this by hypothesizing that the various territorial governments passed laws requiring total decentralization. This enabled me to discuss in some detail the various ways in which investment and actuarial control might be exerted from the parent company without infringing the supposed laws. In particular, though, this scenario made play with the need for System Four at the territorial level (especially if the centre could not evolve a corporate System Four), because local circumstances might offer special threats and opportunities. As an example, the commentary returns to the issue of microcomputing 'in the light of the Fourth Intervention—the most persistently ignored piece of cybernetics ever written!'

Next came a Regulatory Scenario. This was based essentially on the Law of Cohesion. It talked at length about the synergy that could be gained from the new structurally autonomous units, once corporate expertise were regarded as a set of 'assets for sale' to the units, instead of an imposition from above. Training expertise, for example, was cited, and a scenario evolved.

Finally came an Algedonic Scenario. It began:

> 'Although the sum of metasystemic territorial variety will be made equivalent to the variety disposed by the corporate System Three, it has to be noted that the sum of System One operational varieties is much greater than this. It is the job of the *unit* metasystems (territorial managements) to absorb this extra variety.
>
> What happens should they fail? . . . This, above all, is what alarms the existing senior group.'

The report goes on to propose an algedonic alarm system as discussed at length in both *Brain of the Firm* and *The Heart of Enterprise*. Of course, it would hinge on the adoption of a statistical filtration system intended to measure incipient instability, and dubbed Cyberfilter. The company officer, my colleague, and his staff had made many experiments with this statistical tool, and they had long since generated titillating examples of how it would work. However the development had been

> '. . . markedly resisted (ever since 1973) while—at the same time—so much concern is expressed over the legitimate problem that it is meant to mollify. Then let us recapitulate:

● The basis for the lukewarm response to Cyberfilter as a major regulatory tool was usually that, since insurance is based on long-term contracts, there could really be no substantive need for rapid-response devices.

● The answer to this always was that you are a *viable business* existing in a rapidly changing environment. Its viability depends upon the socio-politico-economic contexts that affect all businesses, regardless of the products with which any is concerned.

● To this argument, the lukewarm always responded that no-one knew what indicators should be measured and filtered in any case.

● The answer to that always was that research was required to 'flesh out' the cybernetic model, and that System Four could be focused to that end.

Might it not be time to break this circle?'

The 15-page response to the tape ended with 'Practical conclusions'. The old Operational Reseach man in me was tired of so much theorizing: he proposed three empirical investigations to explore the three scenarios. The first would involve a stay at the English company headquarters. The second would require the investigation of a branch office in the USA. The third demonstrate an algedonic system on a microcomputer in Britain.

The first and third proposals seemed acceptable, but were not yet implemented. As to the second: 'We are not quite comfortable with this.' Or in English translation: 'No'.

The write-up gives a text for the Eleventh Intervention: 1 *Corinthians* 12, 14–21. Anyone who looked that up would read an early account of my Law of Cohesion—one which responds to just those questions that were being bandied about in the company. It goes:

'For the body is not one member, but many. / If the foot shall say, Because I am not the hand, I am not of the body; is it therefore not of the body? / And if the ear shall say, Because I am not the eye, I am not of the body; is it therefore not of the body? / If the whole body were an eye, where were the hearing? If the whole were hearing, where were the smelling? / But now hath God set the members every one of them in the body, as it hath pleased him. / And if they were all one member, where were the body? / But now are they many members, yet but one body. / And the eye cannot say unto the hand, I have no need of thee: nor again the head to the feet, I have no need of you.'

Twelfth Cybernetic Intervention

As events moved into 1981, it became increasingly evident that the cybernetic team (which included a senior vice-president and a few other senior stalwarts) was becoming essentially a coterie of the CEO. We had all been in this game for six or seven years, and were by now speaking a language of great sophistication. Most of the senior management still knew the basic Five–Four–Three language of the VSM; but, as changes occurred and people came and went, it was noticeable that newcomers learned just enough of it to get by. It was as if they had just enough French to fail to impress the head-waiter. Moreover, the original intention to spread 'basic VSM' throughout the company by means of tapes, videos and pamphlets (a plan more-or-less consciously modelled on the bilingual 'immersion' courses that the Canadian Government ran in French) had lost any potency it had ever had. As for the coterie itself . . . we were indeed speaking a foreign language of our own invention. And it was a language that the President had not learned. When we met, he and I spoke together in courteous English.

The Twelfth Intervention was founded in an intense interaction on audio-tape. As I review the transcripts seven years later, it is clear that we had implicitly agreed to continue our conversations in a private language and an esoteric domain—in order to reshape a strategy conforming to the postures of the previous Intervention. In fairness, we were not trying to use these materials to influence anyone else. But that is just as well: they would have been incomprehensible, and the audience uncomprehending.

Meanwhile, the initiatives that had led to the formation of the Corporate Planning Committee (which was earlier described as a surrogate System Four) had somehow lost their cybernetic form and momentum. I could barely contain my anguish when, after all these years of joint work, the President addressed the new committee with the words: 'I feel that we should have a better idea of our future direction before we really get serious about our organization.'

At any rate, the CEO was undaunted. Out of the new tapes, mentioned above, and subsequent meetings with him in London and Montreal, was forged a new set of intentions. These included the visitation to the British company and the demonstration of the power of microcomputers to search out algedonic signals, that had been proposed before. Above all, it called for a final attempt to design the company as a true 'multinational', on the bases that the coterie understood, and to find a way to present it to the rest.

In the event, the planned activities in Britain pushed matters ahead quite fast. Effective Operational Research really does need 'hands-on' involvement—getting dirty if necessary, getting the feet wet. Since the days of WWII and its foundation, OR has slid into a morass of academic respectability and routine. Streams of publications refer to techique-oriented 'applications' of long-standing theories; it does not seem that a totally new approach in an unexplored area of contemporary concern has been made in many years. Who could guess the origins of OR had to do with addressing life-or-death problems that no-one knew how to tackle? Or that management needs to anticipate what is likely to happen?

In order to illustrate these contentions, I had for long fostered the development of Cyberfilter in the company, and had talked endlessly about the microcomputer revolution—as already recorded. There was also the question as to whether the industry was alert to changing societary habits and needs. My friends were proud to inform me when they arranged special life rates for non-smokers, which is fine. However, that is within the conspectus of insurance as we have always known it. It is little more than a display of actuarial finesse.

To illustrate what was meant by a System Four activity as coloured by such considerations as those just mentioned, I drew on some American work dealing with the sour relationship between women and the life insurance industry. The contention was that the industry and its marketing strategies are geared to a model of the family which is now barely, if at all, relevant. This model says, essentially, that father goes out to work to provide for mother, who stays at home with the children. We do not have to forecast: statistically, today, this system is a myth. Figures vary, but this set-up applies to only 10–20 per cent of families in the USA: that is a small minority.

Insurance markets primarily to the man, working on his 'duty' to protect this dependent wife and children. The hype exaggerates both need and dependence—even for the minority to which it might apply at all. As to wives: the male chauvinistic model expects to pay her burial expenses, and to cover the replacement of her domestic services until the husband should remarry and a new wife would assume them. Alternatively, a wife's policy is automatically written for half the value of her husband's—regardless of her salary.

The investigation found strong alienation of the working woman (now of course in the majority) to the industry, in that it disregards her contribution to the life-style of a two-income family, and her responsibility for her own support and that which she may share for children and other relatives. It has, moreover, antagonized many politically knowledgeable women by donating millions of dollars to defeat the Equal Rights Amendment in the United States.

It is not necessary to push one's thoughts much further down this route before wondering whether the concept of insurance as it exists will be relevant to our developing society for much longer. Even if it is relevant, it may not look too attractive as an investment compared with other—and innovative—ways of using money.

My insurance friends would of course listen to such arguments as these, and discourse on them—although I suspect that they found them rather quaint. Business, after all, is good. But as to justifying, or even exemplifying, what is really meant by System Four . . . it did not work.

When the company finally *did* acquire another (it was not the one recorded earlier as lost), and whether or not it paid far too much for the holding (as financial commentators alleged), the absence of System Four thinking was stark. Why was it acquired? Where would lie the synergy? It was evident that the CEO was driven by the ambition of sheer size: he wanted his company to dominate the scene. So it was a requirement of the deal that the two companies merge within a year. On the other hand, the President went on record with the press that the two companies 'have no present plans that would lead to any sort of cross-fertilization of each other's business': a remarkable admission.

Thirteenth Cybernetic Intervention

This was the final major effort of the consulting episode. In a document dated May 1982, I put forward the detailed cybernetic plan for turning the company into a genuine international organization. The phrase contrasts with the reality we knew: a Canadian company with offices oversea. The plan set out to define autonomy-with-cohesion in operational terms.

The principles for doing this were by now common ground. The practicalities had been under discussion for the last several years—in 'political' terms. For instance, the CEO valued his trained sales force at a billion dollars: how would *they* react to the various regulatory alternatives? The word 'lemmings' had achieved the status of a technical term in these conversations. . . .

The 1980 document *Prospects* had reported:

> 'It is now being openly said that there exist strong pressures for territorial autonomy—not all deriving from the company's staff (i.e. from self-image) but from the outside (i.e. the market-image).'

To this end, I had for a long time argued that the *Canadian* company should be detached from the Chief Office, which would then be seen by both staff and clientele as a proper international HQ. This was accomplished at about this time. A move was made into makeshift offices. But these six years later a fine new building has been observed some miles from the down-town HQ: good. It is amusing that the reason why this happened was that the merger required a continued presence in the township concerned; the implementation of my recommendation was virtually accidental. But such is *real-politique*.

The difference between this ultimate plan and those earlier presented was that by now we had full agreement as to what it was intended to do—that is *which* viable system called the company had the form that the company wanted. The weak term in the laudable claim is the little word 'we'. 'We' were consultant-cum-client, certainly; but there is no doubt that the coterie, much as it was in charge, would not be able to pursue the plan in the plan's own terms. The plan would have to be filtered through the management system in whatever terms were acceptable to the current sub-culture.

By this date I had for some time been using a picture of the VSM in which the next lower recursion (that is, the infrastructure of System One) is presented at a 45° angle (see page 23 of this book, Chapter 1, which also explains the evolution of this pictorial topological algebra). This May 1982 report was, however, the first occasion when I traced through the full interrecursive topology in a detailed set of rules for a client using this precise rubric. It may be interesting to note that I left Toronto after the submission of this report to spend most of June and July in Montreal. The seminars at Concordia University used this topology extensively; and the third book of the VSM trilogy, *Diagnosing the System for Organizations*, was the result.

A second innovation of this report is published here for the first time. It is often difficult to enter as much detail on a VSM chart as one might wish—even when these are enlarged (as often happens) to a height of two metres. But the nine-year enquiry behind us had generated a truly daunting supply of interactions which a systemic model was needed to depict, but which would swamp any reasonably sized depiction. In February 1982 we had worked hard to make a long list (reductionist, what else?) of activities, issues and relationships that we wanted to account for within the homeostatic loops of the VSM. The list was too long.

The answer was to divide the model, like Gaul, into three parts. But surely the problem of putting the whole system back together, of relating the three parts, would be insoluble? The answer was to nominate each part so as to be an *orthogonal dimension* of the whole. Here is how the report expressed it:

> 'The term dimension refers to a condition of existence; the expression of an activity in one dimension tells us nothing about another dimensional activity, to which it is perceived as orthogonal. But when all three dimensions have been mapped out, any point in the viable system may be considered as the intersection of each of the dimensions in which it occurs. The three managerial dimensions we propose to consider are: financial management (to include all professional aspects, such as actuarial management and accounting management); operational management (to include product management and money management); people management (to include both agency management and management development).'

The report undertook this task for the corporate recursion, and went on:

> 'It is suggested that the map of the viable system, the Main Figure, should be taken in three copies by someone concerned with each of the projected national subsidiaries, and developed for the three dimensions proposed. It would then be possible to consider the wisdom of forming an amalgam of each set of four plans, as depicted in three dimensions.'

Figure 12 reproduces the diagram supplied, standing for three VSMs set at right-angles.

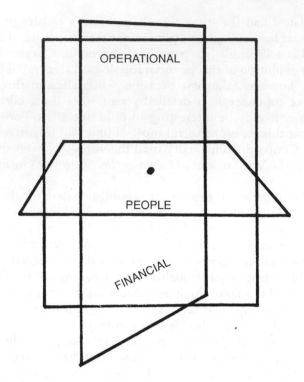

Figure 12. Three dimensions of management. Any point on the viable system diagram can be considered as coterminus with the central dot. Then its financial operational and human meanings can be studied on the three separate charts

This report ended with a technical appendix discussing the possibilities of *measuring* cohesion in the organization. Taking the function for cohesion discussed in *Heart* (p. 355) as monotonic-increasing over time, it should be possible to concoct measures of entropy, as I had done in *Brain* (Chapter 14) for measuring the process of decision. There was no response to this: it was the same fate as befell the mathematical ideas of Intervention Four. Of course, no-one would expect busy senior managers to bother with these matters; but this was an unusual company, full of actuaries, and I had unrequited hopes of capturing their interest.

Fourteenth Cybernetic Intervention

The end of this project must count as an intervention, insofar as it had the usual characteristics: a visit, a report, consequences. As was said at the start, the feeling was that matters would now resolve themselves within the internal dynamics of the company. The official conclusion of my assignment came after the usual discussions of the May report, in November 1982. However, letters were still being interchanged with 'the coterie' well into 1983. I still meet nostalgically with two who have since retired.

After so long and so intimate an involvement, I personally experienced a kind of a trauma: it was like losing a family. It was strange not to be thinking continually about their affairs. Readers will be making their own evaluations of this story (and I hope that they do not forget the Introduction), so I have no compulsion to conclude with conclusions. But I do have a couple of thoughts, offered in *l'esprit d'escalier*.

More serious than any other of their problems as detected by the cybernetic diagnosis was the absence of an effective, that is *focused*, System Four. The whole of the Seventh Intervention had been devoted to the matter. No-one disputed the problem, but there was no acceptable solution. I have seen this situation in other places, and indeed had the personal experience of being parachuted in from outside to be Development Director of the (then) mammoth International Publishing Corporation in 1966.

It is a very difficult role for the newcomer: for instance, it takes him at least a year just to bury his parachute. He never really 'belongs'. Because I was not on the staff in the insurance company, I was not nearly such a threat—and lasted ten years instead of five! However, I had in fact become their System Four myself. I took the developmental lead on the organizational issue, persuaded others to be involved, and managed—by dint of the visits, which averaged three or four a year—to seize attention and focus it in a way an in-house man would have found difficult: he can be put off, whereas I was leaving town.

At the time, being a Temp Development Director seemed a good compromise—just until they sorted themselves out. It kept looking as though they would. Four of the top men in turn were given responsibility for some aspect of planning, but the effort leaked away into the sand each time. Since it was evident that the culture would never tolerate a parachutist, I several times pressed the strongest personality to take on the job himself. 'What!', he would say, 'and lose my power base!'. Quite so.

It is easy to see with hindsight that my long sojourn was counterproductive in respect of this corporate role. As long as I was doing the job, what was the hurry? And yet it is likely that otherwise no new thoughts would have been stimulated or metabolized. Certainly I knew likely candidates who were at the time too junior to take on the part. My guess is that they have been getting steadily older. Another guess would be that the perception of need for System Four will have faded along with the language.

Tailpiece

From the Fourth Intervention, early 1976:

> 'These simple expedients introduced the design for an insurance variety generator which would use descriptive rules for the prospective client, transformation rules for the actuarial expertise to be applied to the descriptions, pick out the statutory invariants that apply in the "second reality", and cement all of this into a box of microprocessing chips. This box could then be handed to the agent, at very low cost (much lower than his use of a terminal in the branch office to speak to the mighty but somnolent dinosaur computer at head office). It points the way to highly sensitive selling (because it handles Ashby's Law by design rather than by historical inadvertence), and to a vast reduction in cost. I do not give any further details here; partly because they were preliminary, and partly because this company is still my client. . . .'

From *In-Flight Magazine*, 1982:

> 'Insurance agents can store rate tables to be called up in the client's home. A California company has developed a software package to individually tailor insurance schemes; it asks the client questions such as: "How much can you afford?" and "How much monthly income would your family need?" and then computes what it considers to be the best plan.
>
> By eliminating the need for extensive tables, the system saves agents hours every day.'
>
> (Bob McElwain in his regular column)

The Viable System Model: Interpretations and Applications of Stafford Beer's VSM
Edited by R. Espejo and R. Harnden
© 1989 John Wiley & Sons Ltd

10

Developing organizational competence in a business

Bengt A. Holmberg

ASSI, 105 22 Stockholm, Sweden

ASSI is a Swedish paper and packaging company with 8000 employees in five countries. Stafford Beer's model has been introduced in the company as a consistent tool to develop organizational competence, and especially to promote decentralization and increase the conscious flow of information. Methods and experiences are described in this chapter.

Background

An old Chinese proverb says that all trips start with the first step. Thus if you want to change something and go to a new point it is very important to identify where you start. And if you have a long way to go you must also be prepared to change the route somewhat when moving towards your goal. This is very true about organizational development.

Ten years ago the ASSI Group was organized along functional lines. The Sales Director was, for example, directly responsible for 80 per cent of Group sales and thus for all the marketing effort to support the production of some 6000 men and women. At that point in time we started a process that is commonly called 'decentralization'. Head Office consisted then of more than 200 people; today that number is 60.

Our Group hit upon hard times during the period 1979–82. Commercial problems created financial problems and resulted in a number of tough decisions. Activities were closed down or sold and the workforce diminished by some 20 per cent.

In 1983 we found that we had to be more conscientious in developing a competent, decentralized organization. How to develop competence in

271

organizational matters became a subject for serious study. Two issues became especially important in this work:

(1) How could we make many people conscious of our organizational philosophy and get acceptance?
(2) How could we balance autonomy and controls?

There was nothing unique in our approach. Many companies follow the same roads when developing their organizations. One could even say that it was easier for us just because others developed in analogous ways. We did not have to invent the wheel all over again.

However, our approach was absolutely unique in one sense: to the 8000 men and women in the ASSI Group.

Why is organizational competence important?

There are some basic factors that make it very important to develop organizational competence.

All social organizations must fulfil certain requirements to survive; that is, to stay viable. There are a number of requirements internally and there are a number of external changes that continuously require reactive/ adaptive ability. It is important for the management of a business organization that as many members as possible are aware both of the objectives of the organization and of what is expected from every individual. Management must aim at motivating all employees to do a first-class job.

In many business organizations there are today two factors which require special attention.

Increasing educational levels of employees

Every manager (irrespective of hierarchical level) is on a borderline between, on the one hand, a customer (internal or external) who expects a quality product and a boss who demands profitability and long-term efficiency, and on the other hand employees who expect job security, to be proud of their jobs, and personal development.

Managers are not merely directors of a play: they are both directors and actors on the stage in front of their employees. They must direct their organizations along commercially sound lines. This is very demanding and requires a lot of conscious 'acting'. At the same time we must remember that the average manager tends to be one generation older, or at least half a generation older, than his or her employees. And employees today tend to be better educated and have values different from those of their managers. Managers have to acquire a high degree of 'social competence'. This is important: most managers have got their jobs because of their technical competence; their 'social competence' is often their Achilles heel.

Internationalism

Most business organizations must develop competence to handle an international environment. This is important not only for marketing people and high-level management. The knowledge must go deep down in the organization to people in production, packaging and shipping.

The ASSI Group

Any company can be described in several ways. A few static descriptions of the Group will be given as background. Other ways of describing the Group could be in financial, historical or cultural terms.

People

The Group employs 8000 men and women, 60 per cent in Sweden and 40 per cent in other European countries. We have production units in Sweden, Denmark, Great Britain, West Germany and Switzerland. There are sales offices in most European countries.

Our 65 workplaces are distributed from the Polar Circle in northern Sweden to Milan in Italy.

Products and market (Figure 1)

Our home market is Western Europe. We produce:

(a) packaging papers and pulp (55 per cent sales):
    1 million tonnes annually;
(b) corrugated boxes (40 per cent of sales):
    800 million boxes annually;
(c) joinery products (5 per cent of sales).

This means that the Group is very capital-intensive. The capital employed is roughly the same as our annual sales (700 million GBP), which is typical of a heavy chemical industry.

Logistic system (Figure 1)

The sheer number of tonnes makes the transport systems and the integration between various products very important. Furthermore, one tonne of

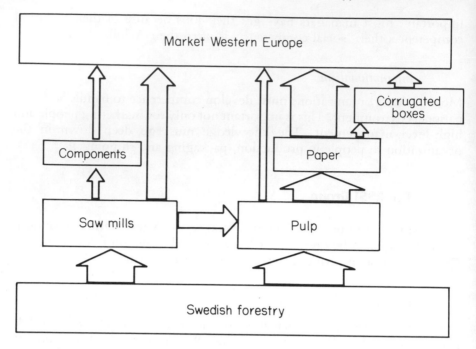

Figure 1. The ASSI logistic system

product to the customer requires 2–4 tonnes of raw material to be transported into the system. The main raw material is pulp wood from the forests.

Organization (Figure 2)

We are organized into 14 separate profit centres. Each one of these has its own profit and loss statement and balance sheet and full responsibility for product development, marketing and production of its product range. Formally they report to a board of directors. In reality, naturally, the system is much more complicated. There are also two service organizations, one for the transport system and one for wood procurement. Finally, the managing director has a small staff with the usual expertise required in a Head Office.

Towards a better description of a viable organization

Traditionally there has been just one very simple way of describing an organization. This picture is the organization chart. It may describe the

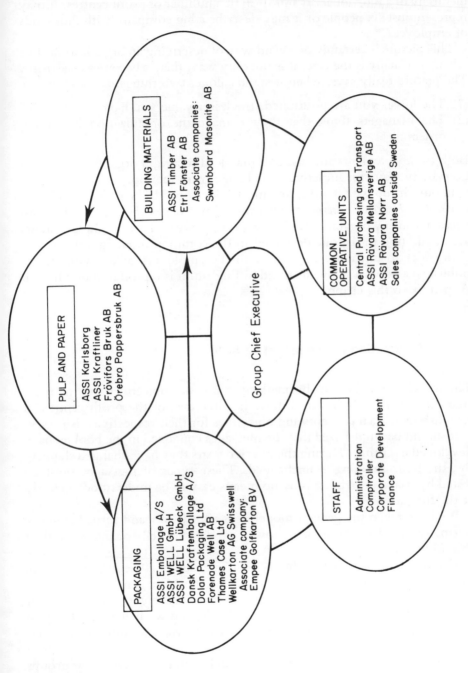

Figure 2. The Group's organization chart

hierarchy of a functional organization or a number of profit centres. It may represent just six people or it may describe a big company with thousands of employees.

This picture is certainly one valid way of describing an organization. But when, as usually is the case, it is the only way, then it becomes disastrous. The picture really says, when it stands alone, two things:

(1) The lower you are positioned, the lesser value you have.
(2) The managers think that they manage the organization by giving orders.

Both these statements are false and may lead to a rigid organization where everybody waits for initiative and change from the top.

In our Group we turned to the model developed by Stafford Beer in his book *The Heart of Enterprise*. This model postulates the systems that are necessary in any organization that wants to survive. Three of these systems are in addition to the one described in the common organogram.

In order to introduce the model to the employees of our Group, we published a booklet in Swedish and in English. It is reproduced in full as an Appendix to this chapter.

Application of the model to ASSI

The booklet describes our business concepts, our objectives and the main rules that everybody has to follow. It is the task of the profit centres to establish their own concepts and objectives for their respective businesses.

It should be emphasized that the messages contained in the booklet have developed over time. During the last two years they have changed slightly, but the basic thinking is unchanged. These types of messages must be tested by trial and error: it is by no means easy to be understood correctly at the first attempt.

We have taken the general model of Stafford Beer and turned it into a normative model. For us it is now a model that tells people how they should behave.

We started by introducing the model to a group of 20 people. They were the top management of the Group; that is, the Management Committee and a number of the managers of profit centres. We devoted the larger part of three 2-day seminars to the subject. At the end we had a fair ability to describe the various systems in which we had to work as individuals. A lot of shortcomings were also detected.

The next step was to introduce the model to all the management groups of the profit centres. These groups also have union representatives as members. We now had a few people from our Training Department arranging the seminars. Today the model is a regular part of our management training programme. This programme covers all people in managing positions, starting with the foremen.

It is comparatively easy to delegate authority to somebody. The manager regards this as an increased freedom of action. But the next step is more difficult: that is, to make this manager in his turn delegate authority further down the hierarchical ladder. We try to do this in various ways.

The direct way stresses our key ideas:

Delegate authority.
Increase *adaptability*.
Formulate and preach *key ideas*.
Show *heart*.

We also to some extent employ the method of *backselling*. In all 'missionary' work by management, especially by the managing director himself, we stress the three words

Quality   Training   Information.

These words are used very frequently within the organization. Thus we raise the expectations on continuous delegation of authority based on better training and on better information.

We also stress the words *information* or *business intelligence*. We talk about business intelligence *from* customers, competitors, research and development, society, employees; business intelligence *to* employees who can act upon it; business intelligence *between* employees or groups who are interdependent, and *between* Group profit centres.

This all ends up with one of Stafford Beer's important statements: 'Information is free.'

The following is an excerpt from a speech at a management meeting:

> 'Information must be free within the ASSI Group. This does not mean that everybody can ask for anything; but it means that everybody shall have access to information which enables him to do his job properly. Information must not be monopolized or hoarded internally. We must always fight those tendencies where some people try to make information into an internal base of power.
>
> We talk a lot of delegating power and authority. Decentralization is also a key word. It is necessary to remember that from delegation of authority follows a requirement to increase the flow of information. Otherwise people will not be able to act accordingly or their job will be done badly.
>
> Decentralization means that decisions should be taken as low as possible in the hierarchy—that is, where there is enough information on a certain problem (but not lower than that!). Employees (for example, foremen) should be encouraged to establish their own information channels between steps in the chain of production. Usually there are many flows of internal information over which managers have no control and usually no knowledge. These flows usually carry

very essential information, so let us stimulate them. This is often called *net-working*.

Finally, another type of general message to all employees is the following: You are the only one who knows your job, you are the professional. Try to reach your objectives, follow the rules and act when you have the necessary information.'

Experiences so far

Have we been able to achieve our objectives through this long process? The answer is still uncertain. We started by talking of 3–5 years to make noticeable progress. After roughly three years we knew that five years would be required to change the culture of the Group. However, some preliminary conclusions can be stated now.

(1) We are profitable: that really is the 'acid test'.

(2) Stafford Beer's model as applied by us has helped in several ways. Foremost, top management has felt itself completely confident that the message is good and consistent. There is a new consciousness of the importance of systematic information flow.

(3) All groups given new autonomy are first happy, then troubled. They cut themselves off from the world for a while when they try to establish their own identity and their own work rules. After one or two years they are ready to cooperate with the outside world again.

(4) It is difficult to make middle management understand that it is to their own advantage to delegate authority. They feel sometimes that their power base is threatened.

(5) It takes an extra effort to make profit centres realize that it is to their own advantage to cooperate with other profit centres to gain synergistic effects. Centrally we assist in creating 'network' groups between key professionals.

(6) The model has not, however, become part of every manager's daily language, although quite a few often think in terms of the systems of the model. It is notable that engineers especially find the model interesting: their technical background in regulatory systems makes it easier for them to understand.

(7) Some union people also accept the model easily. Skilled workers are proud professionals used to solving their own problems.

(8) When introduced to the model people start asking 'Where am I?' It requires a lot of effort to make people understand that they have to work within more than one system.

(9) We stress very carefully that decentralization does not mean independence. It means that more people have to take interdependence into

account on their own. The basic concepts, objectives and rules are to be followed.

(10) We also stress carefully the nature and the significance of our control system. However laudable autonomy might be, it has to be embedded in some overall cohesive context for the corporate identity to emerge. As one of our German managers put it; *Vertrauen ist gut, aber Kontrolle ist besser.* (Trust in people is all very well, but some control mechanism is necessary.)

(11) Finally, and perhaps most important, we can see how a new generation of middle managers is taking on new tasks and creating new opportunities. They are not waiting for their 'marching orders'.

APPENDIX: ASSI—The Employee and the Organization

INTRODUCTION

In the ASSI Group, we must establish a shared perception of our organization. If we start with this assumption, it should be easier for us to discuss and to develop our way of working together. All this is with a view to achieving our primary objective: survival through the ability to adapt to a changing world. If we are to achieve this, everyone must be prepared to play an active part.

This paper presents one way to describe an organization and those demands placed on all employees to ensure survival. The description is essentially a general one. It can just as well be applied to the Group as a whole, as to a work group with a common problem that needs to be solved.

The description is based on ideas developed by Stafford Beer, presented most fully in his book *The Heart of Enterprise.* Stafford Beer uses the human nervous system and its way of functioning as his starting point. Our nervous system follows a number of key principles to ensure its survival, by processing an excess of information. Our brain has to regulate a large number of variables (such as temperature). It must learn, adapt and develop. Parallels can also be drawn with the control and regulatory techniques applied to our manufacturing operations.

Our nervous system is, like the ASSI organization, a complicated network designed for evaluating information and for choosing alternative modes of action. Following the evaluation of both external and internal information, thousands of decisions are made within the ASSI Group daily, all of which should contribute to the continued health and welfare of the Group as a whole.

This sounds complicated, and it is. However, it is only by becoming aware of how we work that we can hope to understand, to change and improve. In this way we become more adept at survival.

Bengt A. Holmberg, Corporate Development

DEFINITIONS (see Figure 1)

Every organization must establish procedures and rules to obtain internal **stability**.

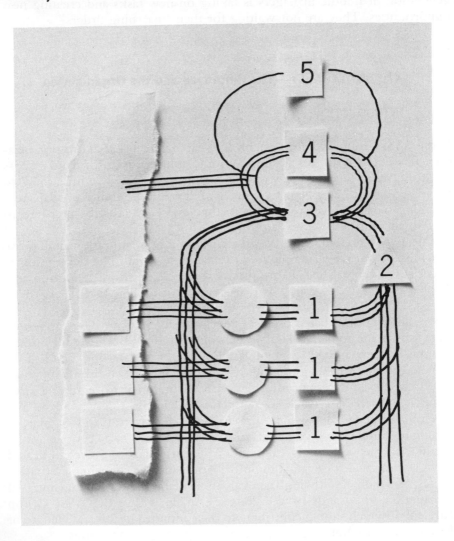

Figure 1. External information

Every organization is dependent on the world around it, and must be able to **change**.

Every employee should have free access to all the information he or she needs to carry out his or her duties: **information**.

Every decision should be taken at the lowest level at which adequate information is available: **decentralization**.

Every organizational unit represents in its turn part of a similar parent organization: **replication**.

System One refers to those units that are to be controlled. These are defined as, for example, departments in a pump mill or subsidiaries in a group of companies.

System Two–System Five are the channels for controlling the separate Systems One.

System Two comprises the information systems necessary to decentralized decision-making within System One and to solving problems which develop between the separate System Ones. This is carried out through formal reporting and through people building their own networks of contacts.

System Three, 'Here and Now', is the channel for orders related to current operations. The requirement for a control system is integral to the right to make decisions.

System Four, 'Change and the Future', handles contacts with those outside the company, and initiates changes and development work. System Three and System Four maintain a continuous dialogue.

System Five completes the system, balancing today's operations against tomorrow's needs, investments.

PART 1: ORGANIZATION FOR SURVIVAL

'We can only control a system in as far as we understand it.'

In the ASSI Group, employees have to make thousands of decisions daily. For example:

(A) There's litter around the winder. Decision: Clean it up.

The decision is simple and direct. The one who notices the litter is also the one to do something about it. But a number of demands are still made on the employee.
— He/she must know that it is important for the litter to be removed quickly.
— He/she must know which safety rules apply to prevent accidents.
— He/she must have access to the right tools.
— He/she must know where to dispose of the litter.

(B) The customer requests delivery in week 520. Decision: We offer to deliver in week 521, or to deliver half in week 518 and half in week 521.

When the sales manager in the Amsterdam office makes this decision, he calls on a considerable stock of knowledge acquired from his contacts with various parts of ASSI. For example:
— The customer's attitude and reaction.
— The actual production schedules for the different paper qualities from a given paper machine, and the current status regarding work on hand.
— The mode of transport, and its availability.

(C) Purchase of fuel oil for Norrbotten prior to the winter. Decision: We postpone purchase for 14 days.

The decision is based on a careful evaluation of the internal and external information available:
— Assessment of current price trends for oil and freight costs.
— Current stocks.
— Rules concerning stocking obligations.
— Implementation of current investment designed to reduce oil expenditure.

(D) It is noted that a bearing in the paper machine is overheating. Decision: Stop the machine, call for a repairman, and contact the boiler plant.

It is obvious that steps must be taken to get the machine operational again. But it is just as important to ask yourself 'Who needs to know about the stoppage?' Initially, perhaps, this means the boiler plant. At a later stage, the sales and despatch departments may have to be contacted.

AN ORGANIZATION IS DEFINED BY ITS PURPOSE

The purpose of each organization is the key factor in determining its structure. Since the main purpose is the survival of one's own organization, it must be structured to achieve that objective. In ASSI, we have defined our business concepts and objectives in terms of products and ambitions. To stay in business, we must meet the demands made by customers, employees, competitors, suppliers and owners.

A viable organization must always be able to
— Make all normal decisions simply and effectively (inner stability).
— Adapt itself to changes in the demands made by the world around it. This 'world' consists mainly of customers, employees, suppliers, competitors and owners.
— Learn from experience.

Demands are also made on the inner stability of the organization. It is often a question of clear rules and rational routines. Channels of communication should also exist which allow the outside world to influence the organization. An organization cannot remain viable if it lacks the means to change and adapt. It becomes out-of-date, and eventually vanishes.

The word 'independence', when used in the context of an organization, is basically an illusion. Everyone is dependent upon everyone else, and upon their surroundings, which are often shared to some extent. But when, at a particular moment in time, a decision-maker feels himself to be in possession of reliable information related to objectives and regulations, the peripheral situation (another System One and the market) and his own unit, he is then able to make an 'independent', decentralized decision.

A DESCRIPTION

Traditional ways of thinking and structuring are inadequate if the demands imposed by such an organizational description and its mode of operation are to be met. For this reason, let us construct our model of reality in another way.

A diagram can indicate each group of people which shares the same work objectives. For instance, this might involve chemical recovery and the boiler plant in a sulphate mill. Another instance could be a profit centre within the ASSI Group, such as ASSI Karlsborg.

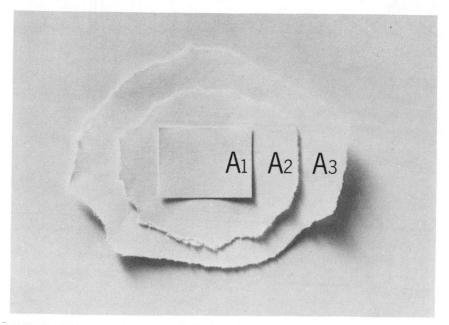

Figure 2. Viable unit: A_1 indicates management, A_2 comprises the entire unit, and A_3 the external environment

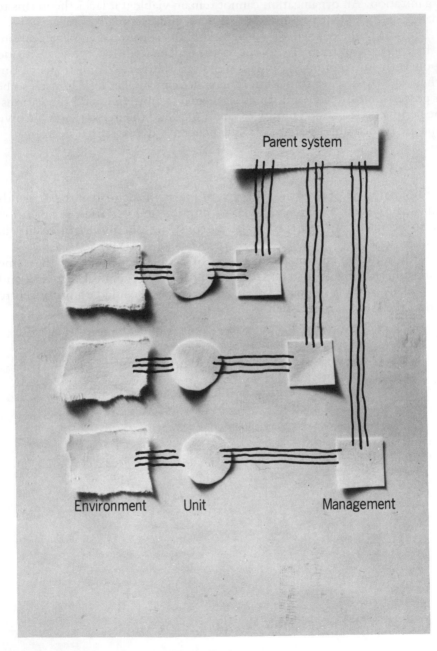

Figure 3. A company

The boiler plant has

— Customers who purchase steam, such as the digester department and the paper mill.
— Employees who have to complete a joint assignment.
— A supplier of black liquor, fuel and purchased chemicals.
— Competitors with whom to be compared, in this case other boiler plants.
— Owners, represented by the parent unit, such as ASSI Karlsborg.

If the diagram is instead used to describe Frövifors Bruk, the ASSI Group fulfils the role of the parent unit.

We can show that each organization consists of a number of units, as demonstrated in Figure 2. If we expand Figure 2 to include several units, we get the pattern in Figure 3.

We now have a company divided into independent units. It could, for instance, be these three units within Frövifors Bruk (the meta-system): pulp mill, chemical recovery and boiler plant. For practical reasons, management, the individual unit and the surrounding environment are drawn separately.

The lines in Figure 3 indicate all types of contact and communication necessary to every organization.

AN ORGANIZATION AS A COMPLEX REGULATING SYSTEM

The thermostat on a heating element is a good example of a simple regulating system. The system is based on the idea that information about air temperature is converted into a variation in water flow. The thermostat strives to maintain a predetermined temperature in the air of the room (see Figure 4).

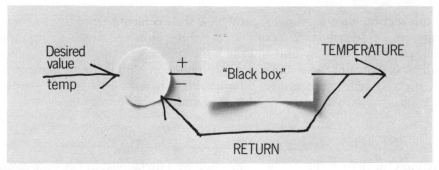

Figure 4. Thermostatic regulation of room temperature. A stable temperature is possible using feedback. Impulses from outside are converted inside the 'black box' into a means of adjusting temperature

Note that even this simple system has a defined objective, that of maintaining a constant temperature within the room.

Each manager deals with information from his own unit and from the surrounding world in the same way. The difference is that there is a considerable amount of information. The manager of the boiler plant receives information from the foremen about fuel availability and the boilers' status. He also receives external information concerning how much steam his customers need, today and tomorrow. Further, he must always be prepared for the previously mentioned phone call, that the paper mill will require no steam at all for the next three hours.

Each manager of an organizational unit therefore receives a large quantity of information. This forms the basis for a large variety of alternative modes of action. Naturally, no manager or management group can systematically evaluate every theoretical alternative. Experience and subjective evaluation act as a filter to determine the alternatives selected for rational evaluation. Consequently, the great majority of decisions are simple and self-evident for an experienced manager. He is therefore able to spend more time on dealing with the few difficult decisions, which often tend to be those related to changes in the demands made by the surrounding world.

Inner stability can be achieved through feedback within the unit. Alternative modes of action and stimuli for change are received from outside. A large number of channels are required, supplying a considerable flow of information, to meet demands related to stability and integration within an organization. These channels of information must flow between manager and employee as well as between individual employees, inside and outside a unit. It is important to develop and support lateral contacts.

PART 2: FIVE SEPARATE SYSTEMS CONTROL THE ORGANIZATION

In this section we are going to study how the management of one or more units can be described. We can take the three units used earlier, the pulp mill, chemical recovery unit and the boiler plant, together with the paper mill, as one profit centre, and study its interaction with the management of the meta-system profit centre. Alternatively, it can be the profit centres and subsidiaries within the ASSI Group and Group management.

SYSTEM ONE

System One defines those units controlled by a meta-system Let's move on from Figure 3.

The three organizational units, which must be 'viable', are called System

One. Seen from outside, they are 'black boxes'. For the moment, we don't need to bother about what goes on inside them.

The three units are directed by a meta–system (see Figure 5). At this point, we can merely note that lines exist between the units which indicate that they have direct contact which each other (lateral contacts).

The meta–system must make it possible to handle
— Management of the existing operation, System Three.
— Integration with the surrounding world, System Four.
— A balance between current needs and future needs (investments), System Five.

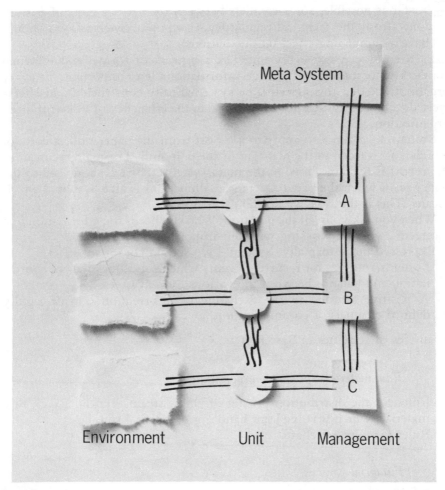

Figure 5. Management, unit and environment for three units controlled by a meta-system

Note that the different systems do not normally define official positions or persons (turn to Part 4 for further details).

SYSTEM TWO

System Two has contributed something new and important to our traditional organizational thinking.

System Two is a system of individual rules and behaviour for coordinating operations in terms of current needs. The system comprises everything that makes us into a good work team. It must

— Enable the various units, System One, to solve their own problems to a large extent. In short, to make decentralized decisions.
— As far as possible, solve conflicts between the various units of System One (from the technical/regulatory viewpoint, System Two should have a dampening effect on fluctuations).

Each employee and every unit has a *network* of formal and informal contacts who supply and receive information. It is important, in every organization, that this network be systematically constructed and effectively developed. System Two is integral to the efficiency of a decentralized organization.

System Two can also apply to a report from the paper mill, a copy of which is forwarded to the manager of the pulp mill. The main recipient of the report is System Three in the meta-system. The telephone call to the boiler plant when the paper machine malfunctions is also System Two. It means 'This is the way we do things in ASSI'.

When you ask yourself the question 'Who else needs the information I've received?', you are working with our important System Two.

Expressed more formally, System Two should comprise

— Systems/routines for recurrent events which solve questions of coordination and balance between the various System Ones.
— A positive attitude to the exchange of information outside rigidly defined channels of communication.

Examples of activities in System Two:

Raw materials—production coordination

— Rules for the distribution/pricing of chips, steam, kraftliner etc., raw materials which involve joint handling or internal supplies.
— Rules for joint purchase.

Financial

— Required return on operations and investments.
— Rules for quantitative and qualitative reporting.

— Plans (budgets and long-term plans).
— Financial demands (debt–equity ratio, cash handling, foreign exchange transactions).
— Investment proposals and cost calculations.

Personnel

— Personnel policy.
— Salary and wages policy.
— Codetermination policy.
— Training, internal and external.

Schedules

— Planning cycles.
— Frequency of meetings.

Resources

— Organization manual.
— Directives from managing director.
— Directives from Group staff and profit centre managers, concerning planning and accounting.
— Unwritten rules.
— Formal meetings.
— Lateral contacts (such as meetings with the Group's market analysts).
— Oral information from individual to individual.
— Workplace get-togethers.
— Local personnel newsletters.
— ASSI Newsletter and ASSI Info.
— Noticeboards.

SYSTEM THREE

System Three is the channel for *orders* within the organization. This system represents the lines we inscribe on a traditional organization chart. System Three refers to the management of the existing operations within the separate System Ones.

The lines in Figure 6 indicate two-way traffic between System Three and System One. This is intentional. Obviously, a smart System Three should utilize his right to make decisions and issue directives only after contact and discussion with the managements of the various System Ones.

A manager, such as one in charge of a profit centre, is working within System Three when he makes a decision about marketing, production,

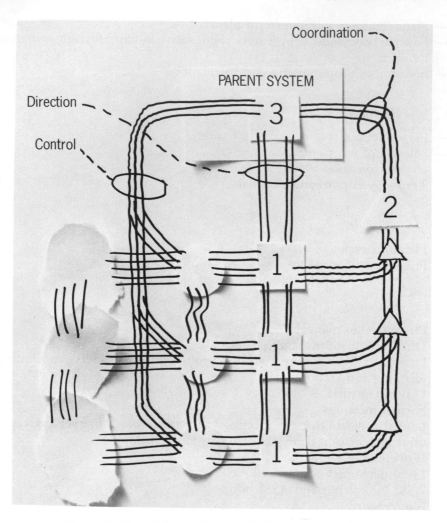

Figure 6. Many channels of communication in one organization

finance or personnel. The characteristic of this activity is 'Here and Now'.

The relationship between System Three and System Two is decisive to the successful interplay between System One and the meta-systems (S2–S5). An organization is increasingly centralized and authoritarian in proportion to the quantity of decisions made about 'Here and Now' in System Three. Fewer decisions reflect a corresponding increase in decentralization.

An important, in fact decisive, condition for maintaining a low frequency of decision-making via System Three is a well-developed System Two. Units within System One can only attain a satisfactory ability to select courses of action and make decisions, if System Two is well-developed and attitudes to lateral contacts within the organization are positive.

Only well-informed and knowledgeable people can reach satisfactory decisions. This applies to all employees of ASSI.

All decisions should be made at the lowest possible level of the organization. It is there that the alternatives and the consequences can best be evaluated.

System Three also includes an imperative *control system* (Figure 6). This often operates according to a separate set of directives. The personnel representing the control system have the right to acquire information from any source within the organization.

Examples of control systems include the following:
— Finance department within System Three.
— Internal accounting department.
— External accountants.
— Work-safety committee.
— Environmental protection authorities.
— Tax authorities.

SYSTEM FOUR

With feedback between System Two and System Three, we have attained a balance in 'internal' handling of information, developing alternative courses of action and making decisions. The world outside the organization has so far affected our individual organization only through its influence on the separate System Ones.

The environment for the meta-system comprises the environments of the separate System Ones and the greater environment around all of them.

If the organization is to survive, we must have a system that can function in its environment, and in future, by determining the potential risks and the potential benefits. System Four must be created as a dynamic function, which develops and changes.

We can now draw a new diagram within the meta-system (Figure 7).

All the work related to change within a company is included in the concept 'System Four'. This means that:
— Every manager is responsible for development within his area.
— Special development functions are to be instituted where appropriate.

A dialogue must be established inside the company between System Three and System Four. They are mutually dependent. System Four identifies needs and potential for change and 'tells' System Three about them. The selected courses of action can then be implemented by System Three.

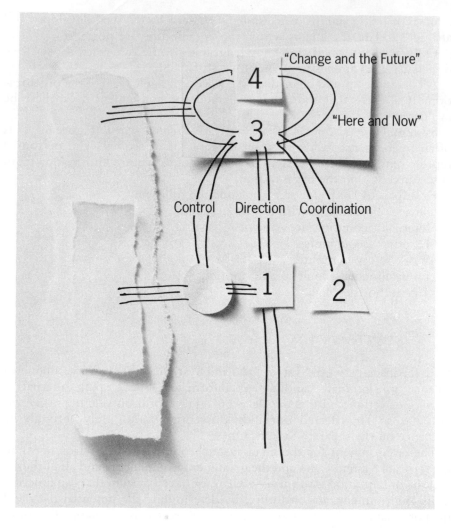

Figure 7. 'Here and now' and System Four: 'Change and the future'

It is important that there be a cross-fertilization of know-how concerning development work between the different parts of the company, and between the functions, such as marketing, technology and personnel development. It is also important that there be an open exchange of information between the separate System Fours operating at different levels within the company.

The following areas have been selected for rational development work to an appropriate scale:

— Society, as an environment for demand and production.
— Economic development.
— Assessing the competition.
— Financial freedom of action.
— Market and product analyses.
— Technical development (product/process).
— Development of management systems/leadership.

SYSTEM FIVE

System Five, the last in the model, exists to complete the whole system by balancing current activities (System Three) against future needs (System Four), the latter normally being in the form of investments (see Figure 8).

System Five generates neither know-how nor alternative courses of action. It simply ensures that a balance is maintained between System Three and System Four. When an imbalance develops, System Five plays the role of judge.

System Five should not be thought of as the 'king of the castle' or the boss of bosses. System Five often comprises a large number of company personnel. The Group's board of directors always forms part of System Five.

PART 3: SAME SYSTEM AT SEVERAL LEVELS

We build up each organization using work groups with joint objectives and joint assignments. Then we bring these together in their turn into new groups. We can then say that each 'viable system' forms in its turn part of a new and similar meta-system.

One of ASSI's independent profit centres can be described in terms of our model. The various parts of ASSI Karlsborg, for instance, therefore form System One, and company management sees to it that Systems Three, Four and Five function. In the same way, ASSI Timber, ASSI Karlsborg, ASSI Kraftliner and so on represent System One at Group level. The Group chief executive, board of directors and staff represent Systems Three, Four and Five.

Note that System Two in both cases should be seen as support and service for the respective System Ones to enable them to fulfil their assignments.

Thames Case Ltd. and all other independent profit centres within the ASSI Group simultaneously function as System One, and as System Three to System Five (the meta-system) within the profit centre. Box plants and speciality plants comprise the ten elements of System One within Thames

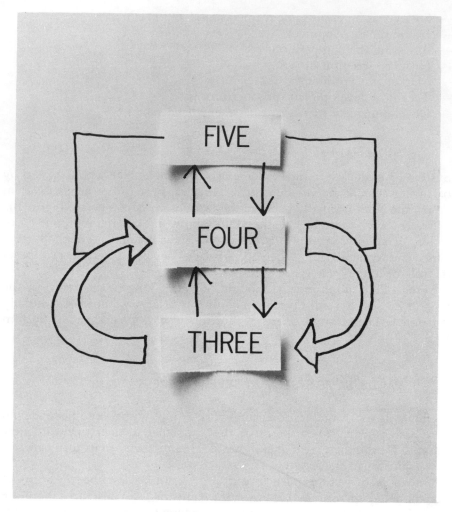

Figure 8. The meta-system

Case. Each plant becomes in its turn the meta-system for a number of System Ones such as log handling, sawing and despatch.

We can thus use the same model within the Group at a number of different levels. Naturally the different systems vary in their formulation, according to which part they refer.

The requirement that all systems should function in a satisfactory and balanced manner is equally important at all levels. To ensure that the organization functions, all employees must make a conscious effort.

PART 4: THE SYSTEM AND THE EMPLOYEE

'Where do I stand, and what is the role of my group within the organization?' This is usually the first question to be asked in a discussion of an organization. In this particular case, the question is 'Which system do *I* work in?' There are a number of answers to this question.

EMPLOYEES SHARING A COMMON VIEW

Each ASSI employee is part of a network of contacts which link people within every organization. The network comprises all these contacts, including the individual work group and other employees in other parts of our organization. Many employees also have some of their contacts outside the company, such as customers and suppliers.

If we are all to 'pull in the same direction' we must accept a number of basic principles which are common to all of us. These are necessary to our survival.

The *business concept* indicates *what* we are to do, which is to concentrate on our customers and prducts.

The *objective* indicates *how* we are to act to ensure satisfied customers.

Our *policies* represent a joint approach to establishing comparability and similar procedures for dealing with important questions such as capital budgeting.

HIERARCHIC ORGANIZATION RETAINED

It is important to recognize that the standard model for our organization remains valid. Traditionally, only the channels of communication which enable imperative decisions (directives) to be reached are described. The approach outlined in the preceding section does not change this. It is equivalent to System Three.

When we wish to describe how the decision-making process functions in the ASSI Group, we may continue to describe it in terms of the traditional model given in Figure 9.

The new approach means that we consciously complement our model with our personal experience of what we all know happens in an organization. Reports are not sent only to the manager—they are also sent to the next manager, and to adjacent departments. Plans and information are spread through the organization by word of mouth, as well as the written word. By bringing this to everyone's attention (especially System Two), we can all make a conscious effort to develop this necessary aspect of our work.

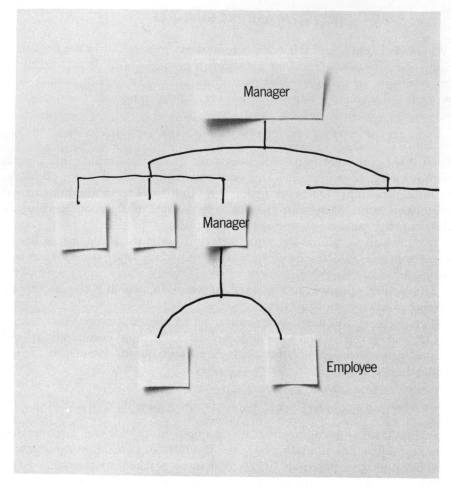

Figure 9. Hierarchic organization

Everyone who requires specific information for his/her work is entitled to get it. This is crucial to our efficiency.

THE ROLES OF THE EMPLOYEE

When the shift foreman at the papermill decides to order the machine stopped because of a malfunction, he is acting within System Three. The telephone call he makes to the boiler house to inform them of the situation falls within System Two. When he makes a proposal the following day, concerning improvements to prevent further stoppages, his action comes under System Four.

	Level	
	Group	**Profit centre/subsidiary**
System One	Profit centres/subsidiaries	Operational departments (sawmills, pulp mills, papermills, paperboard mills)
System Two	Group policies and guidelines Meetings/seminars Internal information Word-of-mouth contacts	Policy and guidelines for the unit Meetings/seminars Local information Word-of-mouth contacts
System Three	Messages and directives from CEO Instructions from Group staff	Memoranda and directives from managing director and PC-managers Communiques from, for example, finance or personnel
System Four	Directives from the CEO Corporate development The efforts made by all managers to achieve change within their area of responsibility	Directives from managing director and PC-managers Development department Market analysis The efforts made by all managers to achieve change within their area of responsibility
System Five	The Group's board of directors Discussions between the CEO, Group staff and all PC-managers and managing directors	Board of directors Management group or equivalent

Figure 10. Examples of System One–System Five at the Group level, and at the independent-profit-centre level

Many ASSI employees have jobs which mean that they operate within several systems. Relatively few operate exclusively within a single system.

At the two highest levels within ASSI, a diagram (Figure 10) can provide a schematic illustration of how tasks are assigned within the various organizational units.

It is obviously possible, and also useful, to do this for several aspects of Group operations.

The Viable System Model: Interpretations and Applications of Stafford Beer's VSM
Edited by R. Espejo and R. Harnden
© 1989 John Wiley & Sons Ltd

11

Strategic planning and management reorganization at an academic medical center: use of the VSM in guiding diagnosis and design

Michael U. Ben-Eli

The Cybertec Consulting Group, Inc., New York

Two consulting assignments involving strategic planning and management reorganization in an academic medical center are presented in this chapter. The Viable System Model provided important guiding concepts for both. The way the model was used in assisting diagnosis and design is highlighted and some of the implications and consequences are discussed.

Serendipitous milestones in one person's journey: a tribute to quality Beer

In many obvious ways, this volume and its various contributions stand as a tribute to Stafford Beer and his work in management cybernetics. This fact gives the book a special aura since, despite the many sources of material and the myriad cases discussed, the spirit of one particular individual provides a unifying thread throughout.

Instead of beginning with a technical commentary it may, therefore, be entirely appropriate to start on a personal note, recounting the circumstances of my encounter both with the Viable System Model and with its author. This will serve as a means for acknowledging the importance that both have had on my own conceptual development. It should also clarify the particular manner in which I have found the model to be helpful in my work, thus setting the proper context for the discussion of two cases of consulting assignments that will follow.

I was first introduced to cybernetic thinking when, in the mid-1960s, as an architectural student in London, I attended a series of lectures given by Gordon Pask at the Architectural Association. I was then involved in issues of urban and regional planning and was immediately drawn to the new material presented by Pask, sensing that although the subject matter was different, there was much of significance in common. I went to work as a junior research assistant at Gordon's uniquely eccentric Systems Research Laboratory, widening my reading in the available cybernetic literature and making my first aquaintance with Stafford's early work.

At the same time I had also become deeply involved with the work of Buckminster Fuller responding, with others, to the challenging vision he was projecting of social evolution, the future of the planet, and the possibility of a peaceful global development through a humane and effective design.

The years were times of great intellectual excitement and much talk about the dawning of a new era. The sense of urgency and anticipation was manifest in student unrest that erupted all over the Western world and it found vivid expression in the arts as well as in the widespread youth movement seeking new life-styles. Bob Dylan, the Beatles, the staging of *Hair*, and, of course, 'Woodstock' became popular symbols of the time.

For me, much of it boiled down to the question of change and thus to the whole issue of constancy and adaptation. Change: how can one sensibly approach it in the enormously complex domain of social systems? To what extent can it be designed? How could it best be 'managed'? Could it be mediated with minimum stress and not too much violence? What under-lying processes are essential to bring it about effectively?

My interest was thus shifting from three-dimensional design, which at the time still dominated approaches to urban planning in architectural schools, to questions of underlying processes. A concept of planning was gradually emerging as the activity of searching and specifying viable options in the dynamic process of managing complex affairs. I use the term 'managing' in its broadest possible sense, for it was slowly becoming clear that whether the context was that of urban, regional, or global planning, whether it was related to a single enterprise, a piece of biology, or a whole ecosystem, there was a similar underlying logic throughout. It clearly related to the problem of balancing various, sometimes conflicting, forces in a coherent, regenerative, self-reinforcing manner, mediated by a pur-pose and a set of primarily qualitative goals, which, depending on the context, are subject to particular constraints.

The concepts of 'regulating for viability', of 'planning', and of 'management' were thus becoming inexorably related, and as I was beginning to grapple with these questions it was, again and again, Beer's work that facilitated making the link between the general cybernetic theory of regulation and a newly emerging view of management processes.

But this was not all. Soon after graduation I joined Fuller's operation in the United States. Among the various exciting projects actually taking place or contemplated in those days, one was of particular historic significance. It was Fuller's concept of the 'World Game'. The idea called for a giant simulation facility, with an emphasis on dynamic visual displays, where data on humanity's world-wide conditions, needs, resources and trends would be made available to teams of players, potentially linking satellite-monitored world-wide information with actual decision-making processes. Members of such teams—policy-makers, scientists, researchers, statesmen, lay persons—would be engaged in a cooperative effort of developing winning strategies for our planet's future viability.

In the face of practical expediency and the short-sightedness that dominates world politics, here was a tremendously ambitious, perhaps outrageously naive, blueprint for a conscious, whole-system, problem-solving approach to real-time management of world affairs for the benefit of humanity as a whole.

Many difficult issues, philosophical as well as technical, were obviously raised by such a vision and novel tools were required for dealing with them. I was searching for such new tools and gradually became convinced that one could look to general systems theory, and particularly to cybernetics, for help. Between them, these fields of inquiry appeared to have developed not only the appropriate language for dealing with complexity, but also an entirely new general paradigm and the conceptual means for realizing its design. I soon found myself, therefore, back in London pursuing a doctoral program in cybernetics under Pask, while commuting to the United States to continue my association with Fuller.

Then came the year 1973 and with it the third Richard Goodman Memorial Lecture and the first public account of the work Stafford was doing meanwhile in Chile. The news was very exciting indeed. Here, to my mind, was an early prototype realization of a World Game operation. It was embodied in an actual physical facility, backed by the appropriate, even if relatively simple, technology, and firmly founded on sound scientific insights. Personally, a lot had come together at this point and I felt a strong sense of intimacy with Stafford's work, although I never had the chance, until then, to meet the man himself.

It was thus with a true sense of awe that I undertook the journey to the remote quarry in Wales, when Stafford turned out to be the external examiner for my doctoral thesis. There, at the old Black Lion Royal Hotel in Lampeter, on a dark and stormy winter night (really), I was treated to a ritual of a rite of passage that, with Gordon and Stafford presiding, enacted in a mysterious and powerful way the story of cybernetics itself.

By now the euphoric sixties were giving way to the more sober seventies. Fuller's World game, in spite of early promises, was never funded on the scale that would have made its realization possible, although the concept is kept alive through university workshops and other similar types of activity conducted by faithful disciples. The story of Chile is known. I was beginning to establish a management consulting activity in New York and was soon entirely absorbed by the surprisingly intense demands of a growing small firm and the needs of its clients.

Our firm has been set up as a general-purpose management consulting firm and the intention has not been to focus deliberately on any specific methodology. Nevertheless, for me, management cybernetics and the ideas inherent to the VSM have always been there, ever sharpening my discrimination faculties in the face of complex management situations. This comment is significant insofar as it will help clarify the role of the model in the approach developed for the two projects that I am about to describe. The scope of the cases involved was such that a comprehensive institution-wide view was essential and in this the model was extremely useful. It provided the conceptual framework for both diagnosis and design. In neither case was the model itself the purpose, nor was the focus on precise and full mapping of existing situations on to its mold. Rather, the model offered a set of concepts which, not unlike a craftsman's tools, were essential, but often remained invisible to those who made use of the products.

I am neither a researcher nor a scholar in the academic sense of the terms and my interest has not been in refining the theory or offering critical comment on its efficacy. Rather, within the limitations of actual client assignments, I have been interested in making use of the useful. I am fully aware of the dangers underlying such an approach and the connotation of superficiality it may evoke. Nevertheless, it is precisely on this level of sharp-tuning and honing one's perceptions that I have found the model most helpful.

The point is this: In approaching systems of exceedingly high complexity we need tools that help simplify without trivializing; tools that can help clarify confusing events, and that are accurate, focused, and potent enough to ultimately effect proper actions. It is precisely in this regard that the VSM is immensely powerful. For above all, it provides a set of concepts that help orient our thinking and guide our actions as we try to come to grips with high-variety management situations.

This compressed meta-comment should not detract from many other aspects of the model for, indeed, the work is monumental and in the whole of management literature is second to none. In addition to its technical merit and logical consistency, it has other important dimensions as well. In fact, I often think about it as about a work of art. For there is passion in this work and poetry, love and desperation, and deep concern for humankind, its 'Actuality', 'Capability' and 'Potentiality' . . .

So on to the more mundane realm of the cases I chose to describe, in which an attempt was made to give a working expression to some of these qualities.

Design for a strategic planning process

Background

From the early days of our firm's inception we have been involved in planning for the health care industry. After undertaking a number of planning assignments of different degrees of complexity and varying scope, the opportunity arose to tackle a comprehensive strategic planning effort at a large academic medical center in New York. The project involved all questions that were pertinent to the organization's future and, from a management point of view, it was conducted in a particularly favorable environment, where careful design for the planning process, as well as a consistent and thorough execution, were possible.

The organization, The Mount Sinai Medical Center, is a proud institution recognized as a leader among the nation's academic medical centers. It began its existence as a 45-bed hospital founded in 1852 by leaders of the Jewish community in New York. The hospital grew over the years, changing locations, expanding its services, and increasing the number of its beds. By the turn of the century, it had settled on its current site on the Upper East Side of Manhattan, gaining prominent reputation among medical institutions as a major medical resource.

By the early 1960s, the institution had reached an important turning point in its history with the realization that, in order to ensure its position of leadership and secure its tradition of clinical excellence, a growing commitment to clinical and basic research, as well as to medical education, would have to be made. A bold decision to establish its own independent medical school was made by the board of trustees and, in 1963, the Mount Sinai School of Medicine was founded. An ambitious program was soon under way to ensure the appropriate nourishing support for the school's development and growth.

At the time we became involved, the center comprised a 1200-bed hospital where some 36,000 inpatients were discharged and some 250,000 clinic visits and 60,000 emergency visits were recorded annually. The medical school enrolled some 460 medical students, offered PhD and MD/ PhD programs in addition to its MD degree, and oversaw a substantial number of active research projects. Strong teaching affiliations were maintained with a number of other medical institutions in the New York area. Operating budgets topped 200 million and 70 million dollars in the hospital and school, respectively. The center had an extremely active board of trustees, a dedicated faculty and medical staff and, at the time, a management intent on progressive and innovative practices. From a purely managerial view-point, it presented, as with other similar institutions of this type, a particularly interesting set of management problems. This was so particularly because of the need to blend and integrate essentially different activities that by their very nature respond to management strategies of a different kind. Many aspects of the hospital's operation, for example, could clearly benefit from a typical model of rigorous corporate-type management control. The more elusive requirements of successful education and innovative research obviously called for a somewhat different approach.

By the late 1970s, the sense was growing among the leadership of the institution that the time was ripe for a process of self-assessment and renewal. A relatively new management team, having just emerged from an effort of resolving financial difficulties and streamlining internal operations, was ready to orient its view to the future. A planning committee of the board of trustees was established and the center's planning department, unusually strong among institutions of this kind, became the focal point for a new and intensive planning activity. The climate was ready to pause and take stock, review the institution's position, establish new priorities, and reach consensus on future directions for growth. Our firm was retained to assist the medical center in designing the appropriate planning process and developing the center's long-range plans.

Key planning principles

Much discussion took place at the outset concerning the planning effort, its appropriate scope, and the most suitable approach. The issues were numerous, as were the opinions, and for a while it was difficult to see the forest for the trees. Fundamental concepts developed in management cybernetics, and particularly those embodied in the VSM were helpful in sorting things out, separating the essential from the trivial, giving events a clear definition as well as a coherent structure, and communicating to others the essence of the task. I kept returning to these sources for guidance, again and again, as the process unfolded.

In viewing the most relevant ideas involved, distilling their essence, and expressing them in the context of the challenges presented by designing and carrying out this strategic planning assignment, a number of principles emerged as essential concepts. They are accompanied here by a set of diagrams that were used originally to convey the meaning and gain acceptance for the approach. They can be summarized as follows:

A whole-system view of the institution

The notion of whole systems, relating to complex identities comprising interacting, inexorably interdependent parts, is fundamental to the view of the world that is inherent to the VSM. It is rarely, however, the view adopted by management, whose attention and energy, more often than not, are focused on specific special cases that are symptoms of current crisis events. It was typical of the early stages of the planning process, for example, that different participants would narrowly focus their view on specific isolated problem areas, as though this or that particular issue was in itself the most important to resolve.

Shifting the emphasis at the outset to a whole-system view of the institution became, accordingly, a vital consideration in laying out the blueprint for planning. This was achieved by developing a series of visual representations—'system diagrams'—an example of which is depicted in Figure 1. The intention was to highlight the key factors that affect the institutional totality and draw attention to the manner in which such key factors interact. The interdependency and mutual effects of the key components was thus explicitly recognized at the fore, and specific planning issues could be placed in the context of the complex whole.

This essentially pedagogic device was extremely useful in scoping out the planning effort, identifying pertinent systemic boundaries, defining the appropriate levels of resolution, and developing consensus concerning the critical areas which ought to come under review.

The commitment to such a whole-system view and the understanding of underlying systems dynamics is important if one is to avoid the potential risk of developing policies which would turn out to be only partial in scope.

Planning as an issue-driven process

At the heart of the systems view of the world there is implicit an 'organismic' view of organization, be it involved in economic enterprise, government, or other aspects of human affairs. This view, which brings a touch of biology to societal systems, has its roots in the pioneering work of von Bertalanffy, and it is clearly in the very essence of the VSM itself. The underlying approach is significant insofar as it replaces a static

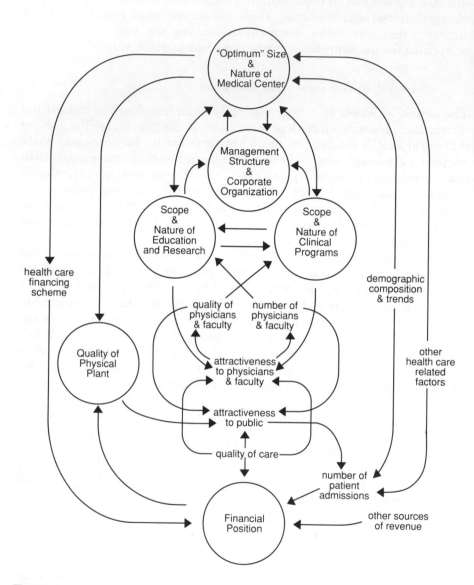

Figure 1. A systems view of a medical center. *Reproduced from Ben-Eli (1988) by permission of Hemisphere Publishing Corporation*

view of organizational entities with an emphasis on their dynamic and, ultimately, evolutionary properties. Observable behavior is linked to a specific internal anatomy, the structure of which involves dynamic interactions that specify the characteristics of the system itself.

There is a legitimate sense in which even a complex and multi-faceted organization, such as the medical center involved, can be regarded as a living individual. It is an extremely active 'living' system with its own personality and style. It has its own tradition, internal culture, many active constituencies, competing interests, and sometimes conflicting internal drives. It has its own peculiarities, idiosyncracies, definite 'ways of doing things', and, not unlike the case of a patient, a therapy prescribed by planners which if it does not fit its temperament and most important critical needs could very well fail.

Above all, perhaps, the momentum of an institution's ongoing activities cannot be halted while a 'master plan' is being developed, nor can all institutional problem areas be dealt with at once. From the viewpoint of strategic planning, therefore, it is important that at any given time attention is focused on identifying the most critical issues requiring long-term resolution, those that are clearly central to questions pertaining to effective survival and future growth.

Accordingly, the planning process at Sinai was issue-driven, and the process of issue identification, although time-consuming, was given special attention. By separating the critically important from the secondary, and sometimes even the merely 'smokescreen' types of problem, the target areas for the planning process were clearly defined, consensus about their content was reached, and much potential confusion was avoided as the process continued to unfold.

The process of issue identification, incidentally, was derived from an exhaustive 'external environment' analysis, as well as an extensive review of various institutional profiles. The key to effectiveness in such a process is in conceptually linking external events to internal activities in a way that has general adaptive significance, but also a clear and specific enough operational meaning.

The primary role of purpose

At whatever level of recursion, the type of organizations that the VSM seeks to depict are first and foremost purposeful systems. As is typically characteristic of social organizations, they are 'ideal seeking' in the way suggested by Ackoff, a fact that is sometimes lost in the confusion and shuffle of daily affairs. Their purpose is often projected on to their external world, as it relates to whatever they seek to accomplish; but above everything else it has to do with the essence of 'selfhood' itself.

Perhaps of all the issues that are relevant to strategy development, the issue of purpose ought to come first. It is, in my view, probably the most important. Around it revolves the whole question of institutional identity, its internal cohesion, and the very reason for its being. In fact, without purpose, the whole notion of planning is quite meaningless.

Too often, however, planning processes are driven by an almost reflexive quest for data. Their early activities are automatically focused on information gathering, without establishing first a clear idea of what it should deny or confirm. Data are obviously essential for backing arguments, for testing, for validating, calibrating, and refining ideas; but it is a sense of purpose, even if vague, that gives it a framework for meaning. It is ultimately purpose that is responsible for fueling the imagination, inspiring commitment, and galvanizing action. From the viewpoint of cybernetics, in fact, whenever variety proliferates uselessly, a purpose acts as the most effective variety-reducing device.

At Sinai, the idea of tackling the issue of purpose was greatly resisted at first. Most participants felt that it was much too obvious a question to be of any use and that it was unlikely to yield anything but vague and, from a practical viewpoint, meaningless results.

'Everybody knows what the center is all about' was the typical attitude. Yet, upon insistence, the issue was faced and soon acknowledged to be of vital importance when it became clear that fundamental differences of opinion and attitude existed among key players about policy issues concerning the institutional mission.

To resolve these differences a process involving trustees, management and medical staff was launched which, after a few intensive and, at times, surprisingly tense months, yielded agreement on a clear statement of purpose. This, in turn, greatly facilitated decision-making later on in the process, when questions of program priorities, resource allocation and the like had to be faced.

The concept of levels of management

The concept of levels of management, differentiating between functionally and logically distinct domains of management concerns, is central to the VSM. For any given recursion, the distinction roughly corresponds to normative, strategic and various kinds of operational questions. The pertinent point is that approaching problems that are related to each such level requires a different conceptual orientation, a different language, a different method of handling, different emphasis, information aggregated at different level of details and, more often than not, a different group within an organization.

In a much simplified form, as shown in Figure 2, these essential

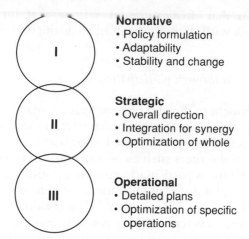

Figure 2. Levels of management/planning. *Reproduced from Ben-Eli (1988) by permission of Hemisphere Publishing Corporation*

distinctions provided an important concept for structuring the planning process. It helped in organizing the content and sequence of issues that had to be dealt with, and in defining the key fora for study, discussion, formulation of recommendations, and review and decision-making, as related to each step. The concept was helpful, not only in sequencing activities for the planning process as a whole, a sequence depicted in Figure 3, but also in helping to dismiss at the very beginning those operational issues which were persistently raised but did not necessarily belong to the proper domain of strategic concerns.

Planning issues were thus sorted out by a two-dimensional matrix: by functional type, such as financial, organization, program-related and the like, as well as each with respect to the appropriate level of management to which it logically belonged. This particular practice was useful in avoiding

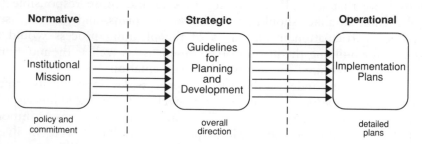

Figure 3. The planning process: conceptual sequence. *Reproduced from Ben-Eli (1988) by permission of Hemisphere Publishing Corporation*

a typical confusion that often results when issues that ought to be logically separated are dealt with without the proper distinction and all at once.

Commitment through participation

A strong commitment to the idea of participatory planning was at the heart of the planning process and, given the character of the institution involved, it would have been difficult to proceed otherwise.

Academic medical centers such as Mount Sinai have been recognized in management literature as particularly complex institutions. They are often highly diversified, pluralistic organizations, where decision-making and authority are diffused and where many key individuals play a number of different—sometimes even potentially conflicting—roles. A clinical chairman responsible for medical services at a teaching hospital, for example, will typically wear the hat of departmental chairman of the medical school, belong to some framework in which private medicine is practised, and be expected to play a central role in management of the center as a whole. The same individual could have a significant function, accordingly, not only at different levels, but simultaneously at entirely different recursions.

Further, a corporate-type management structure is normally superimposed on the practice of medicine, while at the same time leaving the question of authority, responsibility and accountability in control of resources largely unresolved. For example, many physicians—'volunteers' as they are sometimes called in America—use the hospital resources for treating their patients but are not individually and directly accountable to management in the usual corporate sense of the word. As a result, management by broad-based consensus is a matter of necessity rather than merely a question of style or choice.

Consequently, it was clear from the outset that in order to be effective, the planning process would have to be deliberately designed so as to encourage a considerable degree of participation. The underlying considerations were that key players, policy-makers and those responsible for implementation alike, should be authors of the plans, and that extensive and open participation in all stages of the planning process would be essential for building the institutional consensus, as well as the individual commitment, required for successful implementation.

Participation expanded as the planning process developed; at its zenith, when implementation plans were being developed, some 200 individuals, including trustees, management, faculty, staff, representatives of important public constituencies, and various specialty consultants, were involved. Orchestrating all the different activities and integrating the different contributions into a coherent whole was, in itself, an important aspect of the planning process.

From the viewpoint of cybernetics, of course, the notion of broad-based participation relates directly to the idea of making full use of an organization's potential variety and enhancing its own self-organizing capabilities. To be successful, however, the process has to be carefully staged. Though popular in recent management literature, the concepts of 'participative management' and 'participative planning' are not easy to accomplish. The energies that such processes release can easily turn things into chaos. They must, therefore, be thoughtfully channeled in order to achieve an effective result. This requires discipline, sensitivity, some skill, and, very importantly, an appropriate structure.

The need for an underlying structure

Perhaps because planning is often regarded as a 'staff' function—merely an on-tap specialized technical support, an adjunct to the mainstream of management activities—the need for an underlying structure for planning is not always obvious. Structure is the vehicle through which processes actually take place, and the need to embody the planning process in an appropriate structure cannot be over-emphasized. In fact, one of the classical contributions of early cybernetics was due to the insights of Wiener and Rosenblueth, who linked the characteristics of observable systems behavior to their underlying structure.

In too many organizations a well-functioning institutional structure for planning simply does not exist. In such cases, if a serious planning effort is contemplated, such a structure must be deliberately designed, put in place, and encouraged to persist. A well-designed structure will clearly define the major components—committees or groups—that are to be involved in the process, their roles, mode of operating, and manner in which they interact. At Mount Sinai, the planning process was embodied in a structure depicted in Figure 4. It consisted essentially of a governance-level steering committee, a senior-management and faculty planning group, a number of specially appointed institutional task-forces, and a core group of planning staff and consultants who provided the technical support and overall management of the process itself. Overlapping membership was used as a means to ensure continuity.

As a formal mechanism this structure was designed to reflect the various functions and tasks required by the planning process in a manner consistent with the VSM's concept of levels of management. In the way that it integrated activities and people it acted, in a sense, like a giant homeostat,

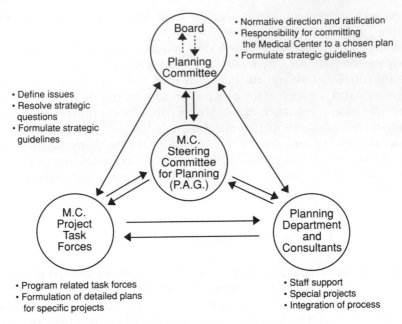

Figure 4. The planning process: structure

mediating the various processes involved and facilitating the resolution of issues, formation of consensus, emergence of decisions and reinforcement of the necessary commitment.

Process overview and summary

The planning process proceeded in a sequence of reiterative steps moving from the general and relatively open-ended to the ever-more specific and precise. Planning issues were identified, sorted out, analyzed, and their implications spelled out. A mission statement was developed while alternatives for future developments were being reviewed and various options studied and narrowed down, until a general sense of the institution's strategic direction emerged, was clearly expressed, and finally ratified. This general 'strategic posture' was then given sharper focus with the development of a strategic plan in which the scope of particular programs was specified and questions of timing, priorities, resource requirements, sources, and allocation were addressed and resolved. Implementation planning in all essential areas then followed.

Two and a half years were required in order to complete the whole effort, from a short but critically important period of 'planning for planning', during which the stage was basically set for all that followed, to the final ratification of the strategic plan by the board of trustees. Work by the various committees and groups was intensive throughout the period, with the institution's leadership clearly on a special footing. This continuously sustained mobilization was modulated by various institutional events that occurred at significant milestones, and included a few special retreats with the full board, designed to review and ratify recommendations for important decision. The circumstance and ritual of these occasions combined with thoroughly prepared presentations and solid analytical staff work to produce an emotive ambience that encouraged the spontaneous emergence of institutional commitment and energized the process for the next required step.

The process as a whole is depicted by Figure 5. It unfolded in a way that, in its dialectic characteristics, could best be described by Pask's Conversation Theory. Like a vast 'entailment mesh', a new fabric of concepts was being woven by groups of different participants. Ideas were sketched out, reviewed and tested, at times to be rejected and then returned to again, until agreement was reached and with it a stable footing secured for moving on to the next relevant topic.

The strategic planning effort at The Mount Sinai Medical Center was completed successfully. It was hailed as a pioneering effort because of its

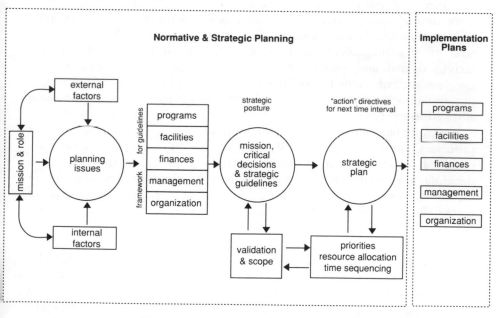

Figure 5. The planning process: overview

quality and unprecedented scope, and it positioned the institution confidently before yet another significant step in its long history. Of all the lessons learned perhaps the most surprising was the time and sustained level of effort that were actually required. There is a definite gestation rate related to major insitutional decisions, and time is needed for common concepts, for a common language, as well as agreement, to emerge. The front-end period dedicated to planning for the planning process was crucial, as was the energy and focused attention devoted to orchestrating and managing the process itself.

A special effort of this kind may well be required periodically, for it can rejuvenate and refocus institutional vision which can otherwise lose its edge through the dulling effect of daily routine. But even more significant is the need to incorporate the concept of planning in the very fabric of management at all levels, so that it becomes an integral and consistent part of every manager's role instead of an effort to be carried out by a specialized team. This concept is inherent to management cybernetics and is important enough to merit a few additional comments.

Cybernetic theory has generalized the concept of regulation and, in an abstract formulation that is largely due to Sommerhoff and Ashby, produced a rigorous definition of regulation that is independent of the particular fabric of the system involved. By making 'regulation' synonymous with 'management' a view is obtained that focuses on questions of process, structure and outcome, de-emphasizing the usual preoccupation with conventional, specific management disciplines, such as marketing, accounting, production, finance, and the like. In the context of this theory, the concept of management can be interpreted as the continuous dynamic activity of matching varieties between organizations and their context. It involves the conscious balancing of variety, amplifying or reducing it as the situation demands, so that coherent identities are maintained and allowed to evolve.

In a more familiar vein, management involves organizing structures and processes, and executing activities such that specific desired outcomes are obtained. In this regard planning focuses essentially on the process of deciding 'what to do' and specifying 'how to do it'. It relates to articulating the context of management activities and involves, for any given level of recursion, formulating and reformulating goals as well as specifying procedures for their attainment and for monitoring the results. The two concepts, management and planning, are sides of the same coin, or better, the fabric of the same continuous loop. They coexist and are inseparable.

In the practice of management, however, the function of planning is often misunderstood. The predominant tendency is to regard it as a remote, vaguely intellectual, even idle, activity, whereas management in a 'hands-on' job. The constant pressure for immediate, 'practical', short-term results leaves little room for thoughtful contemplation; for the typical manager, planning is a time-consuming luxury regarded with some degree of suspicion, even with scorn. In fact, a deep dichotomy exists in our management culture between the notions of 'doing' and 'thinking'. Planning and managing are seen as two entirely separate types of activities. A definite premium is put on quick, 'resolute' action, whereas planning is delegated a secondary role and is all too often compromised.

It is because of this attitude, perhaps, that management, at all levels of society, often finds itself in a difficult vicious loop. Since effective planning is not done, management must continuously stagger from crisis to crisis, and alas, there is no room for planning when one is constantly operating in a crisis mode. . . .

A case of management reorganization

The existing situation

As the development of the strategic plan was drawing to a close, attention was given to organizational issues and future management needs. Since it was felt that this obviously sensitive subject could disrupt operations and interfere with the smooth progress of planning, it was delegated to a small task-force of three trustees, including the chairman of the board, and was dealt with outside the mainstream of activities of the planning process. By then I had spent considerable time reflecting on the medical center's organization and was asked to assist this group in its work.

Primarily because of its long history and the more recent addition of the medical school, there were aspects to the existing organizational structure which were somewhat confusing. Various important features represented a sequence of makeshift measures that, in response to legitimate needs, were added piecemeal over time without the blueprint of a coherent and comprehensive design. Recently added features overlapped and coexisted with organizational aspects that did not change from an earlier time. This is not an unusual circumstance, of course, and just as is the case with their mission, or other aspects of their operations, organizations can obviously benefit from a periodic consolidation of their management structure. Thus, as new plans were being developed at the medical center, attention was directed to the question of future leadership and of simplifying the overall structure so that it could effectively meet future needs.

From a management viewpoint the existing situation was briefly as follows. The medical center consisted of three separate, legally autonomus corporate entities, none of which was a legal subsidiary of another. The three were The Mount Sinai Hospital, the older and initially dominant force; the Mount Sinai School of Medicine of the City University of New York; and The Mount Sinai Medical Center, Inc., a non-operating entity, conceived as a means for facilitating the interface between the school and the hospital and raising and distributing charitable funds for both.

There were a number of ways through which integration between the three entities was effected, although officially there coexisted three different organization charts, each with its own board, president and other typical functions. There was a great deal of overlap in management roles, with individuals assigned particular corporate responsibility and holding a similar title, vice-president for finance, for example, reporting to a different person depending on which entity one would choose to focus on.

While some such instances of overlap were not particularly harmful (as for example, in the case of the separate boards that had overlapping memberships and met conjointly), other cases of overlapping functions contributed to some degree of confusion and tended to aggravate potential conflicts that would have existed anyway; for example, the typical tension over resources that existed between the hospital and the school.

In a nutshell, key management-related features of the three individual corporate entities can be summarized as follows.

The Mount Sinai Hospital

The hospital was the primary operating unit. In addition to patient care and education, it provided to the rest of the organization all basic maintenance and support services, such as financial services, payroll, housekeeping, materials management, security, administrative services including data processing, and the like.

The key officers of the hospital were the chairman of the board, the president (a position which was then vacant) and a director who acted as chief operating officer. In addition to the hospital's board of trustees and its standing committees, other important governing or management bodies included the Medical Board, a policy-making body of the medical staff responsible essentially for issues concerning patient care, and a Senior Management Group consisting of vice-presidents and heads of various management support services responsible for hospital resource management and operations.

The Mount Sinai Medical School

The school was the second operating entity discharging all academic and administrative aspects of medical education and research. It had an unusual contractual arrangement with the City University of New York which made it officially one of the many colleges of the university, although the latter had no significant fiscal or management role in its operation.

Key officers of the Mount Sinai School of Medicine were the chairman of the board, and the president and dean (one position as specified in the bylaws) who acted as the chief administrative and operating officer. In addition to its board of trustees other governing bodies at the school included the executive faculty and the academic council, responsible for academic excellence and for advising the dean on issues of policy related to academic affairs. In a somewhat similar fashion to the Senior Management Group at the hospital, a Senior Dean Group was responsible for policy implementation and day-to-day operation of the school.

The Mount Sinai Medical Center, Inc.

As mentioned, the Mount Sinai Medical Center, Inc. was a non-operating entity, established in order to provide coordination and act as a fund-raising body to channel resources for the hospital and the school. Key officers consisted of the chairman of the board and chief executive officer, a president and chief administrative officer, an executive vice-present, senior vice-presidents for academic and clinical affairs, and vice-presidents for various administrative functions. A President's Advisory Group consisting of departmental chairmen and key managers developed recommendations for the president on center-wide policy issues.

Imagine, then, three different organization charts, of the familiar conventional type. They would look roughly as follows. *For the hospital* there would be a box representing the board of trustees, under which there would be shown a position of a president to whom a hospital director reports. The latter would have directly reporting to him the various administrative and support functions and, in a matrix fashion, he would interact on matters concerning control of resources with chiefs of clinical services officially, through the medical board. *For the school*, the president and dean would be shown reporting to the school's board of trustees; and in turn, to the dean, there would be shown reporting three types of groups: the chairmen of the school's departments, various deputy deans and associate deans for different areas of academic concerns, and a group of vice-presidents and directors responsible for corporate-type management functions. Finally, the organization chart *for the center* would show a president reporting to the center's board; and to him, in turn, there would be reporting an executive vice-president in charge of all key administration and management support functions.

A significant degree of functional overlap existed across these three organizations which could be summarized as follows:

- There was an extensive overlap in membership among the three boards of trustees, and the three entities had the same chairman of the board who acted as chief executive officer.
- The president of the Center Inc. was the same individual as the president and dean of the school, and the executive vice-president of the Center Inc. was the same individual as the director of the hospital.
- The chairmen of the departments of the school were the chiefs of the corresponding clinical services in the hospital, and all medical staff of the hospital were members of the faculty of the school.
- Almost all of the senior management staff (e.g. vice-president for planning, vice-president for finance, vice-president for personnel) held the same or equivalent positions in all three entities.

Organizational diagnosis and design

Behind the apparent complexity there was a subtle but enormously significant aspect of organizational reality: namely, that the institution was caught in the midst of a natural evolution, toward full integration of its autonomous and for a long-time completely independent hospital, and the recently created but separate School. The fact was that a concept of the 'Medical Center', distinct from the existing non-operating legal entity, the Center Inc., was used organizationally and functionally to describe Mount Sinai's total activities as an academic medical center. Formally, however, it still did not fully exist, although some aspects were functioning as though it did.

For example, in a partial attempt to strengthen the 'center' concept, two new positions were created approximately at the time that the strategic planning effort was launched: a senior vice-president for clinical affairs (held by the chairman of surgery), and a senior vice-president for research and education (held by the deputy dean). Both were supposed to provide the executive vice-president with input on clinical and academic matters. This in itself was not a sufficient move towards full integration and, in fact, superimposed on things as they were, added some ambiguity to existing relationships.

From the viewpoint of looking at the center as a total entity, a number of other significant ambiguities related to the existing status of things. For example, the president of the Center Inc., who was also the president and dean of the school was not an officer of the hospital; the director of the hospital had a direct reporting relationship to the chairman of the board and the board of trustees and also reported, both as hospital director and as executive vice-president of the center, to the president of the center. In spite of his position as executive vice-president of the center he was not an official of the school; the directors of various support services for the school (e.g. finance, personnel) reported directly to the dean, but in their hospital and Center roles they reported to the director of the Hospital and executive vice-president of the center. Equally significant, the Senior Management Group, which reported to the executive vice-president and saw itself as a center-wide function, had no membership from the academic component.

These and other type of ambiguities tended to accentuate problems of divided loyalty, constantly required various organizational adjustments, and consumed energy which could otherwise be clearly focused on the progress of the center as a whole.

Many of the related problems could clearly be given an interpretation by the Viable System Model. In fact, it was precisely the use of the model which made it easier to untangle the complex web of relationships as presented by official documents, personal interviews and observation of the behavior and practice of key players. It thus became clear that the language, concept, and practice of management at the medical center confused different recursions as well as different functional levels, mixing them up in one not-sufficiently differentiated view.

Many of the existing organizational difficulties were there, it appeared, because the 'center' concept was emerging as an operational and psychological reality, but was not yet given complete and unambiguous expression in a clear organizational form. Put in the language of the VSM, the first recursion, that of the institution as a whole, existed in an incomplete embryonic form, and it was mixed up with the legitimate functions of a second recursion, related to each of the major individual operating entities which were resisting the need to give up a measure of their respective autonomies in order to reinforce synergetic qualities of the whole.

With this in mind, the subsequent organizational intervention focused on Systems 5, 4 and 3 of the VSM. The emphasis was on the functioning of the first recursion, underlining the center concept as an operating total whole, with the hospital and the school as its two primary operating units. Note that in this regard at least three different recursions could be of relevance. The first, the one dealt with in this case, pertained, as mentioned above, to the academic medical center seen as a whole. The second would relate separately to the hospital and the school, and the third would focus on single departments as elements of a framework within which patient care was rendered, research conducted, and medicine taught.

With attention focused on corporate management, even a quick sketch-mapping of the VSM on to the existing reality, as roughly depicted in Figure 6, revealed important deficiences. These related primarily to the functional distinction between the integration of operations on the level of System 3; the appropriate mechanisms for formulating strategies at the

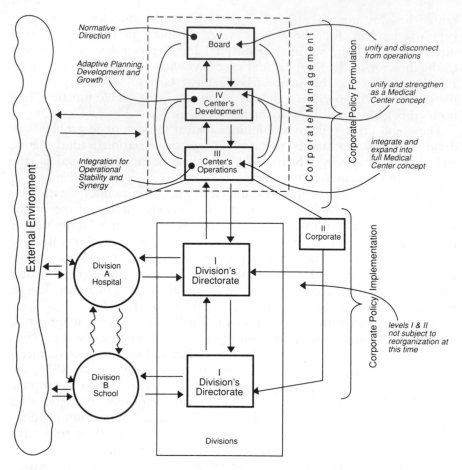

Figure 6. First recursion: medical center view; intervention at Systems 5, 4 and 3

level of System 4; and the role of the board as the embodiment of System 5. Even from the viewpoint of a functional concept only, and without considering titles, positions, individual roles, or questions of executive overlap, it was clear that the appropriate mechanisms, both for integrating operations and for policy formulation, ought to be better defined, strengthened, and institutionalized in a clear and recognized form.

As these issues were reviewed there began to emerge a 'center' concept as captured in the structure shown in Figure 7. It identified the need for a distinct center-wide policy formulation function, balanced by an appropriate mechanism concerned with the integration of center-wide operations. The first would be embodied in an office of the center's president and it would be concerned with questions of effective adaptation and the continu-

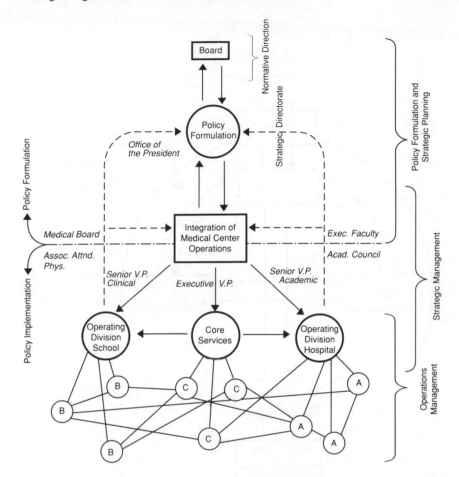

Figure 7. Consolidation of medical center structure

ous balancing of external events with the institution's own aspirations and needs. The second, embodied in the office of the executive vice-president, would be responsible for integrating internal operations, and, by overseeing the core management services, integrating the needs of the operating units for well-balanced results.

Both functions would incorporate clinical, educational and administrative considerations in a way appropriate for their respective levels of concern. Thus, for example, the executive vice-president would preside over a senior management group, expanded to include the appropriate clinical and academic inputs, as required for the smooth management of the center's operations. At the same time, the office of the president would be designed to incorporate the institutional mechanisms that were recently

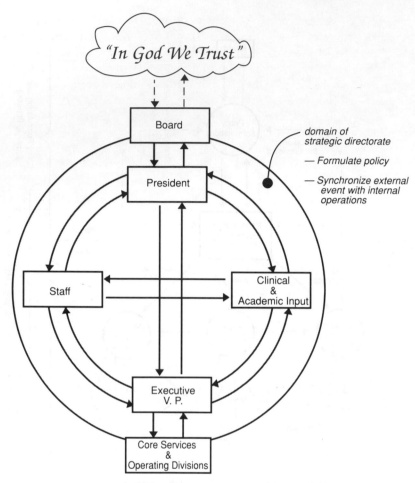

Figure 8. Design for a strategic directorate

put in place for the needs of the stategic planning effort. These would be institutionalized in a manner depicted in Figure 8 and, under the president's leadership, would constitute a permanent 'strategic directorate'. As a general function it would include the executive vice-president, an expanded version of the President's Advisory Group, with its key clinical, academic and basic science members, as well as other selected administrative and corporate planning staff.

From a center-wide viewpoint, these mechanisms would each provide a specific forum for an essential management function corresponding to Systems 3 and 4 of the VSM. At each level they would enhance integration of requisite domains of activity and ensure the broad participation that was advocated by those concerned as the appropriate approach to management.

One particular issue concerning the role of the board, and thus, related to the functions of System 5 had to do with the fact that, at Sinai, the board had a long tradition of active involvement and it sometimes occurred that members of the board's standing committees issued instructions and requests directly to lower-level line managers, thus not only circumventing normal management channels, but also involving board members inappropriately in an operational role.

These kinds of general considerations, together with a review of organization and management structures at other medical centers, provided the

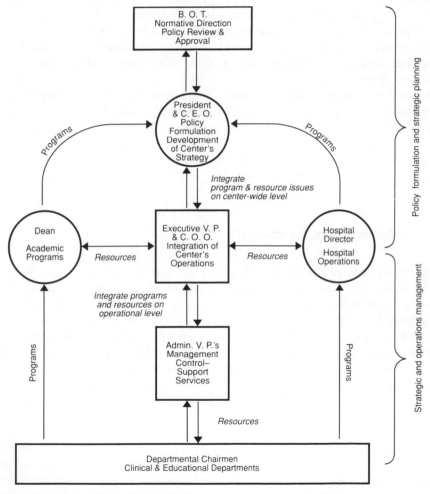

Figure 9. Design for management reorganization

general background. Ultimately, two major considerations fueled the deliberations on the need for reorganization. One focused on perceived weaknesses in the existing structure. The other pertained to the organizational and management implications of the strategic plan, particularly in light of the major commitment it made for an ambitious development effort that included a major rebuilding program for the Hospital.

In its final report to the board, the task-force recommended full implementation of the center concept. As a general framework it adopted the structure shown in Figure 9, which gave a clear expression to the idea of the center as an operating umbrella entity providing for distinct but organizationally integrated medical school and hospital components. It called for the appointment of a new president who would act as a chief executive officer of the center as a whole and provide the authority and focal point for institution-wide leadership. It recommended separating key management responsibilities (president/dean; executive vice-president/hospital director) and restructuring these functions with emphasis on effective integration.

In order to alleviate some pressing management problems, particularly in relation to the need for managing the hospital's rebuilding programs, the report also called for implementing a transition phase in the reorganization while a search committee was established to select a new president. The responsibility for implementing the full center model and finalizing all the specific organizational details would be left for the new president.

Comments on the aftermath

Implementation of the proposed reorganization did not fare well, in my judgement. The board of trustees adopted the recommendations for management reorganization in their entirety and set about looking for a new president. The task took longer than anticipated and meanwhile there ensued a period of some obviously predictable unrest. Unfortunately, things did not quite stabilize when a new individual was finally recruited. In fact, while the crucial decision of vesting in one office the authority and responsibility for unifying the medical center was being implemented, the spirit, concepts and key structural features of the proposed center model were being compromised.

For ultimately, unlike the open participative approach underlying the planning effort, the work on management reorganization was conducted in relative isolation. The conclusion did not convey the full richness intended. Even the language in which recommendations were expressed, although essentially correct, had to use familiar management terms, thus suppressing the variety of important underlying ideas. In retrospect, insights which drove the design were not shared, they were not part of the motivation that fueled implementation, and thus they simply could not succeed.

There was also involved the question of management conviction and personal style. In prior years, a progressive outlook of management had been introduced at the medical center and its practice was slowly evolving. It was committed to an open process, to participation and shared information, to planning, to team building, and to thorough professionalism. With all its possible faults, it was fundamentally positive in orientation and nourishing and constructive in attitude. It showed an interest in innovative management concepts, as well as in operational results, and was championed by the executive vice-president and hospital director, and a small group of associates. Together, they led the center to a period of management renaissance, which, among other things, made the strategic planning process itself possible.

Now the mood was changing. The new president (a surgeon) brought with him, in my view, a fundamentally primitive concept of management. As he took office, he moved decisively and quickly to consolidate power and, in a series of steps that were welcomed at first as signs of purposeful rigor, proceeded to reduce management variety that was previously distributed, deliberately, throughout the organization.

A fundamental conflict of personal outlooks and management styles was inevitable, and the executive vice-president, who by now was acting as president of the hospital as well, had to bow out. As he departed, other parts of the management structure, which had taken a long time to cultivate and develop, were dismantled and participation in policy formulation and decision-making was greatly reduced. Before long, a senior management structure emerged with a flat chain of boxes reporting to one source.

An autocratic management style was firmly imposed, which, while in itself is not necessarily always negative, in this case was intrinsically destructive of the collaborative management framework to which many had become committed. Its power–centred mode of operation fostered a climate that did not bring out the best in people. The management process was starved of essential variety; and all the symptoms of poor leadership, as manifest in widespread dissatisfaction, suspicion, mistrust, and non-cooperative behavior, rapidly proliferated.

Thus it was that my naive eagerness to contribute to enhancing the organization's creative viability was dealt a sobering blow. Given the idealistic intent, the personal emphasis is perhaps appropriate. For a while I blindly believed that the coherence and logical argument provided by using the VSM for guidance would dominate events, simply because they were correct. But logical correctness and managerial integrity are obviously not what institutional politics of power is about, and slowly but surely the destruction of a fragile management fabric went on.

Medicine continued to thrive at the medical center, but its management processes retreated, almost tangibly, to a dark age. Now, an observer, I watched with apprehension as basic management principles that I advocated, together with others, were being persistently compromised. Some vindication, perhaps, finally came, when as acts of managerial arrogance continued to mount, the new president was suddenly dismissed in a manner, which entirely uncharacteristic of the institution itself, was as brutal in its abruptness as the style of this short reign.

Reflections from the meta-level

In closing we come full circle to face the Viable System Model again, for there have been many lessons derived from its use. Where do they lead?

First, perhaps, a brief comment on a question that is often raised with respect to the efficacy of the model and its ability to address the complexity of human behavior, and the fact that, by necessity, it is of significantly lower variety than the organizations it seeks to describe.

The model recognizes structurally distinct levels of management as abstractions of functional prerequisites for viability. In real-life institutions, such a structure is embodied in something more like an amorphous cloud. Humans constantly move around. They come and go, appearing and disappearing at various levels, at different times. They constantly agitate, follow their own self-interest, and are motivated by ambitions that do not always align with neat, well-specified organizational demands. The immediate impulse is therefore to charge that, while the model could perfectly fit an ant-hill society or a system of a well-programmed automatons, it is simply not rich enough to deal with the evident 'mess' of human behavior, which by necessity transcends the model's tidy design. To argue thus would be to miss an important point; for as a meta-statement about viability, the concepts inherent in the model are invariably comprehensive enough to contain the notion of such dynamic behavior and, as it turns out, it is precisely this kind of redundant untidiness which underlies the ability of complex systems to fulfill the conditions for viability that the model prescribes.

Notwithstanding financial constraints and accounting convention, the fact is this: ultimately the stability of complex organizations is embodied in their very complexity. Their fundamentally redundant structure, with all their imperfections, makes it possible for them to absorb tremendous punishment. In fact, they can even survive bad managers! This inherent complexity is important to bear in mind, for when we map the VSM on to an organization we usually address the formal management structure. But it is not through that structure alone that management for viability, in its broadest sense, is mediated. It is actually achieved by the numerous interactions between individuals and the enormously complex web of communications that ties them together in a common, self-actualizing process. The VSM offers an excellent metaphor for this process. It provides categories of thought that are immensely potent in diagnosing management pathologies and guiding organizational design, but the model's tidy structure should not be mistaken for the actual 'real thing'. Ultimately, as long as an organization exists, it is in some sense viable and thus, by definition, it must in some way fulfill the necessary conditions for viability even when their functional embodiments are not immediately apparent, or when they exist only in partial form.

Complex living organizations are necessarily richer than any single specific attempt to portray them. But in the realm of management applications, the critical lack of sufficient variety is not so much in the model itself as it is likely to be in the way that the model is applied. It is the model–user combination that is crucial, and it is to this combination that Ashby's Law of Requisite Variety must apply.

The process by which application takes place is, therefore, important. To succeed, it should itself be constructed to embody the principles of viability and, in any organization, become an integral part of the management processes that the model describes. Creativity and innovation in application are thus essential for amplifying variety, while too literal and pedantic an approach is likely to prove brittle, perhaps even harmful, ever evoking the image of the sorcerer's apprentice's fate.

The focus on the user begs the question of the individual, again. It is precisely in pointing to the role of the individual as a participant in organizational life that the VSM offers one of its more important and challenging insights. For in every human organization, the basic unit of autonomy, the ultimate recursion, is the individual himself. It is the individual human who is the source of initiative, creative variety, and the constant unbalancing, agitating and unpredictable acts which ensure that the door to future possibilities is kept open. To this primary force, constantly seeking the viable route, driven instinctively to explore, sometimes blindly, for the next adaptive move, the rest of an organization can only respond. Whether such a response is designed to enhance this creative variety or suppress it is at the root of the question of liberty. Thus is closed an important loop that ties the individual to the communal and specifies the requisite conditions for the continuous viability of both. Here, the underlying spirit is more significant than the technical details. For the technical details can help us correct and refine the existing, whereas the spirit behind the model can help us find the way to the next important plateau.

In this regard, there is inherent in the VSM, and the tradition from which it emerged, a number of crucially important ideas: that the way we manage our enterprises, institutions, communities and societies has an inevitable impact on the well-being of the whole, as well as the identity, integrity, fulfillment, and self-realization of the parts; that in a true evolutionary sense, management processes have a great significance to human affairs since they constitute a loop through which humans can participate in guiding their own destiny; and finally, that the whole complex process is accessible to understanding and can be made a subject to conscious design.

These are very significant ideas, especially as we face a world of growing complexity, where critical questions of threatened viability are being constantly raised all around. The need for wise and effective counsel, for a fresh conceptual reorientation, and new tools for approaching complexity, is becoming increasingly apparent. Ultimately, the future of our planet is at stake and ensuring its lasting viability is, perhaps, the greatest current management challenge. To this very end, the Viable System Model has already made an important contribution.

Acknowledgements

Special acknowledgement is due to Alfred R. Stern, then Chairman of the Board of Trustees, Thomas C. Chalmers, then President of the Medical Center and Dean of the Medical School, Samuel Davis, then Executive Vice President of the Medical Center and Director of The Mount Sinai Hospital, and Raymond K. Cornbill, then Vice President for Planning, who with their vision and persistence made the planning process described here possible.

References

Ackoff, R.L. (1970). *A Concept of Corporate Planning*. New York: Wiley-Interscience.

Ashby, W.R. (1962). *The Set Theory of Mechanism and Homeostasis*, Technical Report 9, Electrical Engineering Laboratory, University of Illinois, Urbana, (1962).

Ben-Eli, M.U. (1988). 'Cybernetic tools for management: their usefulness and limitations', in *Science of Goal Formulation* (eds. Sadovsky, V. and Umpleby, S.). Hemisphere Publishing Corporation.

Beer, S. (1975) 'Fanfare for effective freedom—cybernetic praxis in government' (the third Richard Goodman Memorial Lecture), in *Platform for Change*. Chichester: John Wiley, pp. 423–51.

Bertalanffy, L. von (1969). *General Systems Theory*. George Braziller.

Pask, G. (1976). *Conversation Theory*. New York: Elsevier.

Peters, J.P. and Tseng, S. (1983). *Managing Strategic Change in Hospitals—Ten Success Stories*. Chicago: American Hospital Association.

Rosenblueth, A., Wiener, N. and Bigelow, J. (1943). 'Behavior purpose and teleology', *Philosoply of Science*, **10**, 18–27.

Part Three
Methodology and Epistemology

The Viable System Model: Interpretations and Applications of Stafford Beer's VSM
Edited by R. Espejo and R. Harnden
© 1989 John Wiley & Sons Ltd

12

National government: disseminated regulation in real time, or 'How to run a country'

Stafford Beer

*Chairman, Syncho Ltd.**

Drawing on actual experience, this chapter outlines the way to implement managerial cybernetic principles at the national level. Some examples of contemporary computer software relevant to such a process are also discussed.

Introduction

The approach of managerial cybernetics (Beer, 1981) to the regulation of very large, complicated, probabilistic systems is based on a number of postulates (Beer, 1979) which are effective in them all (and see Chapter 1 of this book). In particular, these postulates apply to the organization of government, to the organization of the enterprises that generate the national income, and to the organization of the human communities that constitute the nation itself.

The response to this contention is often incredulous, because the systems just listed seem at first sight to have nothing in common. In what sense is a ministry of education comparable to a ministery of ecology? How can small businesses be comparable to large, or industrial giants to public utilities? Are villages and cities and metropolitan centres comparable in any sense that matters? Even more: what do the organization of government, enterprise and community have in common between themselves?

*Visiting Professor of Cybernetics, Manchester University Business School.

Here are the fundamental answers to these questions. All of the systems mentioned (large, complicated, probabilistic as each is) have a powerful investment in their own identity. Each seeks to define its identity, to maintain it, to flourish out of a commitment to itself and a confidence in its selfhood. Each has an organization as we saw; and the primary purpose of that organization in each case is to preserve identity—in a word, to *survive*. Survival, moreover, is not a concept of stasis. Identity must change—and be gradually modified—as the world changes (it is called adaptation); otherwise there will be no survival.

Managers and ministers, however, regarding themselves as 'practical', as 'down to earth', as 'hard nosed', often seize on a single or overriding constraining feature of survival-worthiness. The chosen feature is genuine enough, but it is only one of many critical features. Those others are often forgotten; and the basic laws of survival are not even considered—except in the merely economic sense of viability.

For example, a democratic government, in order to survive, must renew its political mandate at the polls; a dictatorship must instead restrain the exuberance of the people. Enterprises must make a profit, or they will not survive. Communities must find ways to survive in balancing their books —between local and federal taxes, between remunerated and voluntary effort, between recreation and rip-off.

Because of this, governments formulate their objectives as 'winning the next election', or 'suppressing subversion' depending on their political colour. Enterprises formulate the objective of maximal profit, while community objectives are of the book-balancing, budgetary kind. Unfortunately, all of these objectives are narrowly focused. All are short-term, and all may be actually inimical to survival when pursued single-mindedly.

The democratic government cannot deliver on its election promises although its objective to be elected was met; while the dictatorship by the very oppression that meets its non-dissidence objective is fostering its own overthrow—sooner or later. The enterprise cannot entertain a new idea because it is so far sketchy and has no demonstrable 'bottom line'; that enterprise is overtaken by its competitors. Communities fail to change, because the book-balancing emphasis destroys adaptability: villages become ghost towns and urban centres decay. Yet, all these systems met their stated objectives.

The outcome is that the intention to maintain an identity (albeit evolving), that is to survive, cannot be encapsulated in mere slogans. Survival is a function of the *total organization* of any system that does survive, and includes its capacity to learn, to adapt, to evolve. A system that does all these things is called a *viable system*. The postulates referred to at the start are the natural 'laws' of any viable system. The third book of the trilogy

(Beer, 1985) shows how to use that model to diagnose structural and informational faults in the organization.

The recursive structure of the VSM

The first demonstration of the VSM is that all viable systems contain viable systems, which are themselves of identical cybernetic organization to the totality and which are largely autonomous. We say 'largely' because autonomy can be exercised only within limits imposed by the cohesion of the whole. The VSM has much to say about this in detail, and devotes a theorem to it.

Thus in government, Education for example contains primary, secondary and tertiary components—all viable systems—and all largely autonomous in the sense explained. These in turn have largely autonomous viable systems contained in them called schools and universities, for instance. In Enterprises, the holding company may have largely autonomous operating divisions, and they in turn largely autonomous companies or plants. In the largely autonomous provinces or states of the Nation, there are largely autonomous cities, each having largely autonomous fire and police departments, each with largely autonomous districts.

Let us take it that we have established 'largely autonomous' as a technical term in VSM parlance, so that the tediously repetitive adjective may be dropped in future. Figure 1 shows the shape of the Viable System Model itself, where any of the wholes I have discussed can be thought of as containing the parts (only two are shown) that belong to the next level of recursion down. 'Down' itself refers simply to organizational containment: the VSM is not essentially hierarchical, it is essentially an interaction of subsystems. For details, reference should be made to the book already cited or, conveniently, to Chapter 1 of this book.

The point in printing the bare structure here is to demonstrate the cybernetic principle of organizational recursion involved: namely that the viable system's components are structurally identical with the whole. So would their components be in turn, if we had sufficient optical resolution to show them. It is for pictorial and not logical reasons that the VSM draws only one pair of recursions at a time. However, every component and every connection to be found in the total picture of Figure 1 will be found reproduced exactly in the included VSMs that stand at 45° to the main axes, as will the connections between each of the subsystems across the two recursions. It is this mathematical property (called isomorphism) that entitles us to talk about 'laws' of the viable system. And we can follow through as many recursions as there are pairs: A & B, B & C, C & D, and so on.

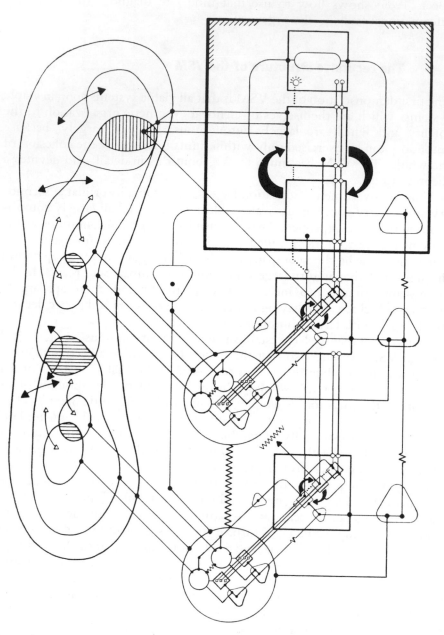

Figure 1. Bare framework of the Viable System Model to demonstrate identical recursions of the whole included in each of two parts

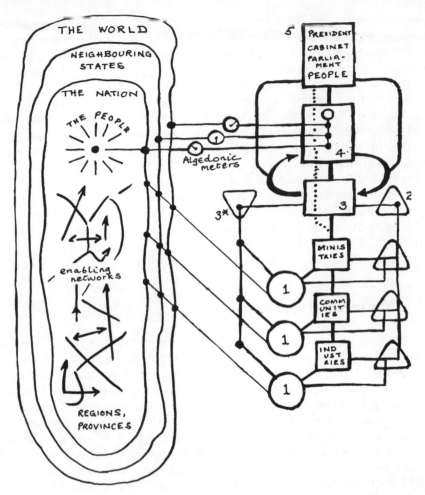

Figure 2. A first mapping of the nation on to the Viable System Model

These theoretical considerations, briefly considered above but precisely exemplified in Figure 1, have practical consequences of immense power when it comes to large-scale applications. Teams of people are trained to investigate the organizations in which we are interested. At the national level, we chose to discuss three viable systems that are components of the Nation itself. They are shown on the right in Figure 2.

Now if we were to enquire into these three major organizations by asking them for their 'organization charts', covering every level of recursion, we should find that we had hundreds of charts, each as idiosyncratic as the 'family tree' of some noble lineage. Such presentations are without coherence. Any one chart of any one large organization, if reproduced entire, would cover the whole side of the multi-storey edifice housing the enterprise, and no-one would be able to review its viability. But if all such charts are mapped on to a standard model—the VSM—this becomes possible, severally, and also in interlocking recursions.

Thus if, for example, we wanted to 'computerize' the Family Tree Organization Chart model, we should need to go into endless differentiations of structure, relationship, regulatory function and managerial need. Indeed, this is exactly what has been happening for over 30 years, and a good deal of chaos is the result. But if each and every organization, at each and every level of recursion, is first mapped on to the VSM, we can handle all of them according to a common rubric. For instance, as we shall see later, the investment made in a major regulatory computer program for the VSM as such can be used for every organization and every recursion without modification, regardless of content. This is because the descriptive organizational language of the VSM is the same for any viable system.

The practical approach

The mapping of actual organizations on to the VSM is a matter both of cybernetic technique and of profound knowledge about the particular organization under study. Thus any given investigatory team must meld together cyberneticians with local people. The latter are folk who, between them, know how the whole place 'ticks'. Then they must include managers and workers, and functional people too—such as an accountant, an engineer, an administrator . . . depending on the type of entity reviewed.

Do not be misled by the recursive simplification of the diagrams. Each of them deals with just one pair of recursions. As we have already seen, each of the three viable systems shown in Figure 2 contains viable systems —and so do those in turn. It follows that a sizeable number of in-house people need training in the VSM language, and in the cybernetic paradigm of management for which it was created, before they are ready to act as confident collaborators with the cyberneticians joining the team. This is not a task for the cyberneticians themselves. Training is not their special expertise. Moreover, if the in-house group do not know what is happening in advance, they will be either hostile to the innovators or overawed by them. In either case they fail to make their vital contribution. We need trainers capable of specializing in the cybernetic paradigm: such people exist already; let us hope they proliferate.

Consider now a practical example of what would happen. The constitutional regulatory system of the Nation is Recursion One. This includes (Recursion Two) ministerial government, communities, and the wealth-producing industries, public and private. Select from Recursion Two, public industry. This includes (Recursion Three) Water Supply, Energy Supply, Mining (perhaps), and so on. Select from Recursion Three, energy. This includes (Recursion Four) the viable systems of oil, gas, electricity, nuclear power (perhaps). A VSM team will need to map each of these industries on to the VSM, and in doing so to visit each of the component companies or plants of each: that will be to map at Recursion Five.

The degree of complexity now adumbrated may sound alarming. It is not. In the first place, the multiplicity of basic activities encountered across the country (in this case incorporating five levels of recursion, but in other cases maybe more or less) have to be managed in any case, and have to be incorporated into the governmental perception of the national weal in any case. The cybernetic approach is already making matters easier in two ways.

Firstly, by using the same model, the same regulatory language, and the same information technology across the board, it becomes much easier to synthesize a view of what is really happening throughout the nation. Secondly, because the recursions are richly interconnected, inside each other, models of the higher-order recursions can rapidly be integrated once the basic (say fifth recursion, in the example) systems have been mapped. The first of these desiderata was once met by accountancy, but see later the question of real-time regulatory processes for today's world. The second could not be met beyond the ordinary numerical device of aggregation: the provision of totals and averages. In managerial cybernetics, the VSM is passing to-and-fro among the encapsulating recursions not merely aggregate numbers, but *Gestalten*—whole and integrated patterns—of viability.

Here is a final point on the size of the proposed task. Because of the way in which accounts are currently prepared, we entertain a managerial illusion that wealth is being produced, or important results are being attained, at each level of recursion. We look at 'regional sales' for instance, or at qualifications obtained in the private as compared with the public sector of education. In the end, as the board of directors or the presidential advisors sit in their top-storey offices reviewing figures especially prepared for them, it might seem that the figures are being passed through the window by angels sitting on the clouds. Although these figures are extremely valuable, they do not relate to actual operations at the levels cited—neither in the regional offices, nor in the clouds. These figures are aggregates of the results at the lowest recursions.

People take plenty of credit for them higher up, because they have been 'organizing' things. Plenty of costly effort is put into massaging the basic data so that this 'organizing' of things is manifestly justified. All of this glossy activity creates the illusion that each level produces. Of course it does not. What it does, if it is effective, is to generate a measure of added-value, deriving from the informational energy of synoptic vision. Even then, things are fine only so long as the basic operations do well; see what happens when they fail or fall short of expectation. The illusion is proven to be such because only credit and not discredit is equally shared. The integration of a set of recursions of VSMs will not underwrite the illusion. It creates the interlocking model fast, as a corporate whole, as a seed crystal instantly crystalizes a super-saturated solution.

Now the output of the teams is twofold. In the first place, we expect a VSM-like version of the organization at each level of recursion. And if that organization has weaknesses (and which organization has not?) we expect that the modelling process will generate a succinct list of them. Because the VSM sets out to give a necessary and sufficient account of the laws of any viable system, it is a tool of intense diagnostic *power*, as many organizations have discovered. (Note: if the VSM language is used loosely and merely descriptively, then of course its power is lost.) So we expect some prescriptive suggestions too. After all, the management is itself implicated in these studies—and so are the workforce representatives whose members will doubtless bear the brunt of any substantive operational change.

The second output from the teams' work is a set of quantified flow charts (QFCs). These are iconic representations of the wealth-producing, or result-generating, parts of each organization. An actual example is produced at Figure 3. Again, the intention is to obtain, display and communicate vital information in the most powerful yet economic way. Flows linking major operations are depicted proportionately to the rate of flow, and stocks or process-rates use proportionality too. The aim is to state in iconic (visual, recognizable, stereotypical) terms what are the key functions of each operation. It is wise to put effort into the design criteria of iconic representation, because it is amazing how much incomprehension, ambiguity and general muddle can be eliminated by the use of ergonomically satisfactory standards. Colour codes, curvatures, proportionalities, radii at corners . . . all these help to provide a perceptual framework, an iconic language, that aids actual people in understanding the many new systems with which they are confronted. A new subsidiary has been bought? Its set of QFCs are much more accessible to manager and worker alike than its balance sheet.

In particular, the QFC highlights major flows, and also process *bottlenecks*. The purpose of so doing, apart from making perceptual rather than numerical data instantly accessible, is to design monitors for the management system.

Most management information systems try to monitor too many items, so that their human regulators are overwhelmed. The cybernetic approach determines to isolate whatever is important to the manager to know at any

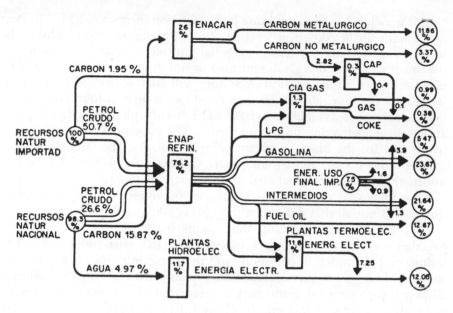

Figure 3. Quantified flow chart (ENAP, Chile, 1972)

given moment. We shall see later that the major tools for this concern real-time technology and the theory of probability. But our teams are here concerned with setting up relevant structures.

The mapping of the organizations on to the VSM retains all the necessary complexity of viability with all the possible simplicity of topological mathematics—the basic diagram. The QFC in turn offers necessary complexity in operational realities, depicted by a uniform, iconic set of conventions. And the key conclusions of the QFC work are the agreements that the whole team reaches as to which major flows and which potential bottlenecks shall be monitored. There are usually about ten to twenty of them at each level of recursion, although some may not be simple measurements but more elaborate ones. This is like saying that if you wish to monitor a succession of circular areas, you had better measure each radius, and transmit this figure squared times pi to the monitoring system.

It is worth emphasizing why the iconic approach to information transmission is regarded as so important. Some senior people take umbrage at the notion of 'pretty pictures' rather than 'hard figures' (which are in any case available, on call). The brain's perceptual machinery is, however, dominated by the visual mode, including amazing computational devices in the retina itself, and flowing back to the huge occipital lobe of the brain at the back of the head. Thus it is that even financially trained people on boards may be observed scribbling little charts on their scrap pads. 'It went up, it went further down, it levelled out, it's going up again'. This is the sort of impression that has managerial impact. Certainly a numerical statement presented to a greater number of significant figures than could be measured in a national physical laboratory will not have so potent an effect. We readily perceive relative size, relative slope, relative colour, and relative movement, whereas tabulations have to be disentangled from their level of arithmetical abstraction into these forms. The cybernetic approach offers to do that for the brain in advance, by automating the tabulations into iconics—and at best, animations. There will be more on this later, so it is not mere cosmetics to emphasize the issue straight away.

How long will all this take? What sort of investment do we need to make? It is not all that difficult if the task is taken seriously.

Take the training. Will you train teams from oil, gas and electricity together or separately? It is a matter of managerial judgement. The industries have much in common, and their technologies are not incomprehensible between them. How many plants are there in the country? That tells us how many teams may be needed in each industry. One week is a reasonable unit of training. If oil needs five teams of six persons each, then a class of thirty is enough. But if each of the energies needs one or two teams, then maybe they can be trained together. Managers and workers active at this level should not be spoonfed: that is patronizing. They should study properly prepared materials in advance, and then the training week will do. After that, we should expect a transdisciplinary team as defined to complete a basic set of QFCs, including the identification of the major indicators to be monitored, within a week.

In the example (Figure 2) we have been considering the three second-recursion units of government, community, and industry. Each would have its own set of trained teams, although there might be only three fully qualified cyberneticians (say one for each set), each of whom is in charge of one of the operations, but who work closely together in the synergistic interests of the whole project. In this case, we might expect the whole VSM mapping and QFC designing to be completed in about a year. This of course assumes that excellent relations obtain between the project and the multifarious organizational units involved. Then the meta-objective to achieve those good relations must be the cynosure of the training programme, within which the team mixture of managers, workers, cyberneticians, and so on, becomes bonded and dedicated to the task.

Obviously the activity of these highly specialized teams is not to be considered as 'research' or as an 'investigation'. It is a strongly focused, well-defined job. Then it can be formulated as an *expert system*. Such an expert system, dedicated to the mapping of actual organizations on to the VSM, is under development by Syncho. The name of the software is Viplan.

On eudemonic regulation

To this point we have been considering how to structure (by VSM) and how to measure (by QFC) the wealth-producing or result-producing components of the Nation—which in VSM parlance is called System One. Systems Two and Three are concerned entirely with the regulation of System One, and are not a special topic for us here (the books that have been referenced deal with this).

Let us turn to System Four, which handles the interaction of the whole viable system (that is the Nation, in our case) with the outside world. Of course, System One deals piecemeal with its own set of environments, as a matter of local adaptation; but System Four acts for the nation as a whole. For instance, the Minister of Education is part of System One, whereas the Foreign Minister is part of System Four. But System Four is especially concerned with an environment that includes the future of its own people, as depicted in Figure 2. Each component of System One is involved with the home milieu; but overall responsibility for the people's future is a regulatory function shared between the people themselves and the government agencies that act for them.

The problem is how to measure people-satisfaction. What is the QFC for 'well-being', which Aristotle called eudemony?

Standard approaches, summed up in public opinion polls, have these demerits:

(1) They are reductionistic: that is, they try to measure the separate 'components' of happiness (satisfaction with the job, income, party leadership, sex-life) whereas happiness in Aristotle's sense is indivisible.
(2) In dividing the indivisible, they are forced to invent categories, which may not be the categories in which the respondents are accustomed to think. Access to great art, to soccer matches, to natural scenery, to pornography, to religion, to television may or may not count with any individual—and will certainly not carry similar weightings with us all.
(3) Having foisted some categories on to the respondent, they next ask for a crude digital measure (agree, agree a lot, agree very much)—say a seven-point scale with 'don't know' causing statistical problems at the mean. How do we distinguish between the ignorant, the illiterate and the over-scrupulous? The objection applies even more strongly to 'yes', 'no', 'don't know' questions.
(4) In attempting to show the respondent what counts as an answer, there is a real risk of training him/her as to how to answer.

The proposed solution is simple, but not simplistic. If people do not always know why they are feeling happy or sad, they do know that they are so. Fact is, they are doing computations on components and subjective categories with nonlinear metrics *inside themselves*, and they do not have conscious access either to the internalized model or to the weighting system or to the process. Let the respondent do the heavy scientific work for us!

An algedonic measure (from *algos* = pain, *hedos* = pleasure) offers no analysis of the eudemonic condition, but only measures it. The algedonic meter has none of the demerits already listed. It has these merits:

(1) Respondents are offered a task so straightforward that it is not threatening.
(2) They are very deliberately told that they will not be asked to explain their setting; the setting itself is the end of the encounter.
(3) The measurement system is analogue, and therefore does not pose difficult distinctions: it calls on a 'right brain', intuitive response.
(4) Nonetheless, it generates a 100-point two-digit index on the reverse side.
(5) It uses vernacular language, rather than an artificial or academic one— as direct a reading as can be got, short of a punch on the nose.

What is the use of this measure, if it is not susceptible to analysis? It is intended:

(a) To discriminate between sex, age, region, education and social class— which are accepted as objective *demographic* categories.

If all the young people are happy, and only the departing are miserable, we are doing well—unless it is a 'seasonal effect' of ageing. We shall eventually find that out: possibly a major discovery. Or if twice as many educated are miserable as compared with the less educated, what then?

(b) To observe trends and to correlate them with managerial options.
(c) To detect incipient *instability* in the sense of any population's self-image of well-being: a vital political input, hitherto created, monitored, and reinforced by the media rather than by the people themselves.

The algedonic meter is shown at Figure 4. The quantified flow chart that it generates is precisely the set of readings collected, broken down to a next lower level of recursion by the demographic categories used. It is too early to quote results from early experiments, except to say that certain pollsters reacted by saying that they could do much better then discover whether people felt happy or not: they could apply a wholesale analysis to discover why. This of course misses the whole point of the above discussion.

Similar arguments apply to the other two algedonic meters indicated in Figure 2. They represent attempts to measure the satisfaction with the country exhibited by near-neighbours, such as is reflected in border disputes and trading opportunities, and by the world at large, as reflected by trade, culture and reputation on civil rights for instance.

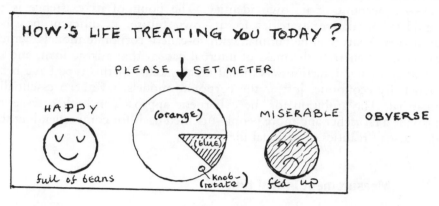

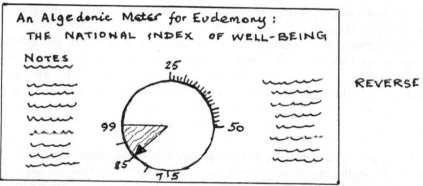

Notes for Interviewer
1. Present meter totally blue. People must then move it, (equal blue/orange may seem 'neutral' but offers in fact a 'don't know' option.)
2. Record sex, age region, education and social class in advance.
3. Say nothing after presenting meter.

4. Answer questions in elucidation only.
5. Ask NO questions yourself, nor any comment on the setting.
6. READ METER TO NEAREST TWO DIGIT NUMBER AND RECORD SCORE.

Figure 4. Algedonic meter ©

Finally, there is the QFC originating in System Four, that represents a model of everything that is here discussed. It is *self-referential*, that is to say—an account of its own identity. The point of its existence is to simulate alternative strategies for its own future, in different possible scenarios. Many will be familiar with so-called 'corporate models' which are able to penetrate the mass of internal data within a large firm, and to examine their interactions. National data systems of this type have been essentially economic, just as the corporate models have been essentially financial. The ambition for the cybernetic approach in both cases is to provide a far richer and more people-oriented basis for comprehending the day-to-day realities of national life.

Measurement in real time

In the last two sections a number of Quantified Flow Charts has been discussed in concept: one set concerns managerial regulations, and another concerns eudemonic regulations. These have been added to our evolving diagram in Figure 5. This vertical line of rectangular boxes represents measurement points for the two sets, which are piped into the two circles that collect the measurements. These measurements are made at the designated points of critical flows and critical bottlenecks for the managerial regulators, and across the algedonic 'potential differences' of societary concern for the eudemonic regulators.

It is a crucial question as to how frequently these measurements should be made. In the inherited system they are made on an epochal basis: each month, quarter, year. It is central to the cyberentic thesis here advanced that they ought to be measured *continuously*. Then the advocacy turns out to say: measure daily. For although a day is itself an epoch, it is sufficiently small as to generate time series that approximate to a continuum. We are effectively in real time.

Critics often argue that government does not need such rapid informational input, and if it had such a thing it would over-react. The first complaint is basically a statement of stereotype: 'everyone knows' that such instant input is not needed because no-one has it, nor can they see how to get it—officially, bureaucratically, that is. On the other hand, everyone knows (without quotation marks) that government is in fact driven, as before a storm, by instant information channelled through the mass media, and often generated by them. This makes nonsense of authenticity. The official bureaucratic information system spends its effort in trying to keep pace, to justify its masters, to excuse the mistakes that may not even have been made. The situation is chaotic. The complaint as to likely over-reaction is merely risible in this context. A properly designed cybernetic

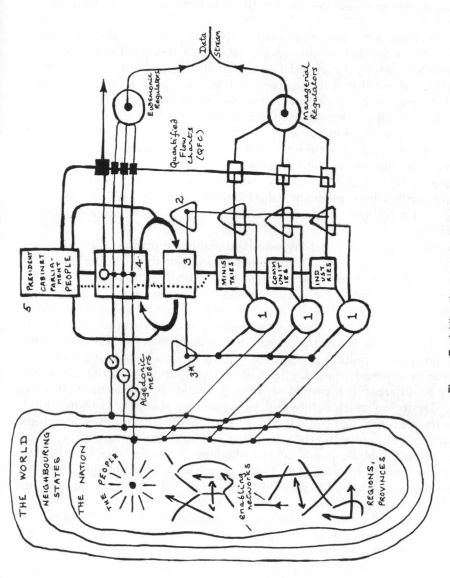

Figure 5. Adding the measurements

system does not over-react, because it has properly calculated feedback functions that smooth irregularities and impose delays that are systemically appropriate. The present instant-response system, which has not been properly designed (nor designed at all), is as over-reactive as could possibly be.

Forgetting then the critics, what is the true case for real-time management? Consider the monthly epoch. Managers are proud if they have last month's figures by the second Tuesday of this month. It is far too late to do anything about any of that, except to learn. We learn from our mistakes; and, crying into our beer, resolve to avoid those particular errors in future. We learn from our successes too, and drink up. But 'the moving finger writes', and nothing can actually be changed. If, however, we are operating today on yesterday's figures (approximating today's, and close to real time), the situation is quite different. It remains the case that we cannot change what happened yesterday. But what we can learn concerns something that has been gradually salted into this text already, in speaking of day-to-day realities. It was even named. It is the recognition of *incipient instability*.

If what happened yesterday, and is probably happening still now, is not so much a triumph or a disaster but a rocking-of-the boat, and if we can detect that at once, then we may be able to restore the equilibrium. The disaster may never happen. The success may be assured.

At last we may return to a concept of management that has the power to manage, that is to say, it may do something now so that the future will be different from the future that would otherwise have been. This definition is my favoured definition of planning, which is not a matter of toying with scenarios (a support function) but of taking decisions—so that the future may be different. It is easy to see how this holds for the future that ought one day to be, which is the topic of normative planning. It readily holds too for the future that could be (if we work hard) fairly soon, whose topic is strategic planning. But the future that will be almost immediately, which is supposedly the subject of tactical planning, is foisted upon us—because our information is so lagged. This 'future' has already happened by the time that its likelihood is signalled, simply because the signal itself is still going through the works.

We may 'return' to the power to manage in the short-term: 'return' is proposed because it was once possible to observe activities under command, dislike the outcomes, and issue new orders instantly. In this way, managers quelled incipient instabilities. The inability to do this today is an artefact of our immensely cluttered, bureaucratic and inept systems—computerized though they may be. Consider the absurdity of a government's employing an army of econometricians in order to forecast (from lagged data) where we already are. It is what happens. And because the forecasts are often wrong, we decide our plans as proceeding from an initial position that we never occupied in the first place.

The point of collecting all the data points daily from the QFCs, and channelling them into a steady data stream, is to be instantly aware of a structured reality. The data stream has to be revitalized within that data structure—provided by the logic of the VSMs and QFCs. That logic is stored in a computer, together with data reference points for every indicator measured. These data points were established when the trans-disciplinary teams agreed their original findings.

For each point identified and measured, the teams established a normative (should be) and a strategic (could be) target. What the tactical result (will be) actually is arrives virtually as it occurs.

Comparisons of these actual results with the stored expectations at each level of planning provides a set of three indices for each arriving data point. Each is expressed as a two-digit number. The task now is to detect incipient instability in the data streams, and this is the task of *Cyberfilter*: a computer software package. As to its criterion of instability, it is not merely picking out exceptions to the norm, and not only measuring variances from means, these being traditional accountancy practices. Cyberfilter has the criterion of discovering instabilities that have import-ance to the manager, in terms of the possibilities of corrective action before any *damage* is done.

The purpose is not so much to worry about the deficiences of the actual results as measured against the planning targets, because these were expected when the norms were set up, and planning decisions have been taken precisely to close the gaps. What matters is the change in the index as compared with recent performance. Each index is represented by its own time series in the computer. Here comes a new reading from the data stream. Let us follow through what happens.

Cyberfilter first examines the new value against its history. By simple statistical tests it determines the likelihood that the input is genuine. If it seems not to fit the statistical parameters of the population to which it supposedly belongs, then its validity will be questioned. Suppose that the new arrival is accepted. Then its planning norms are consulted, and it is turned into a set of two-digit indices.

Now take just one index, newly calculated, and set it into its own time series. The program now uses a technique (Harrison and Stevens, 1971) to estimate four probabilities. How likely is it that this point is merely a chance variation? How likely is it to be a transient (a bit of 'noise' in the system)? How likely is it to be contributing to a change of slope? And how likely is it to represent a step function? All four of these estimates are made for every newly arrived index, using Bayesian probability theory. The device is known to control engineers as a Kalman filter.

It is in the statistical nature of things that most arrivals are either chance variations or transients, and these fluctuations are of no importance to a manager. S/he is not told about them. But if a slope change or a step change seems likely, then this may signify incipient instability. Thus we have an importance index to send to the manager's desk. This manager will be in System One of the viable system using Cyberfilter for operational indices, or in System Four in the case of eudemonic regulators. Let it be clear that this result does not wait around to be included in an epochal report. It goes straight to the responsible manager's desktop computer screen. Because of the rules of local autonomy built into the VSM, no-one but the responsible manager has access to this message. Of course, the whole Cyberfilter system is fully automatic—and it is extremely complicated in its mathematical-statistical entrails. But once the nation or the company commands it, then the importance meter for all viable systems, at every level of recursion, is instantaneously in place.

Cyberfilter is part of a larger software 'shell' called Cybersyn. This has several functions, the main one being to examine the joint effects of the notifications of importance going to managers, as they impinge on the tactical, strategic and normative plans. The software also has a special facility to generate VSM graphics, to aid the original teams, and later reviewers and improvers, in designing the models: it incorporates a methodology to analyse needs and synthesize outcomes.

Brief mention must be made of the algedonic signal the Cybersyn may generate. Part of the original design process of the QFCs within a given VSM was (it was not then mentioned) to estimate the elapsed time that the manager would need to correct a damaging instability in one of the chosen indicators. It is this person's own responsibility to determine this, under advice. Of course if s/he sets the figure too indulgently, it will threaten the viability of the next higher level of recursion, and the senior manager to whom our manager reports will want to argue the matter. But once the reasonable 'repair' time has been agreed, Cybersyn will monitor the consequence of sending an importance signal notifying dangerous incipient instability. If the trend is not corrected within the agreed time, an algedonic signal goes to the next recursion upward. This is in the interests of corporate cohesion that delimits local autonomy. The algedonic 'cry of pain' automatically indicates that help is needed. In theory, the same signal—if not dealt with effectively this time—will be passed on, after the appropriate delay, through higher recursions until matters are in order. This is an organic cybernetic corrective to a common malaise in contemporary management: a loss of control feeds on itself to become worse and more widespread—because algedonic information is missing at the top.

In the case of processes that are inherently dangerous—such as nuclear power stations should they become unstable—the algedonic delay times may be set at zero. In this case, all recursions would receive the alert simultaneously.

Cybersyn as presented here is an integral aspect of the workings of management: it articulates the management process. Managers have always needed to assess the significance of the information presented to them by their staff, and now they have the technological facility to refine that mental process scientifically. To call Cybersyn (in the current jargon) a 'decision support system' would be to miss this point.

The current version of Cybersyn, which is marketed by Syncho as a software package, does not include a VSM generator. There are many versions of this in being around the globe. The most powerful is called Viplan (referred to earlier).

The Management Centre

Provision has been made to feed back importance signals to the responsible manager as soon as they are detected: s/he does not have to call for them. After an absence from the desk, messages may be found on the screen. And if some of these are algedonic signals, deriving from a lower recursion, they will be highlighted.

Management, however, is a collegiate process. Thus although in this cybernetic design each manager is 'on-line' to important signals that will alert him to questions that concern his own responsibilities, it is still necessary to feed the 'college' of management with suitable information. This has always been done by way of the epochal report, to be followed by a management (or typically a board) meeting. People attend such meetings with solemn ritual. They have special rooms, special rubrics, special arrangements for eating and drinking, even (sometimes, as in universities) special clothes to wear. The papers they are carrying are also special: they are typically mendacious, or at least they are irrelevant to urgent needs.

Social anthropologists may reflect upon the reasons for this archaism in an electronic age. The fact is that the information under consideration is out-dated. In the public domain a year's lag is common, and nine months' usual; a ten-year lag for a major study in the USA was recently observed, ironically dealing with societary information; in India, a cabinet minister complained of a quarter-century lag (because of census procedures) on statistics relating to educational procedures for children! Internationally, of course, decision processes often drag on for decades. Given current technology, not only in weaponry but in such activity as produces acid rain, these delays threaten us with planetary extinction.

The 'environment of decision', as I have called the boardroom, has to be changed. 'Paper is banned from this place' I proclaimed in 1972 (Beer,

1981). What we need is a roomful of electronic screens controlled by computers. In the 'sixties, the place was called an Operations Room—because World War II invented the idea of real-time control for management, and provided such environments under that very name. Today the term ought to be discarded from general use: its bellicose connotations are by now too often too severely felt. Perhaps the term Management Centre will prove acceptable.

At the end of the eighties, the idea is just being learned by innovative managements around the world. Even now, however, vital aspects are generally misunderstood. Some Management Centres, under whatever name, are no more than elaborate displays of background facts, such as may be found—and rightly so—in museums. But boardrooms ought not to be museums, be there never so many oil paintings of bygone chairmen decorating the walls.

Here is the key conception. The Management Centre is the environment of decision in which the board or college of managers reaches out into the processes for which it is responsible. The Centre is like a corporate brain, in that it extends the nervous system of the whole organization into the world, integrates the results of its findings, and takes motor action on this synthesis of sensory input. *All this happens in real time.* Under conventional management protocols, the closing of this loop—realities to reports to analyses to presentations to deliberations to decisions to instructions and back to realities—simply takes too long.

To create this Management Centre, we shall need the Cyberfilter device, contained within Cybersyn as already described. In Figure 6 we see how this software assembly feeds the Management Centre with instant information, judged of course by its importance to the highest level of recursion. The importance signals and the algedonic signals are generated and sent out of the system to the responsible managers (at the bottom of the chart). They are re-assessed for the top recursion that uses the Management Centre, according to the QFCs that define its VSM, and are also filtered through the strategic and normative plans—to gauge novel effects. The first set of notifications appears on specially designed screen displays, while the second set passes into the simulator (at the top of the chart). Records of all that is happening are kept in a data bank, unusual for its terseness and relevance. Remember that most of its data will never be displayed directly to managers: it is needed to generate importance-filtered information alone. Finally from Cybersyn, as Figure 6 shows, is the output of the VSM generation already discussed.

Before describing the details of the Management Center itself, let us review the 'other software' to be made available. Many models of the national weal already exist, although most of them are purely econometric. A most impressive and more socially robust an example has been developed

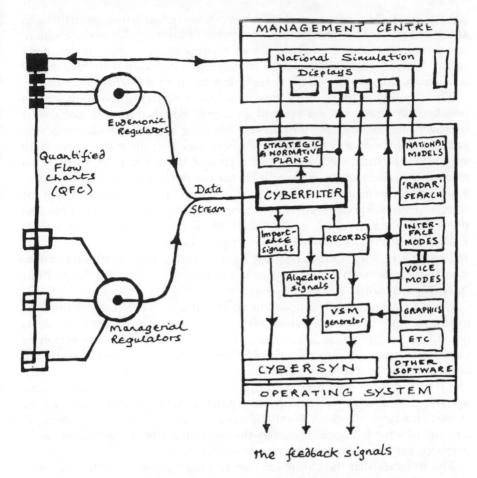

the feedback signals

Figure 6. Handling the daily data stream

by Statistics Canada called SERF (Socio-Economic Resource Framework). This is a massive system, which took 15 years to formulate: it ought to be in wider use. By contrast, a 'mini-system' called Stella comes from the United States. This enables small, experimental *ad hoc* simulations to be designed quickly so that alternative solutions to problems may be examined on the spot, albeit on a small scale. These are the kinds of software, large and small, that support the National Simulation of the Centre.

The software dubbed 'Radar' Search is able to investigate the multi-dimensional array of national statistics with the intention of isolating problem areas. Certainly these should have been notified via Cybersyn. It is inescapable, however, that the QFCs selected to give measure to any organization are simplifications of reality, however expertly they may have been created. The search package, having recognized that something is amiss and yet not so far revealed, is capable of hunting down the cause of the trouble through the whole set of recursions. The British company called Metapraxis markets such an approach through software called Resolve (it markets the physical apparatus for such a boardroom too).

The ergonomic question of man–machine interfacing has long been a problem. We have not yet reached a stage at which ministers and senior managers are comfortable with keyboards. The interface modes indicated in Figure 6 refer to software of an intensely user-friendly nature, and these specifically include modes of input activated by the human voice. This technology is now sufficiently reliable for use in the Management Centre, where a fairly limited vocabulary is adequate to the commanding of screens. Even without this desirable accessory, the Centre does not have to use full keyboards. Such systems can be driven by an infrared switch exactly similar to the now-familiar remote control used to operate the television set. This is made possible by an interrogatory computer mode, which asks the viewer to make selections from a series of single digit 'menus' of options.

Graphics are especially important to the design. Our main channel of information is visual, as explained earlier, and every effort must be made to provide good images. In many cases it is better to use back-projection of slides in a computer-driven carousel than to create electronic images by computer. Photographs of places, plants and people, maps, and especially of QFCs, are very well presented thus. Time series, which need constant updating, look very well when computer-generated, and presented in colour on a large screen. A graph plotter is a useful adjunct, if people wish a record of what has been created by the meeting (rather than tendentiously prepared for it in advance, as convention has it).

The software that the computer uses to juggle all these facilities constitutes a 'shell' that must exert great power. In particular, it needs to be structurally organized in a recursive mode, and to be programmed in an advanced computer language (fourth or fifth generation)—otherwise it will never be completed. But such systems do already exist. The Canadian approach called CATALYST meets these requirements, for example.

There are other useful adjuncts. Inputs to the Centre's screens can be switched in from real-time TV broadcasts, from satellites, by telephone line from screen-displayed public information services, and of course videotape machines. The whole plan is based on an extension of a nervous system to reach out into the whole world. This is stressed for a second time: unless the understanding of this is clear to all concerned, there is a real risk of concluding with a meretricious PR exercise.

As to the design of the room itself: how adventurous are people willing to be? If the room is an environment of decision, served by a nervous system that is countrywide; if the management team is in close discussion, served by electronic artefacts as described; if paperwork is not needed, and is even banned . . . then there is no need for the boardroom table and its matching chairs set out as if for a banquet rather than serious work. The organization that lets go of these talisman accoutrements of power will demonstrate that it has really understood what real-time management is about. The design of room that is then accepted will look more like a clubhouse than a boardroom. Remember that each comfortable armchair can be furnished with its own electronic screen as back-up for the communal high-resolution screen, say four or five feet square, on which attention is focused.

The final question here is perhaps the first question that management needs to address. Is the Management Centre, like the old boardroom, commissioned for use only once a month? Or is it indeed a clubhouse, open to members all the time? In my own incarnations as a senior manager, I used my own office as a clubhouse for my colleagues from the time that the building officially closed for the next two or three hours every day. Drinks were served and news was exchanged. This happened throughout the 'fifties and 'sixties. If the electronic facilities we now have had then been available (although even then we had daily computer printouts), my office would have been the Management Centre here described. As it was, it simulated one. Other managers have certainly done comparable things. Then surely it is feasible to consider the practical fruits of such 'informality' as better products of the formal board than 'the minutes of the last meeting'.

Next: if real-time messages can be directed to the desk-top screens of the responsible manager, as has been shown, then real-time messages for the collegiate higher recursion can be sent to the Management Centre—as they arise. And if this Centre were being used as a clubhouse, then (in a sense) the cabinet or the board would be in continuous session. This idea is commended for further exploration. When we look in at the empty boardroom of today, showing it to our visitor, and lowering the voice out of respect, we might well stop to consider how we ourselves would fare if our brains switched on for just one afternoon a month. Of course and in fact, the cabinet and the board are corporate brains that are switched on all the time. But they could be better served with sustenance than by gossip: that is how they tend to work today.

Completing the cybernetic scheme

Perhaps the major cybernetic notion here, apart from the Viable System Model as a particular tool, is Ashby's Law of Requisite Variety. This asserts that only variety can control variety: in other words, that management must somehow match the variety—the measure of complexity—that it faces. In any ultimate sense this is impossible. All legislation leaves loopholes, because no-one can legislate for every case. Then judges are appointed to determine the law in a given case, and a manager uses skills and techniques to find the cases that are not obeying his laws (known nowadays euphemistically as 'guidelines'). The scene now completed in this statement offers means of reducing variety to manageable proportions —by the use of uniform models, the VSMs, uniform techniques, the QFCs, and uniform computer programs, especially Cybersyn. It goes on to provide ergonomically sound interfaces with managers, individually and collegiately.

The next most powerful cybernetic notion involved is the principle of error-controlled negative feedback. If something is going wrong, modify the input to the process that is producing the unpleasing output, so that the output comes back into a satisfactory state. This is better than wading in to the operations themselves wielding a sledgehammer. Such a process lacks Requisite Variety: there are not enough sledgehammers to go round. Then this is why states who do not understand cybernetics, intent as so many are on controlling would-be autonomous operations, amplify the two-few sledgehammers by upgrading them into guns. Police states abound as a result.

In Figure 7 we have the completed scheme. The error-controlling feedbacks, having been identified by Cybersyn, return to System One. They are not generally available: they go to their respective manager in each case. The objection mentioned earlier that the system will over-react in the face of instant, daily importance filtration is met by the tuning of Cyberfilter. This is a statistically complicated matter. Suffice it to say that Cyberfilter damps down its reactions to the point when the manager needs to know, by a learning system. The manager is always in control of the sensitivity of filtration that concerns her or him.

At this national level, all three Systems One impinge necessarily on the whole nation. When a feedback arrives, its relevance to whatever aspects of national activity are concerned is notified *ipso facto* by its identity implications within some QFC. The manager will want to examine the implications. But in particular, if local autonomy is to exist in the sense and the spirit of this whole approach, it may be necessary for the affected zone to consider its relationships with other zones. This particularly applies if strategic and normative plans involving other zones are affected.

In the evolved diagram, the infrastructure of the nation is sketched by a hodge-podge of intersecting arrows. Across this is written 'enabling networks'. The Ministry component of System One has such a network in the federal–provincial arrangements that are in force in any nation. Some

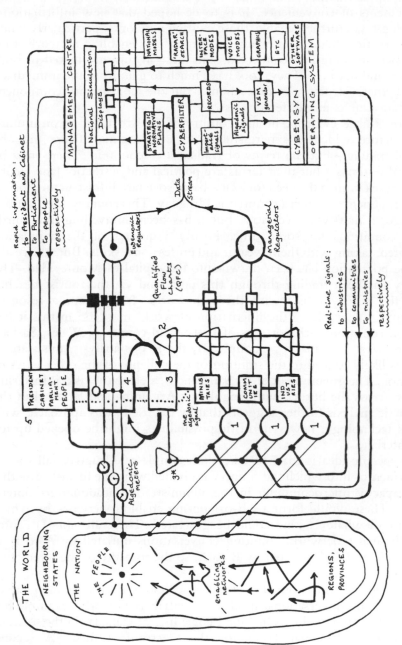

Figure 7. The completed cybernetic scheme

of this infrastructure is in place by law, some by budget, some by arrangements of convenience. It is to be hoped that new informational technology will further 'enable' these existing networks, particularly when the cybernetic scheme is in being. Industries have enabling networks too, their operations are interconnected, and these in turn are linked to both suppliers and markets. They, too, have much to gain from the immediacy and cybernetic structure of this scheme. But the Community component of System One is in a different class.

There is little evidence in any country of the world of a community infrastructure which is synergistic in societary terms. There may be internal trading—but the terms of trading are economic. There may be fairs and festivals—but their terms are political and aesthetic. Jane Jacobs (1984) has examined the need for a city-based societary infrastructure, based on the experience of such a country as Uruguay. This country was, not too long ago, a first-world country, but it has been unravelling into third-world conditions for some time. Nearly a quarter of the population emigrated, taking with them wealth and professional skills. But, above all, half the population has been drawn into the capital of Montevideo. The country has been trading through this port, and via the tourist beaches along the nearby coast, with the rich world. It is, then, a resource economy, vulnerable to the fluctuation of world markets. In this it has much in common with portions of first-world countries: Wales, Saskatchewan, West Virginia, for example. By the Jacobs analysis, this state of affairs is likely to lead to the stunting or the collapse of the other cities of the nation. The government of Dr Sanguinetti, alert to the risk, is fostering activity across the border with neighbours Argentina and Brazil. But the cybernetic analysis further suggests that by the use of this approach, with current technology, new synergistic possibilities might be opened up for the hinterland.

The specific idea is that, to take the example of Uruguay, all the 19 Departmentos into which the country is divided woud be furnished with a large-capacity microcomputer in the administrative headquarters (intendencia). These would form a self-organizing enabling network by which enterpise of whatever kind could be generated. There are $n(n-1)$ routes through such a network, counting conversations initiated by A to B as different from those initiated by B to A, so the measure of the interactive possibilities is 342.

If the network is to flourish, software facilitating a strongly interactive, user-oriented, conversational mode is essential. A powerful example has already been presented by On-Line People Inc. of Canada, in the form of PROTEO (Professional Tools for the Electronic Office). A second example, based on stating the degree of commitment that the conversation entails, is the package Coordinator, presented by Action Technologies of California.

Finally, if the communities involved should wish to set up any kind of collaborative project, as projected by their conversations, they would be encouraged to formulate it as a viable system, and to describe it in VSM terms, as quantified by QFCs. In that case they would be able to monitor the project's viability through the cybernetic system here described. The outstanding example of how this is already done in contemporary society, though without benefit of the cybernetic tools, is evinced by the cooperative movements all over the world. (The artisan cooperative, Manos del Uruguay, is a fine example.)

In some places, such developments are supported by the development of a 'green money' system. Under this, a high degree of networked bartering is available. A paints B's house; B takes meals to C, who is disabled; C designs furniture for the new community centre; D, who runs that centre, does accounts and taxation for A. That example looks at a single loop. In real systems, all sorts of subloops operate to create an enabling network. The prices 'charged' are calculated in 'green money' which exists only inside the computer network.[1] In complicated schemes, prices may be agreed in some ratio of green to federal money. Folk need to eat, and some products will have to be imported for hard currency. In any case, evidence accrues that poor neighbourhoods can do that, perhaps a country can. Certainly if third-world countries are not soon liberated from the traps of international financing, as currently malpracticed, the whole edifice of the world economy will collapse.

Finally we should take a look at the top return loops in Figure 7. The information to be spun off from the Management Centre to System Five needs careful consideration, and cannot be fully discussed here. It is easier to determine what the president and his political colleagues need and wish to hear, then it is to know how to inform the sovereign people. We know what happens to governmental information, both in official press releases and in inspirational 'leaks', when it passes to the public through the mass media. A cybernetic system's analysis of the matter is made elsewhere (Beer, 1983). Meanwhile: television broadcasts from the Management Centre would possible be more interesting than those made from parliament in some countries. . . .

Footnote

The first cybernetic regulatory system of this kind at the national level was commissioned by President Allende of Chile, and was the subject of intense effort between 1971 and 1973, when, on 11 September 1973, he was

[1]Michael Linton, Landsman Community Services, 375 Johnston Avenue, Courtenay, B.C., Canada V9N 2Y2.

murdered and his government was brought down. In these two short intervening years, some three-quarters of the social economy was brought into the scheme. The whole story is described in some detail in the last five chapters of *Brain of the Firm*. As far as it went, it seemed to work.

Currently, this approach is under development in Uruguay, through the United Nations Development Programme. Funding under Project URU/86/004 is hereby acknowledged. A pilot scheme under the direction of Ing. Victor Ganón is working in Montevideo, where President Sanguinetti has an embryonic Management Centre *en suite*. Preparations have been made, and preliminary studies undertaken, for a similar development to serve the government of Venezuela under President Carlos Andres Perez. Professor Manuel Mariña is the local director.

References

Beer, S. (1981). *Brain of the Firm*, 2nd edn. Chichester: John Wiley.
Beer, S. (1979). *The Heart of Enterprise*. Chichester: John Wiley.
Beer, S. (1985). *Diagnosing the System*. Chichester: John Wiley.
Harrison, P.J. and Stevens, C.R. (1971). 'A Bayesian approach to short-term forecasting', *Operational Research Quarterly*, **22**, 341–62.
Jacobs, J. (1984). *Cities and the Wealth of Nations*. New York: Random House.
Beer, S. (1983). 'The Will of the People', *Journal of the Operational Research Society*, . **34**, 797–810.

The Viable System Model: Interpretations and Applications of Stafford Beer's VSM
Edited by R. Espejo and R. Harnden
© 1989 John Wiley & Sons Ltd

13
A cybernetic method to study organizations*

Raul Espejo
Aston Business School

This chapter offers a methodological discussion on how to apply Beer's Viable System Model to social organizations. Its main tenet is that there is no one organization 'out there' waiting to be described by the model, but that different viewpoints, as they ascribe different purposes to an organization, define different criteria of effectiveness. It also argues the need to establish whether the purpose of the study is diagnosis or design. The methodological implications for each mode of study are discussed throughout the paper. Finally the paper touches on the practical use of the method in a variety of situations.

Introduction

This contribution offers a methodological discussion on how to apply Beer's Viable System Model (Beer, 1979, 1981, 1985) to social organizations.

The VSM is a powerful tool; it permits one to study, and establish the adequacy of the strategies used by an organization to cope with the complexity of its tasks. The VSM is a model of the web of regulatory mechanisms that are needed in an organization to cope successfully with the inherent large complexity of real-world tasks.

Though the VSM may be seen primarily as a tool to diagnose the effectiveness of an organization's structure, it also offers other possibilities.

---

*This paper was presented at the Society for General Systems Research Annual Meeting, Budapest, June 1987.

From the point of view of information systems the VSM offers a conceptual model of the organization's management information system. From the viewpoint of policy analysis it is a tool to assess the organizational implications of alternative policies. Where the concern is the contribution of several institutions and/or institutional parts to one enterprise, as in the case of large-scale projects, the VSM offers the possibility to study and design flexible structures and, as a consequence, reduce the chances of costly errors (Davies *et al.*, 1979).

The versatility of this model is due to its abstract nature. The model is relevant to any viable system, whether biological or social, artificial or natural. Yet, because of this generality, its use is not straightforward.

The discussions that follow assume that the reader has already read the chapters 'The VSM revisited' and 'P.M. Manufacturers: the VSM as a diagnostic tool'. The ideas of this paper should be seen as complementary to those presented by Beer in his book *Diagnosing the System for Organizations* (Beer, 1985).

Methodological issues

Social organizations are perceived differently by different observers. Therefore, as will be made clear below, in general it is necessary to name several systems, from different viewpoints, to capture the purposes that relevant observers may be ascribing to them. Moreover, it is necessary to establish not only the purposes that relevant viewpoints ascribe to the organization, but also the purposes that the investigators may be ascribing to the study itself. Diagnosis and design are offered as two alternative modes of study.

The discussion will be focused on the following two main issues.

1: Establishing the organizational identity

The application of the VSM to social situations needs to assume the possibility of several viable systems in the apparently unquestionable 'reality' of one social institution. While establishing the system in focus is not a straightforward task, any attempt to apply the VSM without a proper clarification of this point is bound to produce inadequate results.

An important methodological problem encountered by analysts derives from their difficulties in seeing that there are many equally valid 'viable system models' for any organization. They fail to appreciate that the structure of the organization can be described at any moment in time in multiple forms by different viewpoints. The result is that the analyst is trapped in one viewpoint, unable to see alternative forms of description. Organizations are not single systems but multisystems, being the outcome of the negotiations of multiple viewpoints; therefore, any attempt to approach their study from a single viewpoint is bound to fail because it lacks in multisystemic variety.

2: Modelling the 'structural levels' of the organization

This issue is concerned with the partition of organizational tasks. It will be argued that, though complexity is not an objective property of these tasks, some forms of partitioning are more likely than others to support an effective management of complexity. In other words, this issue is that of establishing the structural levels that contribute to the implementation of the organization's tasks. In Beer's terms the problem is in defining the levels of recursion for the system in focus. In general, it is inadequate to approach this problem by mapping the formal structure of the organization (i.e. its organization chart) on to the VSM; it is necessary to use more subtle criteria in establishing recursion levels.

The above two issues are discussed in two different modes:

Mode I relates to existing organizations and is *diagnostic* in character. Its outcome is, in general, structural adjustments aimed at improving control and communications processes in the organization.

Mode II relates to organizations undergoing a fundamental change in identity, or simply to new enterprises. Its outcome is a *prescriptive definition* of the control and communication processes likely to support an effective implementation of the organisation's agreed missions. Thus, its aim is organizational *design*.

In the diagnostic mode many of the difficulties in applying the VSM stem from confusions about the purpose of the study: is its purpose to model the organization as it actually works in the eyes of the analyst? or is its purpose to model how the organization should work based on VSM criteria? Are the discussions in either case done with reference to what the organization is currently doing (in the eyes of the analyst), or with reference to what particular managers espouse the organization should be doing? Or, are these discussions done with reference to what the business plan might suggest the organization will be doing in the future? Indeed, if the study is done with reference to new missions the mode may not be any longer mode I but mode II.

All these distinctions are subtle but important. If they are not worked out correctly, the study is likely to confuse different forms of description and make very difficult a useful comparison between 'reality' and the 'VSM model'. To a large extent success in using this method depends on establishing clearly, at the outset, the purposes of the study. It is only when these purposes are clear that the relevant 'organizational identity' can be established.

Organizational identity

What are the purposes ascribed by the stakeholders to their organization? Which are the organizational activities that they want to make viable? Answers to these questions permit us to establish the identity of an organization.

However, in practice the stakeholders have hazy ideas about the identity of the organization. For instance, in many cases people in manufacturing companies are not clear whether their purposes are just manufacturing or are both manufacturing and distribution. The structural implications of one or the other identity are, as it will become apparent later on, significant. Haziness makes it more difficult to state uncontroversial criteria of effectiveness for particular organizations. Indeed, the effectiveness of an organization depends upon its ability to make viable its organizationally agreed identity, but the very problem may be that the stakeholders want to keep their options open.

Then, how do we establish an organization's identity?

The main methodological tool for this purpose, in line with soft systems methodologies (Checkland, 1981), is to name the organization of concern; that is, to name the primary transformation(s) of the organization. Indeed, as implied in the section above, depending on both the purposes of the study and the purposes ascribed to the organization and its parts, a number of names are possible in any particular situation. Debates about conflicting names may be 'forced' at an early stage of the study, or may be left to later stages, after the application of the VSM has produced new insights. Most likely, in any serious study, these debates will take place several times, at different stages of the study.

If there is evidence that there are important differences in the identity ascribed to the organization by different people, then it might be useful to force an early discussion about this identity. This was the case in Parker Ltd., where there were two camps, each supporting very different identities for the company:

Name 1. A traditional owner-managed engineering company which manufactures fully assembled switchgear for supply to industrial end-users at the lowest cost consistent with high product quality. (This is the name supported by the Parker-centred camp: it emphasizes the manufacturing tradition of the company.)

Name 2. A company which uses its expertise to advise users of electrical switchgear about the best equipment for a particular application and to provide that equipment either as a manufacturer or supplier. (This is the name supported by the market-centred camp: it emphasizes the marketing problems of the company and the need to develop synergistic interactions with other producers.)

In general, the study of an organization's structure is facilitated by a coherent definition of its business policy, including an agreed definition of its identity. However, if this agreement were not possible, then a study of the structural implications of the alternative viewpoints might help these discussions.

Indeed, discussions about the intended identity for the organization are important. Each identity implies a particular effective structure. In the above example, while the first name implies only the need to make viable the company's manufacturing, the second name implies the need to make viable both the company's manufacturing and services to clients.

Another instance to appreciate this proposition is given by the likely evolution of a 'Management School' under different identities. This evolution is likely to be very different depending on whether the School's identity is perceived as that of an establishment to achieve academic excellence in specific management subjects, or as that of an establishment to excel in the subjects of 'management and organization' as a whole. While consistency with the former identity is likely to lead the School to a 'departmental' structure (with autonomous marketing, finance, economics . . . departments), consistency with the latter is likely to produce a 'business school' concerned with the viability of multidisciplinary teaching and research programmes. Criteria for organizational effectiveness are very different in each case.

Another form of haziness is the case where the espoused view of the organization's identity is in conflict with the 'theory in use'. For instance, if the espoused identity of a company were that of a manufacturing company, but in practice the analyst had established that it was operating in manufacturing and non-manufacturing businesses, then it might be convenient to establish at least two names for the company's identity, one focused in only manufacturing and the other in both primary transformations.

For instance, in the study of P.M. Manufacturers, while managers perceived its identity as that of 'a company to manufacture standard and non-standard electrical generators for foreign and local markets', I could observe a different identity for the company. For me it was clear that the company was also striving for the viability of its non-manufacturing services (i.e. engineering services and spares procurement); and, therefore, that any structural improvements would need to be based on the agreement of an identity like 'company to produce services for the autonomous generation of electrical power in local and foreign markets'.

How does it help to establish the identity of an organization?

In mode I, the diagnostic mode, analysis is done with reference to the tacit missions of the organization (as perceived by the analyst). In general, analysts should aim at making apparent how the organization appears to work. They should hypothesize, as a platform for debate, their views about the way the organization appears to work. The study could alternatively be based on what managers claim the organization is doing.

In this mode it is accepted that there *is* an organization, so that the study will be descriptive and any criteria of organizational effectiveness will be used with reference to tacit and/or espoused purposes.

It is only in this mode of study that analysts can make apparent mismatches between an 'actual' structure and the 'effective' structure as suggested by cybernetic principles. Figures 1 and 2 help to appreciate the meaning of this mismatch. Figure 1 is a description of how the organization works, as perceived by the analyst. Figure 2 would be a description of P.M.'s organizational structure if its current business areas, as defined by its tacit purposes, were effectively organized (according to the VSM). On the other hand, had the analysis been made with reference to the com-

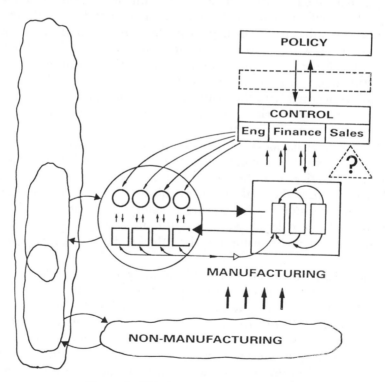

Figure 1. P.M. Organization: theory in use

pany's espoused view of its identity—that is, of a 'manufacturing' enterprise then the non manufacturing activities would have had to be considered as anomalous. Discountinuing these activities would be equivalent
to collapsing the whole organization on to the manufacturing division of
Figure 2.

In mode II, the design mode, studies are done with reference to one or
more statements of identity as defined by relevant actors. This mode is
prescriptive and applies when the purpose (of the analysis) is to design an

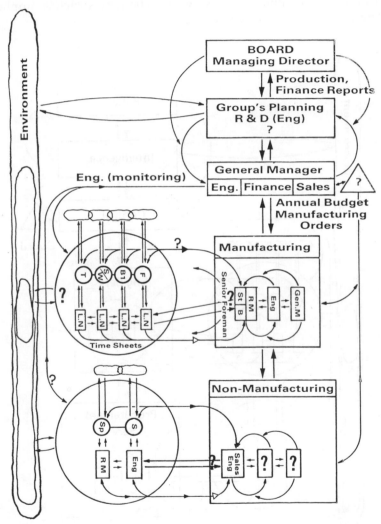

Figure 2. The cybernetics of the organization (mode I)

effective structure consistent with the identity agreed for the organization by the relevant actors.

While the same criteria of effectiveness apply when the study is done in mode II, in practice, if the explicit purposes ascribed to an organization are significantly different from those currently ascribed, then the implied VSM for that organization will naturally have to be different. Figure 3 is an instance of a 'possible' model for P.M. Manufacturers, given that the relevant managers agree to a shift of the company's identity from its current hazy manufacturing and services identity to a new identity only related to its non-manufacturing services. The new structure implied by this identity would need to be designed.

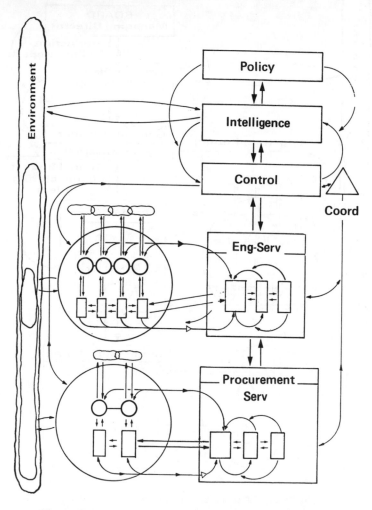

Figure 3. The cybernetics of the organization (mode II)

While it helps to distinguish between modes I and II, in some cases it might not be clear whether the analysis is being done in one or the other mode: indeed, it might not be straightforward to see how the current organizational weaknesses relate either to a hazy identity or to a new, yet unexplored, opportunity. For instance, for P.M. Manufacturers, non-manufacturing activities can be perceived either as already having a degree of self-organization and therefore already defining the tacit identity of the company, or as a new opportunity which lends itself to a mode II analysis.

Modelling structural levels

Naming the system is a first step towards modelling the complexity of organizational tasks. However, in this process, it is also necessary to partition the named primary transformations into activities that fall within the regulatory capacity of particular managerial levels. Establishing activities and their structural levels is one of the key strategies used by organizations to cope with the complexity of their tasks.

The complexity that managers 'see' in their day-to-day activities is strongly influenced by this partition. However, in any organization—that is, in any multisystem—different viewpoints will see different partitions. But, not all partitions are equally effective; the connectivity of activities in the real world suggests that, from the viewpoint of the management of complexity, there are partitions that are more effective than others. This is something that the VSM permits us to appreciate.

However, it makes no sense to impose any such partition on the managers concerned. Methodologically, the problem is to separate different forms of description from the outset; is the analyst modelling the complexity that individual managers should see, or is he modelling the complexity that they appear to see? Conflating these viewpoints is a typical pitfall in the application of the VSM. Analysts have to pay due attention to the views of the organizational clients. Indeed, it is by comparing the views of individual managers and the criteria of effectiveness as provided by the VSM, that useful improvements may become apparent.

In what follows I offer considerations on how to study partitioning of tasks.

Modelling technological activities

The activities necessary to produce the transformations named by the organization's identity are called technological activities. In this sense a model of 'technological activities' is either a conceptual model of the activities necessary to produce the named transformations, or a descriptive model of the activities producing the named transformations. The activities of concern are only those producing the transformations; any other activity, facilitating, servicing or, in general, regulating them, is not part of the model.

The boundaries of an organization are defined, in the diagnostic mode, by those technological activities that the organization actually performs, or, in the design mode, by those technological activities that it should perform.

Indeed, there are a wide range of 'technological' possibilities to produce any transformation. In the manufacturing industry this range may vary from full 'vertical integration' to almost no manufacturing in the case of an 'assembly' plant. In the former case, the organization's strategy is to produce within itself almost everything, starting from the 'nuts and bolts'; in the latter case the emphasis is in the very last stage of manufacturing. Thus, the same identity may imply different levels of task complexity. While the extreme cases may be easy to recognize (names like 'company to assemble electrical generators' or 'company to fully manufacture electrical generators' are clearly different), the cases in between may only be distinguished with reference to their technological models.

Figure 4 is one possible technological model for the 'manufacturing of electrical generators'. A number of alternative models could have been produced. In this example the model takes the form of a 'quantified flowchart' (Beer, 1975). The purpose of this quantification is to measure the complexity of the activities. In practice, most of the time, only proxy

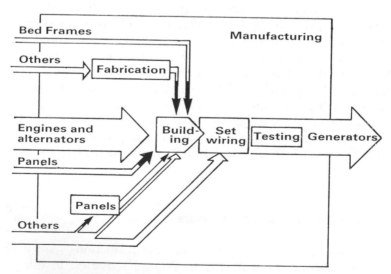

Figure 4. Manufacturing flowchart for P.M. Manufacturers

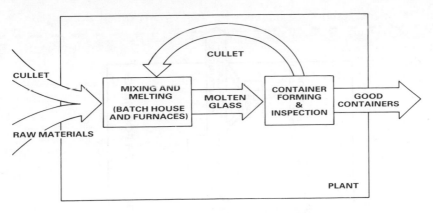

Figure 5. Glass Ltd. plant: flowchart level 1

measurements like assigning money values to inputs and outputs and to the value added by each activity, are used to do this quantification. The purpose of the quantification is to model, at any one level of resolution, activities of a complexity roughly within the same order of magnitude.

In a number of cases it may make sense to 'open' each of these activities to produce a technological model at the next level of resolution (see Figures 5, 6 and 7 relating to 'Glass Ltd.'). Producing these models is a form of variety engineering.

If information about inputs and outputs is not available, or if the interactions between activities is too complex to be described properly by a simple flowchart, then this type of modelling can be done, partially, by using boxes of different sizes, and boxes within boxes as described in Figure 8.

In a mode I study, these models can be produced with reference to the 'technology-in-use' in the organization. By simple observation it should be possible to produce a descriptive model of the organization's transformations. Whether or not these models, and the aggregations of activities implied by them, are a good example of variety engineering, is a judgement that should be left to experts in the transformations. In general, technological models in a mode I study are likely to be useful, simply because they are a summary of the organization's expertise in managing complexity. But, they can also be dangerous; they may be the blinkers that constrain seeing other possible, and perhaps more effective, ways of handling complexity.

In a mode II study these models can only be produced with reference to expert knowledge. In this case good technological models are essential for an effective organizational design.

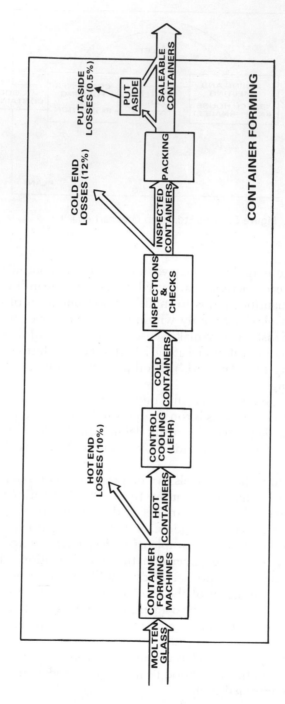

Figure 6. Glass Ltd. container forming: flowchart level 2

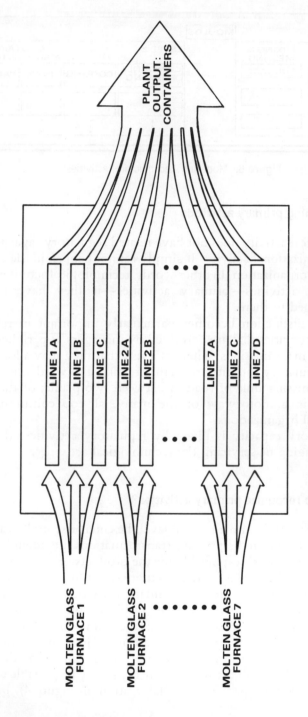

Figure 7. Glass Ltd. container forming for individual lines: flowchart level 3

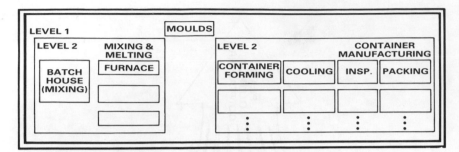

Figure 8. Modelling 'technological activities'

Modelling primary activities

If a technological activity does not have related regulatory capacity, then it is not an organizational activity. It simply cannot happen in the context of that particular organization. In other words, technological activities become organizational activities—primary activities—if they have regulatory capacity attached to them.

This subtle distinction becomes particularly important in relation to technological activities that are perceived by managers as peripheral to the organization's main missions. There is a risk that they may allocate no or inadequate regulatory capacity to them, creating awkward situations; for instance, the missing mile of road in between two plants; or the missing vital services in a motorway; or the many, all too common, similar situations in all organizations.

The regulatory capacity is given by regulatory activities, that is, by activities managing or servicing the technological activities.

How to recognize primary activities?

In any enterprise, at the most general level the enterprise itself is a primary activity with reference to its primary transformations (i.e. identity). At the next level the 'divisions' responsible for the products or services on which its viability depends are the primary activities. Within these divisions the 'sections' producing them are the primary activities, and so forth (see Figure 9 relating to 'Paper Holdings Ltd.') In other terms, primary activities are all those activities which, in the framework of the currently agreed identity for the enterprise, have a transformation of their own. If hived off they would not lose the content of their transformations.

For instance, in a 'manufacturing' company it may be possible to hive off parts of the production process (like 'fabrication' in Figure 4), but it may

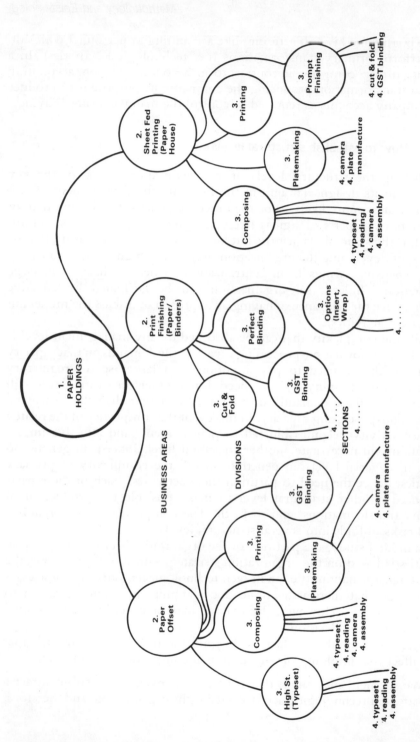

Figure 9. Paper Holdings Ltd.: unfolding of complexity

not be possible to hive off activities like accounting or personnel which do not perform primary transformations (i.e. technological activities). In a 'manufacturing' company, in contrast to an 'accountancy' firm, accounting is not a transformation producing the company; its purpose is to produce the company accounts as required to control the manufacturing activities.

How to establish structural levels?

Establishing the structural levels of an organization is one of the key decisions in its design. Though in the long run the development of any organization is more likely to be the outcome of self-organization than of blueprints by experts, design is likely to make this process less painful. Indeed, if the structure is just the outcome of trial and error, as it often is, the cost of developing the organization may be too high, as proved by the often expensive swings from centralization to decentralization in large corporations. Indeed, self-organization can be facilitated by effective design, hence the relevance of having an approach to discuss the modelling of primary activities.

While the complexity that can be absorbed by any management level is limited to its information-processing capacity, the demands on that capacity increase with the need to pay attention to a larger set of regulatory variables. Oversimplifying, the need for another structural level will emerge when these demands are perceived as larger than the managers' information-processing capacity. In this event the complexity of the related primary activity becomes blurred to management, and effective implementation will necessitate another structural level. Indeed, in general no single managerial level can penetrate in full the complexity of primary activities: hence the need to partition these activities. Each of these parts may need another structural level to make possible the production of primary transformations, and so forth. The complexity of the organizational tasks unfolds into several structural levels.

In a mode I study, the partition of the organization's tasks into primary activities is based on the organization's strategies-in-use in producing the named transformations (i.e. on its tacit technological model). In some cases it might be that there is a one-to-one mapping of technological on to primary activities. However, this overlap may be upset by organizational arrangements and decisions.

There are a number of factors, beyond the technological model, that may affect the unfolding of organizational complexity. For instance:

(1) Management may think that it is more convenient to sub-contract a particular technological activity outside the organization, and therefore,

by choice, decide not to have the related primary activity. For example, the 'division' related to the technological activity 'manufacturing' in Figure 4 may or may not include within its primary activities the activity 'control panels' (Figure 10). The latter case, in which the panels are sub-contracted outside the company, implies not only different boundaries for the division but also a simpler manufacturing activity. Therefore, as can be appreciated in Figure 10, while the technological model has five activities the primary activities model has only four.

(2) It may be desirable to organize work in shifts or in different plants. In either case, from a regulatory point of view, there are one or more additional structural levels. This is the case in Glass Ltd., where the company's complexity is unfolded into two plants, and each plant's complexity is unfolded into four groups, each responsible for a shift. It is only within the shifts that the technological activities are made apparent (Figure 11).

(3) Management may consider that it is better to structure primary activities in forms that are different from those suggested by the technological models. For instance, a range of product lines could be manufactured either individually, under different management, or clustered under the same management. It is not difficult to appreciate that each option suggests a different organizational arrangement for the same technological model. In this case a number of technological activities may be clustered under one level of management, collapsing two technological levels into one primary activity.

In mode I, it is important to keep in mind that the actual decomposition of the organization's tasks may not be apparent in the formal organization structure. For instance, in P.M. Manufacturers the 'non-manufacturing' primary activity was not obvious from seeing the company's organization chart. However, the fact that the transformation was taking place was made apparent by observing the company's outputs.

Indeed, the modelling of primary activities should be based on the model of the technological activities and not on the organization chart. If a technological activity is taking place in the organization, then it is a primary activity. Its structural position is defined *de facto* by its relationships with other primary activities, as implied by the technological model, and not by its position in the organization chart. An instance of this situation was the case of 'testing' in P.M. Manufacturers: while the organization chart put it within engineering, the technological model made apparent that it was part of manufacturing.

In a mode II study, the modelling of primary activities should be done with the support of expert advice. Alternative decompositions of the

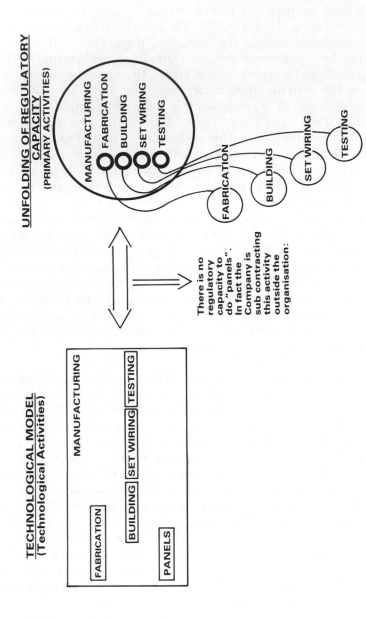

Figure 10. Technological and primary activities

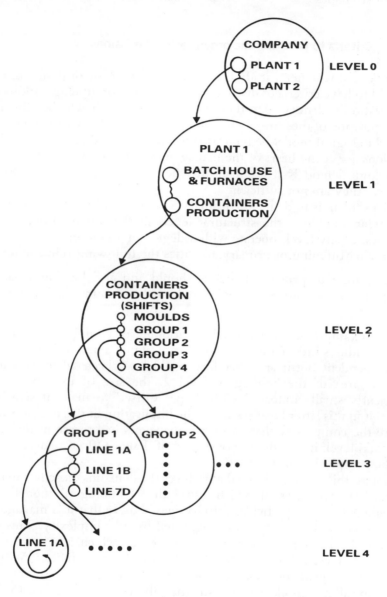

Figure 11. Glass Ltd.: modelling primary activities

organizational tasks will depend upon both the technologies-in-use and the control strategies.

Criteria to partition the organizational missions

Regardless of whether a study is done in the diagnostic or design mode, it is useful to have criteria to discuss the aggregation or disaggregation of the organization's identity. Are there criteria to establish whether the actual decomposition of these missions is adequate or not? To what extent should the technological model be replicated in the organization structure? These questions are at the heart of the management of complexity. While good technological models (e.g. quantified flowcharts) are necessary for this purpose, they are not sufficient.

The problem is to define the primary activities within primary activities and, as an outcome, the structural levels in the organization. Primary activities, at any level, operate with a degree of autonomy.

In the partitioning of primary activities the following rule applies:

Partitioning of primary activities should aim at achieving a balanced distribution of complexity along each of the lines in which complexity unfolds.

A good example of this balanced partitioning of complexity is given by Paper Holdings Ltd. (Figure 9). In this case, while 'High St.' in level 3 was an independent outfit and therefore could have been taken as a fourth business area of the holding in level 2, the size of its activities was sufficiently small compared with 'Paper Offset' to make it sensible to embed it in this latter business area. The implication of this decision was to reduce the complexity that corporate managers had to see in the second structural level; it was the responsibility of 'Paper Offset' management to see the complexity of 'High St'.

Though this is a sound rule that reduces the demands on senior management, there are cases in which either the technological model or the strategic nature of a particular activity may require that it is managed at a higher structural level than that suggested by its complexity. This is the case for the insurance company in Figure 12, which has two primary activities at level 1: the insurance and the investment activities. While the complexity of insurance is large enough to require four structural levels to absorb in full its complexity, investment only requires two levels to absorb its complexity. It should not be difficult to see that a technological model of the insurance activity would make apparent that investment is a global activity, independent of specific insurance types.

Another common case of imbalance in the distribution of complexity is

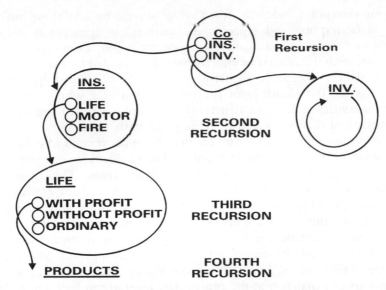

Figure 12. Unfolding of complexity for an insurance company, showing the imbalance between two lines of unfolding

provided by small new strategic units, hanging at the corporate level. Often they become problematic: on the one hand they should not be constrained by structural rigidities, something which is likely to happen if they operate at too low a structural level; but on the other hand they should not become a burden to senior managers—a clear possibility if they operate for too long at a high structural level. Changing the structural position of these units, as they grow and mature, is in itself a strategy that managers should be aware of in order to manage their complexity effectively.

Suming up, in mode I the modelling of primary activities is a description of the postulated 'unfolding of complexity' in the organization (Figures 9, 11 and 12). In mode II this modelling is the modelling of the designed regulatory responses, design which, as suggested above, should be supported by expert advice and/or by any previous organizational experience.

Conclusions

This chapter has offered a methodological discussion to facilitate the application of Beer's Viable System Model to any kind of organization.

The simple, but key idea, on which the use of the VSM is grounded is that organizations are multisystems. Hence the relevance of having a tool to name systems. Different names permit us to separate, from each particular viewpoint, what 'is' from what 'should be'. As these different forms of description and representation are made apparent it becomes possible to compare the actuality of an organization, as interpreted by a viewpoint, with the criteria of effectiveness as provided by the VSM. This is the basis for diagnosis. Equally, the names articulated by relevant viewpoints for the identities of the organization as a whole and of the primary activities at different structural levels, give the platform to design the structure of that organization. This is the basis for design.

The other key idea of this chapter is that the measurement of complexity in a situation depends on the viewpoint. The problem is thus reduced to finding the appropriate viewpoint for each situation. The appropriate viewpoint is the one that copes with the 'complexity' of concern; the measurement produced by this viewpoint is as significant and valid as can possibly be produced. In a diagnostic mode, the method permits us to make visible mismatches between the complexity that particular managers appear 'to see' and the complexity that they should see for effective discharge of their duties. In a design mode the method permits us to focus the concerns of a manager at the appropriate level of resolution in order to avoid unnecessary information overload.

Finally, once the relevant viewpoints have both defined the systems of interest and developed an appreciation of their complexity, then it is my contention that Beer's ideas, as developed in *Diagnosing the System for Organizations*, offer a powerful method to study and diagnose regulatory mechanisms.

References

Beer, S. (1975). *Platform for Change*. Chichester: John Wiley.
Beer, S. (1979). *The Heart of Enterprise*. Chichester: John Wiley.
Beer, S. (1981). *Brain of the Firm*, 2nd edn. Chichester: John Wiley.
Beer, S. (1985). *Diagnosing the System for Organisations*. Chichester: John Wiley.
Checkland, P.B. (1981). *Systems Thinking, Systems Practice*. Chichester: John Wiley.
Davies, C., Demb, A., and Espejo, R. (1979). *Organization for Program Management*. Chichester: John Wiley.
Espejo, R. (1987). 'From machine to people and organisations: a cybernetic insight on management', in *New Directions in Management Science* (eds. Jackson, M. and Keys, P.). Aldershot: Gower.

The Viable System Model: Interpretations and Applications of Stafford Beer's VSM
Edited by R. Espejo and R. Harnden
© 1989 John Wiley & Sons Ltd

14

Outside and then: an interpretative approach to the VSM

Roger J. Harnden

Doctoral Programme, Aston Business School

This chapter considers the modelling facility embedded in System Four of the VSM, from a hermeneutic perspective. Models are viewed as means of orienting about a consensual domain, rather than as a way of mirroring or reflecting some objective domain or truth.

Introduction

That this volume contains a number of applications of a particular scientific model, the Viable System Model, indicates that certain individuals find the model useful. However, insofar as the model has been taken up and applied it has tended to be interpreted in a 'hard' light, as an algorithm to enable rigorous quantification of complexity alone. This has led to a disregard of the model's value by the 'softer' (use as in Checkland, 1981), more hermeneutic paradigm. I want to consider why this should be the case, in the hope that this will encourage a deeper understanding of the model.

My way into the problem will be to focus on just one particular function of the VSM—System Four. Where System Three is held to give closure to the internal workings of the system-in-focus ('Inside and now'), System Four faces towards the environment and future problems which might confront the system ('Outside and then'). In order to fulfil their roles effectively, Beer presents these two Systems as functioning in terms of models. Specifically, System Four 'contains' a model of itself, a model of the total viable system, and a model of the environment in which the viable system is perceived to realize its identity.

As the VSM stands, this modelling function inherent in Systems Three and Four tends to be interpreted according to a particular set of cultural predispositions. Thus, the modelling is generally seen to focus along two distinct lines:

(1) From the *functional perspective*, the system is held to require an accurate picture of the world in order to be able to focus upon and to highlight those needs relevant to it in order to be able to function effectively in the particular environment—in other words, in order to be able to adapt.

(2) From the *perspective of its own organization*, in order to be capable of functioning effectively in an environment, the system is seen to need an accurate picture of its own internal workings—its potential, ambitions, degree of flexibility, intentions and so on.

The above interpretation of the VSM is inherently dualistic and representationist. In simple language, it splits up the world into one particular notion of 'inside' and 'outside', which takes as given that the adaptation of inside to outside is achieved in terms of the accuracy of a mirroring or reflective capacity. Implied is that there is a little observer 'in the head', who makes judgements and decisions in terms of the set of representations passed up to him.

In this scenario, for instance, the eyes of an organism act as a sort of camera or window via which the 'image' of the outside is 'passed to' the inside, for some inner viewpoint to observe and comment on (i.e. interpret). Interpretation itself, from this perspective, is distanced from 'simple' sensory/motor correlations in terms of which an organism finds itself in its coherent dance with the world, and instead becomes an inner perspective able to comment on such correlations. In respect of the VSM itself, this is akin to saying that System Five does the interpretation, at one remove from 'the action'.

The above probably sounds eminently reasonable until you try to make it compatible with contemporary insights that the nervous system is *organizationally closed*. This point will be returned to in the course of the chapter; but in general terms it means that our nervous system functions coherently in terms of ourselves, allowing us to function coherently in terms of what is 'outside', while being closed to, or 'blind' in respect of anything actually going on outside itself (including the rest of the physiology of the organism and the total environment of that organism). From the perspective of hermeneutics, the implications of this last is that 'interpretation' is not some 'higher function', separated from the rest of the nervous system, the organism and the universe, but is a function of the whole unified complex of nervous system, organism, universe, in the course of their ongoing structural coupling with one another. This is what will be later referred to as an 'observing system'.

In the rationalist or positivistic paradigm, one might say that System Five provides the interpretative function by commenting upon and allowing the emergence of decisions by the allocation of competing resource demands by Systems Three and Four, with System Three peering into and assessing the demands of the organism itself while System Four peers out at and assesses the needs for functional success in terms of the environment.

What System Five is actually held to do in terms of the VSM, is to 'monitor the Three–Four homeostat'. It does not monitor System Three and Four themselves, together with their respective models. System Five does not itself 'peer into' these models, in order to comment on them from a 'higher' vantage. System Five enables the closure of the viable system's organization by damping oscillatory tendencies caused by the competing demands of 'inside' and 'outside'. It neither itself has some 'open window' to outside and inside, nor does it have an open window to the so-called 'models of' outside and inside.

In the present approach there is a crucial change in the way we conceive of the functioning of models—the modelling facility. Whereas hitherto, System Four (or some other non-cybernetic category and term of discourse, for example 'I' or 'intelligence') was held to 'peer outside' in order to be able to outguess and outmanoeuvre threats, and to anticipate future changes (i.e. to survive or succeed), it is possible to interpret 'success' in a quite different light. Provided that System Four (or 'I' or 'intelligence' or whatever you choose to call it) performs a function that coherently enables the ongoing self-production of the organization of the system in an effective 'fit' with the environment relevant to its survival, then the system *is* viable.

However, such an 'external' fit cannot be distanced from the 'internal' organisation itself (i.e. System Three model). Interpretation, leading to effective viability (i.e. matching both internal and external constraints and possibilities), is the *closure* of one whole complex of interconnected and interdependant relations, in the course of a flow or dance between distinct phenomenal domains. This closely concerns Beer's own insight into real-time management. To put it another way: instead of viability being indicated as a victory over a hostile, alien world, viability is indicated as a satisfying embrace of, or a coherent dance with, that world.

Viability is not some term in a zero-sum game. It does not concern a neo-Darwinian criterion of sucess 'at the expense of' others. Pointing at coherence in terms of both human individual actors and the ecosystem they inhabit, viability is a function of the persistence of maximal variety of distinct phenomenal domains. Viability is the mark of a non-zero-sum game.

We can view a game of chess in two lights, indicate its significance in terms of our own satisfaction as witnesses of it, from two perspectives:

(a) The best game is the most powerful and decisive win, indicated by the quickest conclusion.
(b) The best game is the most evenly matched sequence of moves, the most persistent coherent dance or pattern.

The first is the mood of zero–sum–games, while the second exemplifies non–zero–sum games.

The fact that modern industrial cultures have tended, for a variety of complex reasons, to adopt as their norm a zero–sum mentality, says nothing about viability. In our common, everyday experience, as observers engaged in a web of relations in the course of which we witness the effects of the interactions of other observers and phenomena, what satisfies and engrosses us is the game of chess that does persist for more than a few hectic moves; the boxing match which does 'go the distance'; the sporting fixture which allows us to witness a 'close finish'; the TV quiz show where the scores are close right up to the end.

To round off these opening remarks, I remind the reader of my intention in this chapter. In point of fact, every and any function depicted in the VSM can be viewed in terms of its own modelling capacity, as it copes with the particular complexity relevant to it in its domain of operation. However, I am going to ignore the manner in which models and modelling reverberate throughout the whole VSM, in order to bring to light certain points concerning the way models are conventionally understood, and specifically concerning how the VSM has tended to be interpreted. Thus, to repeat, I shall focus on System Four itself as my own system-in-focus, and consider the implications of approaching it as an interpretative mechanism rather than a 'window' opening on to 'Outside and then'.

A note on models

A great many people experience major problems on the subject of *models*. This is largely a debt of our rationalist, objectivist tradition in which it is held that knowledge, especially scientific knowledge, is about something 'out there' which is either uncovered or discovered. In this view, human beings move through existence and through history as if through a park on a Sunday morning stroll, noting this particular flower, these grasses, the sounds of that bird. The observant or alert human is held to be s/he who has noticed most about his/her surroundings. Such a human subject has captured best what is 'out there', can 'represent' such a reality, and may in the general course of things be expected to be best fitted to function in this particular environment. This view, for instance, underlies our education system.

Such an insight into the order of things is in the most general terms backed by a particular insight into the function of models. Models, if thought about consciously at all, are what individual subjects are held to build to *reflect* or mirror reality, and, depending on the intelligence or alertness of the individual, the model will be more or less accurate as a representation of reality. Such accuracy can be tested formally in things like aptitude tests and examinations, when an expert who has 'proven' that s/he has an accurate model of reality, compares the student's model with his or her own and declares it to be a lesser or greater reflection of reality. To be clear on this point—it is not a modelling *function* or *facility* that is conventionally being embraced, but a particular stance that implicitly holds that models in some manner represent, or should represent, some phenomena 'out there' in the 'real world'.

Now, on the subject of the VSM, I have often found myself involved in discussions where confusion and misunderstanding have arisen as a result of one or more of the protagonists holding some version of the above in respect of models and their use. Useful conversation then becomes a non-starter. I want to explore why, and in the process hint at some ways forward in our understanding and applications of the VSM.

A model of . . . ?

In spite of Beer's own continuous protestations to the contrary, the VSM is sometimes taken to be the 'model of the human central nervous system' in a fashion that sees the VSM as representing, or setting out to represent, phenomena that actually exist biologically, in the human corporeal frame —in a particular nervous system. When interpreted in this light the model decomposes into a functionalist organic approach, and such approaches tend to reduce all the variety and complexity of things to one particular objectivist biological dimension. Applications of the model then become understood as a sort of translation of this original source phenomenon (i.e. a viable nervous system) into some other domain (e.g. the social domain).

One of the revolutionary breakthroughs in the emergence of *cybernetics* as a discourse, something which set it apart from the main tradition of scientific knowledge, was that it gave up the objectivist dream of being able to 'capture reality'. At the same time it refused to slip into solipsism. Its novel stance and its peculiar strength crucially concerned the status attributed to models—their ontology, so to speak. Now of course cybernetics does not merely have an *attitude* towards models—it uses them as its bread and butter, as the seminal paper by Conant and Ashby (1970) makes clear. It is precisely such *usage* that appears to have given rise to problems.

The Conant–Ashby Theorem demonstrated that the degree of control available to the regulator of a system depends on the accuracy of the model that the regulator has of that system, as defined from the perspective of an observer of the interaction. Thus, for me to kick around a football with some semblance of control, I might be said to have a model of the likely behaviours of the ball, consequent on previous interactions with it. If I am lacking such a model, or if my model is poor or innaccurate (e.g. the case in a very young child), control will be observed to be absent, or minimal.

The paper was argued mathematically, and confusion perhaps ensued, because the notion of a mathematical model is all too often treated as being synonymous with the notion of a visual representation. Now any mathematician would no doubt pooh-pooh this, and treat as naive such confusion; but the fact is that most human beings are not mathematicians. It is only over the last 15 years or so that the full problematic concerning representation (and consequently, models) has been brought to light, largely owing to work gathered around the person of Humberto Maturana (Maturana, 1978, 1980, 1983, 1985; Maturana and Guiloff, 1980; Maturana and Varela, 1980, 1987).

Maturana's 'Biology of cognition' arose out of insight into the workings of cognition as an organizationally closed system—or, as was later put in identifying 'autopoiesis', a system that is 'information tight' (Maturana and Varela, 1980). One of the many complex results of this approach, the crucial one in respect of the present chapter, is that cognition is severed once and for all from any notion that it might involve some form of representation of the phenomena around us.

Just to clarify this point, and bring to light what is involved: whether explicitly a dualist or not (a person separating mind or soul from the material world), it is generally taken for granted, as the Conant and Ashby paper appears to endorse, that to be in a position to be able to 'handle the world'—to walk around, flick peas with a spoon, build space rockets, converse or in fact 'do' anything at all worth the name 'doing'—the doer must have some sort of a *picture* of what is happening in the world within which s/he is doing the doing. This appears to be most clear in the case of our memory, when it seems obvious that we have such a picture, though the picture appears maddeningly hazy and subject to fluctuations at times! And although we might tend to separate doing from our reflection on what has been done, the point about the Conant–Ashby Theorem is that it suggests that even in our dynamic doing we have to have such a picture (recall my above example of the football).

It is this 'obvious' point that Maturana's work throws into doubt. Not only do Maturana and colleagues suggest that such an insight into how cognition works is mistaken, but they go on to demonstrate that it just could not happen in such a fashion (see, for example, Maturana, 1983). This is not so much a refutation of Conant and Ashby, as a criticism of the way in which the Conant–Ashby Theorem has been interpreted in subsequent discourse.

Maturana suggests that the way we orient ourselves in the world is not by direct perception that grasps the features of the environment we find ourselves within, but that ourselves and our medium are coupled in a manner analogous to the coupling between the airline pilot flying instrumentally, and the environment the plane is in. In other words, he suggests, cognition is not akin to the pilot looking out of the aircraft windscreen, picking out visual features, noting condition changes, and acting accordingly (as, it is held, 'I' look through my eyes, and decide to act), but in reality functions more like the pilot who is closed off from the external environment which is buffeting the plane, and instead is *structurally coupled* to a medium by means of his instruments. Thus when congratulated on a marvellous landing in the face of terrible and fluctuating conditions which the observers on the ground have been all too aware of as they anxiously stared aloft, the pilot shrugs and says that all *he* did was to keep the instruments aligned (see, for example, Maturana, 1970, in Maturana and Varela, 1980, p. 26; Maturana, 1978, p. 42).

And this is what 'organizational closure' implies for our own common-or-garden cognitive processes—how we get about in the world—whether for everyday purposes, like mowing the lawn, or for more esoteric processes, such as doing philosophy or composing music. Maturana, in a succession of provocative and stimulating publications, goes on to build up an insight into our interaction with the order of things as a process of our *finding our way around our own cognitive space*, and only incidentally (if crucially) finding our way about the Universe, in the course of the persistence of our structural coupling with it.

In sharp contradistinction to the behaviourists, Maturana and colleagues proposed no causal arrow in the link between observer and the environment s/he inhabits. In point of fact, it is not strictly true to say that an organism inhabits an environment at all. The two cannot be separated into two clear-out components—organism, environment—except from the perspective of a meta-level, by an observer outside the interplay. In the domain of their interactions, however, they are functions of one another, regardless of what some observer distinguishes. Such an observer infers particular categories and system boundaries according to his/her expectations as they emerge in a particular consensual domain (e.g. 'organism' and 'environment').

The actual domain in which organism and niche are in coherent interaction—including the case where the organism is an observer—is a domain in which organism and niche form *one medium*. Such a medium consists of the region of all structural couplings available to the organism. In effect this *is* the cognitive space of the organism. Cognition is what demarcates such a domain, from the point of view of any particular organism—as such, naturally, each cognitive domain is particular to the unique individual, though particular phenomena with the potential for similar organisms to be structurally coupled with, may be, and often are, shared. In this respect, second-order cybernetics would go on to suggest that, strictly speaking, it is not the observer as an individual who affirms distinctions. Rather, distinctions emerge within some observing system.

Models and interpretation

What are the implications of the above for the notion of model as used by Beer; indeed, not merely used by him, but also generalized as a recursive function in System Four of the unfolding VSM? We return to the issue raised at the start of the chapter, and ask whether we discover that the insights emerging from second-order cybernetics (from the work of Maturana, Varela, von Foerster and others) are compatible or incompatible with those emerging in the course of the development of first-order cybernetics? Maturana and colleagues characterized this shift in perspective as being a move away from studying as problematical an observed system, towards studying as problematical the system in which the observing emerges (e.g. von Foerster, 1981).

If we take seriously the insight that coherent coexistence of distinct phenomena is maintained in the Universe without the need for some mechanism of representation as a cognitive aid; if we accept that beings can orient one another and themselves in respect of their cognitive domain, and be involved in a controlled dance in the Universe as a result of structural coupling rather than as the result of some abstracted function such as a space inside our mind where we represent what is happening around us; then as observers, what quality—what *ontology*—do we *epistemologically* ascribe to the processes going on in the organizationally closed space that demarcates, that constitutes the limits of identifiable unities? As regards these issues, which among other things concern the way we think of 'intelligence' and 'creativity', the interested reader should look at David Bohm's notion of 'natural intelligence' (Bohm, 1968), or the papers 'The mind is not in the head', and 'The quest for the intelligence of intelligence' (Maturana, 1985; Maturana and Guiloff, 1980).

Let me put the question another way. I am enquiring about the way we conceptualize 'identity'. How is it that an observer distinguishes a unity as a particular unity, and what precisely are the mechanisms by which such a unity is separated out from a background? How is it that 'unity' emerges for the observer—how does the observer *identify* such a unity? This is important for one simple reason. Control is not something that is asserted over some objective phenomenon that has been isolated and identified.

Instead, control emerges concomitantly with identification. As far as I myself am concerned, I am what I cognize myself as doing. As far as another person is concerned, I am what s/he distinguishes me as doing. Perception is the bedrock out of which control may be affirmed to be effected, for the effects of my doing reverberate in respect of the perceived quality of control in respect of some observer. This quality of perception has a peculiarly machinic character. We cannot but make judgements as to identity in terms of the actions we distinguish as relevant, however much we sometimes like to think that we can 'capture' the 'whole person'.

This does not mean that there is some objective, linear scale of quality involved in the domain within which we find ourselves making distinctions. It may well be the case that I indicate a dustman and compare what I have abstracted with the chairperson of ICI, attributing some qualitative standard to my comparison. But this will be determined by the particular consensual region I find myself making my distinctions in terms of, and will not imply some neutral or 'natural' scale. Let us keep in mind Ashby's note that 'Everyone is world champion of some game (even though some of the games have not yet been recognized)' (Conant, 1981, p. 426). The point is that my activities resonate in the domain of observable activities, consequent on the unique control I am observed to effect in the course of my coupling with other systems. In the case of football, it is not strictly true that I control the ball less well than Pele, except in one particular dimension of consensus. But, as a result of the particular quality of control that I am witnessed to effect, my interaction with the ball manifests itself in a *different* fashion than does Pele's. How myself or some other observer *interprets* this is another matter again.

One of the main thrusts of the work of Maturana and Varela is to introduce into discourse the theme of a non-reductionists materialism, to replace the reductionism of vitalism. The crucial import of Maturana's notion of 'non-intersecting phenomenal domains' is the demonstration that when autonomy is granted a place in the material world, the reduction of one phenomenon into another, or the conflation of two distinct phenomenon, becomes impossible. A representationist view of cognition, or a correspondence theory of truth, demand some form of dualism and its implication of vitalist reductionism (see, for example, discussion in Gregory, 1987).

The problem with a conception of a model as providing the template for comparison by mapping, is that it may all too easily be interpreted as bearing some correspondence relation with truth, suggesting some form of *instructive interaction* between distinct domains—in other words, it may imply that two distinct phenomenal domains might converge towards unity (in this instance, entropy). Conversely, the notion of 'fit' between autonomous unities entails not some synthesis (which is always in the mind of an observer), but a *structural coupling* in the course of which, along a historical trajectory, there are compensatory matchings between organizationally closed systems, as in a dance. This dance entails the process of each of the coupled systems producing its own self in the course of perturbations by the other, rather than either converging towards the other.

From the perspective of the VSM, the viewpoints functioning as System Four should hold a model *of* the environment, the total viable system, and a model *of* System Four itself. On the surface this flies right in the face of the demands of second-order cybernetics. But I want to make the case that such a contradiction is in truth a mater of *interpretation* alone.

We cannot neatly separate out epistemology from ontology. In the consensual region in which we find ourselves embedded, as individual observers we constitute the ontology we find ourselves always historically situated within. Maturana's work throws light on the problematic highlighted by Thomas Kuhn and pushed to the extreme in Paul Feyerabend's 'epistemological anarchy' (Kuhn, 1970; Feyerabend, 1978). We need to realize that the dense play of semantics *binds us* to the order of things as we experience them, rather than tackling, grappling with, or pointing at such an order. Thus, in the unfolding of my own interpretation I cannot but constitute an ontology which I am constructing in terms of the particular consensual domain I am coupled to.

Let me clarify this. It is not correct to say that I am 'just writing about' models. Or rather, that is true if we abstract meaning from a consensual domain—from the interactions of individuals. However, once ontology is granted to the universe *brought forth* in the course of such consensual interactions (including languaging), then we find that what is communicated (i.e. what takes its own place in a particular consensual domain as reality) is the *material* use of models. What I am writing about only emerges as significant (i.e. is effective) in terms of the material reverberations of the use of a model in this or that manner. This has methodological and not merely theoretical interest. Thus it has radical implications for the way we conceive of and design computer systems and computer software (see, for example, the similar point made in Winograd and Flores, 1986, and the ideas incorporated in Action Technologies' *The Coordinator*; also Espejo and Garcia, 1984).

Beer, as has been mentioned, takes great care to point out the indeterminate nature of the sort of cognitive processes that we refer to when we speak of ourselves 'having' this or that model (e.g. Beer, 1986). In other words, though the word 'model' usefully points towards a cognitive procedure, or a procedure in our cultural dealings (e.g. my 'model' train-set), the map is, indeed, never the territory. Model and phenomenon modelled are in two 'non-intersecting phenomenal domains'. From Maturana's perspective, it is not that the one is even the *source* of the other (as hinted at, when we actually think about 'map' and 'territory'). And this is where trouble arises, notably in respect of the VSM.

In the tradition of the Conant–Ashby Theorem, Beer is interested in pointing out *invariances that come to light, when we 'map across' different phenomena* (i.e. when we map across two non-intersecting phenomenal domains). From the perspective of our *interpretation*, what matters is whether we understand the invariances to exist in the *observed* phenomena, or whether we understand them to emerge in the *act* of observing phenomena. Perhaps this is the most fundamental division between those who firmly identify with one cybernetic 'camp' or the other. However, any interpretation is embedded within the horizons of a consensual domain, and brought forth in language—whether that language be the written word, mathematics or art. Interpretation is 'languaged' into being. The thing about the linguistic domain is that we cannot 'get outside' language in order to comment on it at some 'meta-level'—at a certain juncture, all we may do, as Jiddu Krishnamurti and Ludwig Wittgenstein put it, is to remain 'silent'.

Facts and patterns

One of the things that fascinated Gregory Bateson and other early 'cyberneticians' was 'patterns' (Bateson, 1973). Wheresoever an observer observes, s/he may choose to distinguish similarities—hints and echoes of other phenomena and other times, however dissimilar the domains being compared. This leads to such philosophical questions as whether these patterns are 'objective' in the sense of 'out there', independant of the act of observation; whether they are 'objective' as universals 'in here', or conversely, whether they emerge as distinctions in observing systems. But such questions are matters of interpretation alone, depend on perspective and context. They are not matters of fact.

Language is trapped within history, just as it is trapped, self-referentially, within itself. Early insights in cybernetics—rather, strictly speaking, the early *discourse* of cybernetics—did not escape this constraint, nor does this paper. It was apparent to Conant and Ashby, as it was to Beer, that, whatever the actual mechanism (hence the usefulness of black boxes), and whatever the source of the patterns, observing involves such coherenence, such invariances. This implied that, whatever else, cognition functioned *as if* the cognizer were able to represent the external world, and perform cognitive acts on his/her representation. The historical situatedness of that discourse meant that the images (i.e. models) used to explain the phenomenon of patterns, themselves depended on a background of understanding.

From our own historical perspective we may note that what had been distinguished was the phenomenon, since made visible in the discourse of Maturana and Varela, of the coherences entailed in the 'structural coupling' of 'organizationally closed' systems. However, when originally approaching this problematic, what cybernetic discourse indicated was the *fact* of coherences, of 'fits', of coordinated activities across distinct phenomenal domains. The existence of such 'fits' became their ontology, so to speak. However, at some point a fact requires to 'find itself' embedded in explanation. Such explanations emerge on the back of inference, even where derived form an original source indication. In the present case, the indicated fact of coherences required a *model* for its facticity to become visible (i.e. to be accepted as a fact). This is an instance of what is meant by finding our way round our own cognitive space.

Modelling facility as hermeneutic enabler

Keeping this historical caution in mind, I locate the *modelling facility* edged to light by Beer and so crucial for an understanding of his insight into economy in real time, on the historical dimension first suggested by Norbert Wiener and later formalized by Ashby's Law of Requisite variety, rather than the dimension laid down by John von Neumann and apparently indicated in the Conant–Ashby Theorem. Where von Neumann placed ever-increasing value on rationality and logical processes, eagerly embracing the advent of the atom bomb and promises of ever-more gigantic and complex machinery to rule the vagaries of ordinary humankind, Wiener increasingly turned his attention towards the problematic of the '*human* use of human beings', gaining mistrust of giganticism and the threat of technocracy running wild (Wiener, 1951). For such a distinction, see the book tracing the lives and careers of both men written by Stephen Heims (Heims, 1980).

Depending on how we 'bring forth' the VSM, we can contribute towards, indeed constitute, one or the other of the two dimensions suggested in the last paragraph, a point I will return to in my closing remarks. This is our own historicity. We can ourselves choose a representationist interpretation and ascribe the same to the model itself, or we can distinguish the model as a *hermeneutic enabler*. The model itself *does not exist* except as each of us brings it forth. To attack it from a hermeneutic perspective falls into the trap of misplaced concreteness, granting it a spurious transcendental ontology which is not being claimed for it.

However, *within* the context of hermeneutics, the VSM takes on flesh and blood. Now, in line with the Law of Requisite variety as well as Maturana's structural determinism, the modelling faculty has a straightforward function, as it captures the image that Maturana suggests when he speaks of 'the states of the participants as they trigger each other in interlocked sequences of changes of state' (Maturana, 1978, p. 50), and specifically of structural coupling within a consensual domain (i.e. ourselves in society).

The tendency, in the representationist mode, is to set up some modelling function which is held before the eyes of an observer with the exclamation 'See how things are!' In other words, we kid ourselves that we are looking at a picture of 'where we're at'. This is one of the perils lurking in forms of reductionism. Thus, our Western objectivist tradition has it that—as privileged subjects—human beings, if they take the time and trouble, are able to peer into the happenings of the 'real world'. This coloured the nineteenth century attitude towards the pure sciences, an attitude which had such an impact on all intellectual disciplines of the first half of the twentieth century and which continues to sow confusion and doubt.

However, today we are able to view this scenario as simply false. Whether in so-called 'pure' sciences or in any other disciplines, we can never escape our own intimate situatedness as observers, a dynamic part of the observing systems which determine our ontology (Maturana, 1978; von Glasersfeld, 1987). It is simply not the case that we are separate from, outside of, untouched by the natural order of things, a point which the increasingly pressing ecological issues of the day sharply remind us of. And we may never look at that which we ourselves are part of, except in terms of specifications which are themselves constrained by the relations within which we are situated and in turn constitute. Our very looking is in part determining precisely those relations we are seeking to make visible.

The *fact* of a model being a true or false representation of how things are is irrelevant. A model does not gain its positivity—its efficacy as an effective heuristic for action—by being an accurate reflection of an external world. A model emerges with structural integrity (i.e. secures operational closure, and emerges as a useful term for orienting about a consensual domain) by being taken up (i.e. interpreted) as a facility or space for modelling within a particular context.

A simple example will serve to clarify this apparently abstract point. Let us think of a 'hard' model—for instance, the physical model constructed by James Watson and Francis Crick in order to demonstrate the structure of DNA—a construction built of metallic and other materials to 'make visible' the double helix formation, held to be central to the problematic of the structure of the gene. On completing the model, both men shouted 'eureka', or words to that effect, and ran off to drag along witnesses to their breakthrough.

Now what is important to recognize is that the 'model' that was built in the laboratory, the material construction which one could walk around, bump into and point at, did not in itself reflect or suggest DNA any more than it suggested the outline of trees against a winter's skyline. Indeed, no-one suggested that inside a cell the inquiring scientist would 'find' any similar construction. The 'thing' built, was not a model at all, but was a thing that might more accurately have been described as a modern art form! The attempt to suggest a particular set of relations, which would be recognized in terms of a particular set of scientific premises, could only emerge within the context of a particular domain of discourse.

The significance of a model *as a model*—in other words, the model's 'structural integrity'—emerges in a particular consensual domain. In this instance, the consensual domain was that of biologists working at a particular historical juncture, with a particular set of conceptual tools and expectations, concerned with 'the problem of DNA'. In such a consensual context—but in no other—it was 'apparent' that the model provided a coherent means of orienting. Everyone participating in this consensual domain exclaimed 'That's it!', and at a later stage many of us lay persons found ourselves following.

What we take on board with this insight into models is not the power to reflect *back* reality for our own eyes and cognition (i.e. a correspondence theory of truth), but the realization of an ontology which includes one particular dimension—call it history—consisting of our reflections *on* reality. Once more, this bears on the thesis of Thomas Kuhn. This dimension generates its own space or ontology, but such an ontology has nothing to do with some objective world, Reality or Truth.

Historicity and identity

In the language of hermeneutics, we may *effect our own historicity*. To effect one's historicity means to realize one's impact on the order of things in the same move that integrates such an order. It is to move coherently in the order of things, to be alert to oneself and one's medium. It is to realize one's medium in one's own awareness. This is not the same thing as saying that in such a process we can perceive or identify some objective Reality. Rather, it is to effect, to bring forth one's medium in one's actions. It is to seize hold of one's own cognitive space. All this, in a domain of organizationally closed systems.

The System Three–Four homeostat of Beer's model is the transducer between two distinct phenomenal domains—the historical trajectory brought forth by cognition, and the structural eruption of physical being. Any viable system can be distinguished as viable by, among other things, the recognition by an observer that there is such a transducer—such a homeostat. Not that a 'real' transducer provides evidence of such viability, but rather that the recognition of the presence of such a transducer can only emerge upon an indication of the coupling of two distinct phenomenal domains—cognition and structure—from the perspective of an observer.

We realize *not* an objective world which, using a map, we may find our way about, but the existence of non-intersecting phenomenal domains in terms of which our own drift is manifest—in terms of which we sustain our autopoiesis. In a way, one might say that we discover ourselves as observers emerged out of the bifurcation between such non-intersecting domains.

What is 'in the model', including the VSM itself, is not some content corresponding to an objective reality, but rather a series of exercises or algorithms, which give evidence of their relevance and coherence by their successful implementation as heuristics in the course of our 'natural drift' (Maturana and Varela, 1987). Such a 'model' is an ongoing *generative mechanism* rather than a description of some actual phenomenon, and gives rise to an insight into observing systems that requires no overtones of a representationist mode.

The VSM takes its place in the consensual domain for which its implication is that, although the phenomenon of life emerges with organizationally closed, 'blind' unities (i.e. autopoiesis), no single unity may be distinguished in isolation. The further implication is that no unity may be managed in isolation. The bottom-line implication is that no unity may consider itself to be a unity in isolation (e.g. 'I'), however much a reductionistic and atomistic culture tends to inform the contrary. This is the implication of Beer's own stance toward holism (e.g. Beer, 1987).

But—and this is crucial to a full understanding of the cybernetic paradigm—the ontology in which we surface as human individuals is constituted on a fundamental level in terms of unities, and the phenomenon of the observing system is the linchpin of such unities. For operationally coherent cognitive beings, there exists a domain of organizationally closed phenomena, of autonomy. Observation itself, whether conceptualized as active or passive, emerges in the first case as its own organizational closure, in the second as the organizational closure of some 'observed phenomenon'. At whatever level we find ourselves as cognitive beings able to make observations—from the molecular to social—the 'motor of life' is the set of individuals distinguished or distinguishable, even though the 'motor' of the Universe may well be a materially and energetically open system.

This motor of life is not the set of individuals as a system 'out there', with systemic properties and rules, what Felix Guattari (1984) calls 'group phantasy' (Guattari 1984); but is the dance of composite unities as they are structurally coupled to one another during the persistence of their autopoietic organization. In their separation from a background by an observer through an act of distinction, both properties of the background and properties of the unity as a simple unity are specified, although composite unities find their autopoietic space (self-production in the unfolding of their own ontology) in the domain they are structurally coupled to.

In other words, I myself am an organizationally closed cognitive system which is structurally coupled to an equally closed autopoietic organization which is myself as a biological organism, persisting in a 'blind' natural drift. My ongoing self-production (i.e. autopoiesis) is maintained just as long as certain structural constraints are not broken in the course of the structural dance of myself as a composite unity with the domain I am structurally coupled to. From the persistence of this structurally determined 'fit' emerges my own identity as an observer, a simple unity in the space of my interaction with other simple unities in a material and a consensual domain, in terms of which I find myself erupted into Universe.

System Four: a term of conversation

The VSM fractures the tendency inherent in systems thought and process ideas to 'drown' the individual in organizational hegemony, just as it refuses vitalist reductionism by demanding a 'boundary' that is no other than active reflection *on* self rather than passive reflection *of* self. System Four says 'Think what you're doing!', and in that process demands that you discover *who* is doing the doing. The doer is not dualistically situated within some system which s/he contemplates from a privileged vantage (e.g. mind, soul), but is the ongoing reflection upon the doing that is being done. As Maturana put it, the mind is not in the head (Maturana, 1985). Such a process of reflection demands acts of distinction (Spencer Brown, 1972) which emerge within an observing system instead of being the prerogative of some observer *in vacuo*. In the context of social enterprises such acts of distinction are what constitutes the ontology of any one social organization in a consensual domain, from the perspective of the observer.

No social organization—or any organization come to that—has an objective existence 'out there'. The firm, family, state, church, college, etc., all find their grounding (i.e. their ontology) in some particular consensual region in which they are distinguished as performing certain functions, doing certain things. These 'things' they do can never in practice be finally defined and isolated. Does a hospital cure the sick, or promote iatrogenic disorder? It does both, of course, and an indeterminate number of other things, depending on the viewpoint concerned ('viewpoint' as in Espejo, 1987). Such a viewpoint is constrained by a consensual domain, but no two viewpoints may be held to correspond (just as no model corresponds with any reality). The purpose of an organization has traditionally been taken as unproblematical. For instance, traditionally a hospital has been claimed to be an organization whose aim is to heal the sick. Today, problems are arising because hospitals are becoming seen increasingly as organizations to provide career structures or to manage social problems (the medical gaze: Foucault, 1973). The VSM bears on such matters.

In this context the VSM is not held to reflect, or model, what a particular social organization does. Compare and contrast this insight with Beer's point that the purpose of a system is 'what it does' (e.g. Beer, 1979, p. 412). The purpose is what an observer ascribes to the system-in-focus in the course of the distinctions emerging in the total observing system (i.e. observer/system). Were the model to be seen to reflect what the organization does it would fall into the representationist fallacy, and inevitably would fall prey to the interests (i.e. prejudices) of the modellers, claiming, in some guise or another, access to a more objective, privileged reality. In the discourse of Maturana, it would hold up the unfounded offer of instructional interaction.

Instead, the VSM becomes a *term* of an ongoing *conversation* (as in Pask, 1975a,b), and in their implementation Systems Three and Four provide the terms of an ongoing conversation in which the other protagonists are the viewpoints constituting this particular organization. The peculiar quality of the 'terms' of such conversation is that they themselves are organizationally closed, although in the medium of the particular social enterprise (i.e. in terms of a particular consensual domain), they are structurally coupled with the cognitive models of members of the enterprise. Thus, Systems Three and Four are found to be 'fitting' or matching these protagonists, as they 'fit' itself and each other, in structurally realizing a particular organization. Were it not for this closure, there would be no cooperative enterprise —no organizational cohesion.

System Four is often interpreted as containing a model of the whole of the viable system, as well as a model of itself. But we must remind ourselves that this model is not a representation or picture of anything. Nor is it a map of anything. What, then, is it in practical terms? The 'viable system' is not a real thing, any more than a model points at anything real. 'Viable system' is a logical construct in a consensual domain, serving the purpose of helping an observer orient about his/her own cognitive space. The notion of 'viable system' in System Four is likewise a logical construct, this time in the consensual domain in which a particular organization is granted its ontology in terms of the cognitive space of its constitute elements. System Four is a *modelling facility* and not *a* model *of* anything. It is a plastic medium permitting modelling—permitting modelling and allowing the emergence of models. In the social domain it is the space for modelling whose modelling terms emerge in the instant of the confirmation of a collaborative venture by those individuals who constitute the structure through which the organization of the enterprise is realized. To the degree that we as designers provide for or omit such a space, we determine a particular quality of consensual domain, and in turn affect the coherence of interactions for those individuals who will be orienting according to the terms of such a consensual region.

The crucial significance of this mechanism and its recursive situation in the VSM (i.e. having a model of the viable system as well as a model of itself) is that it demands, from the user of the model, the recognition or acceptance that a consensual domain is no other than a *play of models*, rather than that we need to model 'a thing' before we can control it. Further, it demands recognition that the public play of models, and the models that we use to orient within our own cognitive space, though structurally coupled, inhabit two non-intersecting phenomenal domains, each organizationally closed.

Language may indeed be no other than language-in-use or action, as Maturana, Pask, Austin, Searle and others have professed, rather than some closed domain of abstractions which represent the world and our place in the Universe. But the elements of such a domain of consensual interaction, the terms of this process of languaging, are models. Our *orienting* via language ('orienting' as used in Maturana), an orienting which constitutes the particular character of this or that consensual domain (i.e. its ontology), consists of a juggling of models along the historical trajectory constituted in the process of such juggling—what Pask calls 'concept building' (Pask, 1975a,b).

We do not find the cognitive models of individuals in the same domain as the consensual models which constitute the terms of language, however. Cognitively speaking, the individual conversants, though structurally coupled to the consensual domain (which in this context means that they are informationally open), are organizationally closed phenomena. Thus, although Maturana is absolutely accurate in reflecting that *human* experience is 'languaged' into existence, this does not imply that speech-acts and cognition form terms of one domain.

Beer's work demonstrates that, instead of treating a social enterprise as some objective thing, a thing we can paint, sketch, photograph or otherwise represent—in other words, an organization that we can analayse down to its constitutive elements—we need to treat it as an indeterminate series of linguistic terms and commitments *in actual use*. In their structural coupling with such a consensual domain, individuals, whether explicitly or implicitly, find their way around their own cognitive space—*this is the modelling facility*. The organizational closure of such a cognitive space, and at the same time its coherent 'fit' with all other domains it is structurally coupled to, is what we apprehend when we pause to reflect, and any cooperative interaction of human individuals (i.e. any viable social organization) will reflect upon its own context and the particular identity which coheres consequent from the acts of distinction which bring it forth as a particular organization.

The representationist in us all will tend to cry out 'That can't be right! Some *thing* must be modelled—a model must stand for something!' Else, we feel, there is no control, no direction, no coherence. But any model is arbitrary (see, for example, Beer, 1986), and further no model in truth can be structurally decoupled from a consensual domain—from history. In other words, it has to be granted the status of 'model' and the only way this can occur is from within the horizons of some tradition or other. Regardless of whether we like it or not, our models are the terms of our own cognitive domain, and are structurally coupled to the terms of our language-in-use. They attain their positivity, they are selected, not as a result of whether they are accurate representations of reality, but as a result of the continued coherence of this closed, cognitive domain, in the course of its disturbance, its perturbation, by material reality, which includes the consensual domain.

This is not to say that existence itself is languaged. Such a view—what Foucault (1970) called 'Anthropological thought'—grotesquely reduces the variety of Universe beneath some cultural blanket. However, a *model*, whether cognitive or consensual, is a term within cultural horizons, and a vast amount of our lived-experience is constrained within such horizons. This does not specify anything about what is within such models (i.e. what they are models *of*), but merely points out, or distinguishes, a particular phenomenal domain which binds the human observer within history. And so it is with any social enterprise which an observer would have persist. Thus Beer identifies 'organizational cohesiveness' as the process of continuous planning in real time, by all elements involved in a joint enterprise (Beer, 1979, pp. 335ff).

Conclusions

The VSM is not an inert 'thing', if indeed anything may be said to be inert. The VSM is languaged into being, and forms an ongoing *term* of discourse within a consensual domain; but it is not apprehended directly in the cognitive space. An observer within the particular consensual domain which gave rise to the VSM may choose to *interpret* the VSM. One person might interpret the model in the light of Turing machines—input/output transformations. And if that is useful for him/her, good luck! Another person might interpret the VSM in the light of Varela's call for the autonomous mode of operational closure (e.g. Varela, 1986). The point is that the interpretation actually effected will depend on whether the interpreter—the observer—has a representationist model of models, or not. If they themselves have such a model, then they will search for the phenomenon being 'represented' in whatever it is that they interpret the model as being a representation of, and further will assert the phenomenon of the model itself as being something that their own cognition is representing.

In the light of Maturana's work, what becomes evident is that there is no longer any need for a notion of accurate representation to be linked to the notion of 'viability'. We do not need to be able to 'picture' something in order to control our relations in terms of it. We neither need to consider the VSM itself, nor the models as they function within the VSM, as reaching out into the world, and somehow capturing it.

If we are talking about viability that suggests 'fit' (von Glasersfeld, 1986) or 'natural drift' (Maturana and Varela, 1987), rather than neo-Darwinian 'survival of the fittest', then we treat the model as an algorithm for a quality of control that emerges out of and in turn enables a valuable *heuristic for structural coupling*, specifically in terms of our own efforts to coherently orient ourselves within our own cognitive space, as we *cooperatively* constitute the consensual domains we are structurally coupled to.

Conversely, if our goal is to preserve the status quo, to gaze over our shoulders towards Malthus and the social Darwinists such as Herbert Spencer; if we wish to reinforce the type of institutional apparatus that developed in the course of the emergence of the industrial paradigm; then we interpret the model as a representation of some 'natural' (i.e. objective) phenomenon. In effect, the System Four modelling function collapses into System Three control mode alone in spite of all the efforts of Beer and Espejo to tease the two apart. In the mood of the Conant–Ashby Theorem, viability then becomes something to be measured by control over some objective world, instead of control of the actual relations we constitute being evidenced by *communication* within a consensual domain.

You pays your money and takes your choice; but in its interpretation as a hermeneutic mechanism, the Viable System Model has no incompatibility with second-order cybernetics. Instead it deepens our insight into the processes that Maturana and others are attempting to bring into contemporary discourse.

References

Bateson, G. (1973). *Steps to an Ecology of Mind*. St Albans, UK: Paladin.
Beer, S. (1979). *The Heart of Enterprise*. Chichester: John Wiley.
Beer, S. (1981). *Brain of the Firm*, 2nd edn. Chichester: John Wiley.
Beer, S. (1986). 'Recursions of power', in *Power, Autonomy, Utopia* (ed. Trappl, R.). New York: Plenum.

Beer, S. (1987). 'Holism and the Frou Frou slander', opening address to the seventh triennial international congress of W.O.G.S.C., Imperial College, London.

Bohm, D. (1968). 'Creativity', *Leonardo*, **1**, 137–49.

Checkland, P. B. (1981). *Systems Thinking, Systems Practice*. Chichester: John Wiley.

Conant, R. (ed.) (1981). *Mechanisms of Intelligence: Ashby's Writings on Cybernetics*. Salinas, CA: Intersystems.

Conant, R. and Ashby, W. R. (1970). 'Every good regulator of a system must be a model of that system', *International Journal of Systems Science*, **1**, 89–97.

Espejo, R. (1987). 'From machines to people and organisations: a cybernetic insight on management', in *New Directions in Management Science* (eds. Jackson, M. and Keys, P.). Aldershot: Gower.

Espejo, R. and Garcia, O. (1984). 'A tool for distributed planning', *Proceedings—Orwellian Symposium and International Conference on Systems Research, Information and Cybernetics*, Baden Baden.

Feyerabend, P. (1978). *Against Method*. London: Verso.

Foucault, M. (1970). *The Order of Things*, London: Tavistock.

Foucault, M. (1973). *The Birth of the Clinic*, London: Tavistock.

Gregory, D. (1987). 'Philosophy and practice in knowledge representation', in *Human Productivity Enhancement—Organisations, Personnel and Decision-Making* (ed. Zeidner, J.). New York: Praeger.

Guattari, F. (1984). *Molecular Revolution—Psychiatry and Politics*. Harmondsworth: Penguin.

Heims, S. (1980). *John von Neumann and Norbert Wiener: From Mathematics to the Technologies of Life and Death*. Cambridge, Mass.: MIT Press.

Kuhn, T. (1970). *The Structure of Scientific Revolutions*, 2nd edn. Chicago: University of Chicago Press.

Maturana, H. R. (1978). 'Biology of language: the epistemology of reality', in *Psychology and Biology of Language and Thought—Essays in Honor of Eric Lenneberg* (eds. Miller, G. and Lenneberg, E.). Notre Dame: Academic Publishers.

Maturana, H. R. (1980). 'Biology of Cognition', in *Autopoiesis and Cognition: the Realization of the Living* (eds. Maturana, H. R. and Varela, F. J.). Dordrecht: Reidel.

Maturana, H. R. (1983). 'What is it to see?', *Archives of Biological and Medical Experimentation*, **16**, 255–69.

Maturana, H. R. (1985). 'The mind is not in the head', *Journal of Social and Biological Structures*, **8**: 308–311.

Maturana, H. R. and Guiloff, G. (1980). 'The quest for the intelligence of intelligence', *Journal of Social and Biological Structures*, **3**: 135–48.

Maturana, H. R. and Varela, F. J. (1980). 'Autopoiesis: the organization of the living', in *Autopoiesis and Cognition: the Realization of the Living* (eds. Maturana, H. R. and Varela, F. J.). Dordrecht: Reidel.

Maturana, H. R. and Varela, F. J. (1987). *The Tree of Knowledge*. Boston: Shambhala.

Pask, G. (1975a). *The Cybernetics of Human Learning Performance*. London: Hutchinson.

Pask, G. (1975b). *Conversation, Cognition and Learning*. Amsterdam: Elsevier.

Spencer Brown, G. (1972). *Laws of Form*. New York: Dutton.

Varela, F. J. (1986). 'Steps to a cybernetics of autonomy', in *Power, Autonomy, Utopia* (ed. Trappl, R.). New York: Plenum.

Von Foerster, H. (1981). *Observing Systems*. Salinas, CA: Intersystems.

Von Glasersfeld, E. (1986). 'Steps in the construction of "Others" and "Reality": a study in self-regulation', in *Power, Autonomy, Utopia* (ed. Trappl, R.). New York: Plenum.

Von Glasersfeld, E. (1987). *The Construction of Knowledge—Contributions to Conceptual Semantics*. Salinas, CA: Intersystems.

Winograd, T. and Flores, F. (1986). *Understanding Computers and Cognition: A New Foundation for Design*. Norwood, N.J.: Ablex.

Wiener, N. (1951). *The Human Use of Human Beings: Cybernetics and Society*. London: Eyre and Spottiswoode.

Part Four
Critical Views

The Viable System Model: Interpretations and Applications of Stafford Beer's VSM
Edited by R. Espejo and R. Harnden
Published 1989 by John Wiley & Sons Ltd

15

Evaluating the managerial significance of the VSM*

M. C. Jackson

Department of Management Systems & Sciences, University of Hull

This chapter attempts a detailed assessment of the managerial significance of Beer's Viable System Model using the method of 'reflective conversation'. Some broad parameters for the conversation are established. The best case for the managerial significance of the VSM is then made. This is followed by the case against. The conversation ends with suggestions for a possible accommodation between the two positions.

Introduction

Attempting to assess the managerial significance of Stafford Beer's Viable System Model is no simple task. The books and articles (Beer, 1972, 1975, 1979, 1981, 1984, 1985) in which the model is developed and described have produced a very varied response from managers and from academics working in the management sciences. In the operational research/management science (OR/MS) community, Beer is held in high esteem and his work is regarded as being amongst the most substantial, creative and stimulating contributions in the whole literature of the discipline. In the related area of organization theory, however, his writings receive little serious attention. In the realm of practice, Beer and a coterie of devotees find ready application for the ideas encapsulated in the VSM. Many equally intelligent managers and management sciences practitioners can find no circumstances in which they could implement, in substantial part, the

---

*A shorter version of this chapter appeared in the *Journal of Management Studies*, vol. 25, no. 6, November 1988, published by Basil Blackwell Ltd.

recommendations endorsed in this model. It is as well, too, to acknowledge that the VSM and its use have come in for some severe and deeply felt criticism (Rivett, 1977; Checkland, 1980, 1986; Ulrich, 1981, 1983). All this should make us wary of embarking on the proposed assessment of the VSM, without first taking time out to examine the grounds on which an evaluation might proceed.

The process of evaluation

It would obviously not take us far if we attempted to evaluate the VSM *solely* in terms of its own cybernetic assumptions. Such a self-justifying approach cannot even, on the basis of Godel's (1962) Theorem, prove the consistency of a system of thought. Still less could it inform us of the nature and scope of the assistance that the VSM can lend to management practice. The VSM must also be considered in relation to the ideas of those who can find no use for this model and/or who vehemently criticize it. But in trying to do this we shall find ourselves confronted with another important difficulty. Adherents of different theoretical traditions frequently seem to talk past each other, failing to engage in any constructive dialogue. Kuhn (1970) describes this phenomenon occurring in the natural sciences and calls it 'paradigm incommensurability'. The word paradigm refers to the set of concepts, assumptions and beliefs that guide the activity of a particular scientific community. Talking about the most fundamental aspect of incommensurability, Kuhn says:

> '. . . .the proponents of competing paradigms practice their trades in different worlds . . . the two groups of scientists see different things when they look from the same point in the same direction.' (Kuhn, 1970, p. 150)

The notion of paradigm incommensurability can with value be employed to understand the gulf in understanding which exists between Beer and his detractors. Beer's work, as has been suggested elsewhere (Jackson and Keys, 1987), demonstrates a 'structuralist' orientation. He argues in *Decision and Control* (1966) that scientific management should not be content simply with discovering the facts but should also seek to know what the facts mean, how they fit together, and should seek to uncover the 'mechanisms' which underlie them. This is wholly consistent with the structuralist endeavour to provide models of the causal processes at work at the deep structural level which produce the observable phenomena and the relationships between surface elements. Some OR/MS people and many organization theorists have great difficulty with this. Trained within a paradigm resting upon positivism, they assume that scientific knowledge in the management sciences, as in the 'classical' account of physics, is accumulated by the empirical observation of reality and the analysis of the results so obtained. Causal connections are visible at the surface level. It is little wonder that they have problems absorbing Beer's message. A similar problem haunts communication between Beer and his critics. The most serious and lucid attacks on the VSM have been based either directly on the idealism of Kant (Ulrich, 1981, 1983) or have originated from within the 'interpretive' paradigm (Checkland, 1980, 1986) which owes so much to Kant's work. Within this tradition, knowledge of social phenomena is gained by understanding the perceptions and beliefs of those involved in the construction and maintenance of social reality. This profoundly different orientation makes it impossible for Beer to engage in fruitful debate with his critics. He attests to the difficulty he felt in dealing with Ulrich's arguments concerning the VSM and its application in Chile, because communication seemed to be needed between two paradigms (Beer, 1983).

Given this situation, it is tempting to set off in search of some independent point of reference against which to assess the competing claims of the rival perspectives. However, the very notion of science as governed by paradigms which select research interests and strategies and condition the perceptions of scientific communities, warns us that the search for such independent foundations is likely to be fruitless. Acceptance of paradigm incommensurability seems therefore to rule out the possibility of comparing the respective claims to knowledge of rival perspectives. Burrell and Morgan take this stance in *Sociological Paradigms and Organizational Analysis* (1979). There,. they argue that progress can best be made in organizational analysis if particular paradigms develop on their own account without reference to alternative positions:

> 'Contrary to the widely held belief that synthesis and mediation between paradigms is what is required, we argue that the real need is for paradigmatic closure.' (Burrell and Morgan, 1979, pp. 397–8)

Now, while we do need to be alert to the problems posed for the process of evaluation by the existence of competing paradigms, we do not have to accept the implications of the doctrine of 'paradigm incommensurability' as framed in its strongest form. There *is* something to be gained by setting down competing perspectives and subjecting them to some sort of comparison. At the very least this should help counter the chauvinistic claims to foundational knowledge made by adherents of some perspectives. It should help reveal the taken-for-granted assumptions that underpin particular approaches and the limitations these impose. This is indeed the modified position Morgan comes to adopt in a later book, *Beyond Method* (1983a). In this volume he even recommends a procedure for making comparisons between different research strategies, 'reflective conversation', and sets down the advantages of its use:

> 'By reflecting on one's favoured research strategy in relation to other strategies, the nature, strengths and limitations of one's favoured approach become much clearer. In seeing what others do, we are able to appreciate much more clearly what we are *not* doing. In this way, we are able to create a means of developing and refining favoured research strategies in a way that makes them stronger, yet at the same time more modest.' (Morgan, 1983a, p. 381).

It is the intention in this chapter, while being fully aware of the problems posed by paradigm incommensurability, to establish a reflective conversation between the VSM and the position adopted by critics of the VSM. To this end the case for the VSM and the case against it will be kept separate. The best case for the managerial significance of the VSM will be made in terms of the tenets and concepts employed in that model. The best case against the usefulness of the VSM, in the managerial context, will then be developed using the wholly different set of assumptions employed by its critics. Although some suggestions will be made about how an accommodation between the different viewpoints might be reached, the reader must be left to make his or her own synthesis—if indeed this is held to be possible.

In order to maximize the chances of a successful 'conversation', it is necessary first to set boundaries for the debate. Even advocates of paradigm incommensurability in a strong form would accept that it improves the chances of drawing conclusions if evaluation is conducted in relation to some specific purpose. In this instance this means formulating more clearly the idea of 'managerial significance' as a basis for the comparison between the position adopted in the VSM and the position adopted by its critics.

The basis of evaluation

Because of the difficulties, discussed in the last section, of adjudicating between different claims to 'truth', it is reasonable to broaden our basis of evaluation to embrace other considerations as well. As Morgan insists:

> 'Knowledge may serve to explain empirical facts, help us to understand meanings, allow us to act more appropriately, empower in a liberating way, reveal links between everyday reality and the structural logic that produces and reproduces that reality, advance specific political interests, and so on.' (Morgan, 1983, p. 403)

This will be even more the case with an activity such as 'management' which has as its essence the planning and control of human as well as material resources. The managerial significance of knowledge will involve far more than the technical matter of whether that knowledge is accurate enough to permit effective action to be taken. The basis of evaluation needs to reflect this broader conception of knowledge while not losing sight of the need for *some* specific focus.

Another necessity here is to ensure that the guidelines chosen for the purpose of evaluating the managerial significance of the VSM are reasonable and fair in relation to the VSM. Beer set out to try to understand how systems were capable of independent existence; to uncover the laws underpinning their viability. As he remarks:

> '. . . the problem addressed did not include juggling with "ISMS", nor did it commit me to a survey of organization theory, nor have I tried to prescribe how organizations should be designed—although I have been heavily censured for failures in all these departments.' (Beer, 1983)

Bearing these comments in mind, what might we reasonably expect of Beer's model in terms of managerial applicability? In what follows attention is drawn to Vickers' writings on the nature of management, to the popular conception of organizations as socio-technical systems and to the sociological theorizing of Habermas. It is argued that these three sources show a consistent concern with certain identifiable features of the 'management' task. From this it is possible to establish a suitable basis from which to proceed to evaluate the VSM.

Vickers (1967) starts from the belief that 'business is a social activity and management a form of social regulation', and asks what concepts can help him illuminate this belief. He identifies two sets. The first lays stress on circular processes of control as found in feedback assemblies, and involving the comparison of 'what is' with 'what ought to be'. The second set emphasizes

> '. . . the part played by human expectations, socially generated, in setting and changing the manifold and often conflicting standards of what "ought to be"—the standards of success—and hence in determining and constantly altering the "states" which human organizations are set to seek.' (Vickers, 1967, p. 9)

This view of the management function as concerning itself with both the social generation of objectives and the efficient pursuit of those objectives, marries well with the contemporary orthodoxy in organization theory— that organizations are best considered as open socio-technical systems (see, for example, Kast and Rosenzweig, 1985). Organizations, from this perspective, are represented as having a technical aspect, being systems which seek to efficiently and effectively pursue goals in often volatile environments. They also possess a human and social aspect and depend for their viability on the establishment of shared understanding among their members about the goals to be pursued.

The sociological theorizing of Habermas (1972, 1974, 1979) offers another basis for understanding the nature of organization as a socio-technical process and therefore, also, the management task. According to Habermas there are two fundamental conditions underpinning the socio-cultural form of life of the human species. These he calls 'work' and 'interaction'. 'Work' enables human beings to achieve goals and to bring about material well-being through social labour. Its success depends upon achieving technical mastery over natural and social systems. The importance of work leads human beings to have a 'technical interest' in the prediction and control of natural and social affairs. This is one of two anthropologically based cognitive interests which Habermas believes the human species possesses. The other is linked to 'interaction' and is labelled the 'practical interest'. Its concern is with securing and expanding the possibilities for mutual understanding among all those involved in social systems. Disagreement between different groups can be just as much a threat to the reproduction of the socio-cultural form of life as a failure to predict and control natural and social processes.

While work and interaction have for Habermas (at least in his later work) pre-eminent anthropological status, the analysis of 'power' and the way it is exercised is equally important, Habermas argues, if we are to understand past and present social arrangements. The exercise of power in the social process can prevent the open and free discussion necessary for the success of interaction. Human beings have, therefore, an 'emancipatory interest' in freeing themselves from constraints imposed by power relations and in learning, through a process of genuine participatory democracy, involving discursive will-formation, to control their own destiny.

Since organizations are at the heart of the socio-cultural life of humans, it seems clear that they will be the primary centres of social labour, social interaction and the exercise of power. We all have a technical, a practical and an emancipatory interest in their functioning.

Putting together insights gained from Vickers and Habermas and from considering organizations as open socio-technical systems, it is possible to envisage a reasonable standard of managerial significance that can be applied to the VSM. An organizational model might serve the technical interest by providing premises which can become the basis for the efficient and effective design of adaptive goal-seeking systems. It could also seek to support the practical and emancipatory interests by facilitating mutual understanding from which genuinely shared purposes could emerge and by permitting reflection on how purposes are actually derived and on the nature of those purposes. As far as possible, therefore, the cases for and against the VSM will be made in terms of its significance in relation to the *premises* that must underpin efficient and effective managerial practice, and in relation to the *purposes* that managerial action serves and the manner in which these are determined. A brief description of the VSM is a necessary prerequisite to this argument.

The VSM: a brief description

Beer's aim was to unearth the laws underpinning the viability of systems so that we can understand how systems are capable of independent existence. In *Decision and Control* (1966) Beer set out the methodological procedures he was to employ in fulfilling this project. Apparent resemblances between system types are pursued to see if they hold up as fruitful analogies under the discipline of conceptual modelling. If they do, the conceptual models are subject to further rigorous analysis to turn them into mathematically based models. When these mathematically based models are compared they may reveal that the different system types can be mapped on to the same scientific model. If so, the initial resemblances have now led to invariances which demonstrably hold across different system types.

In *Brain of the Firm* (1972) Beer pursued apparent resemblances between the way in which the human body and the firm are controlled and organized. It seemed sensible to search for the principles underlying viability by studying a 'known-to-be-viable' system—the human organism as controlled by the nervous system. In *Brain*, therefore, a neuro-cybernetic model of the workings of the human body and nervous system is set out. A similar model, Beer demonstrates, can be used to understand how firms must operate if they are to be viable. From this comparison of 'brain' and management structures, therefore, Beer is able to construct a scientific model of the organization of any viable system—the VSM. As a further step towards establishing the generality of his model, Beer succeeds in *The Heart of Enterprise* (1979) in building the VSM from cybernetic first principles—though with special reference to organizations. In a more recent publication, *Diagnosing the System for Organizations* (1985), the VSM is presented in the form of a 'handbook or manager's guide', the intention being to aid application of the principles to particular enterprises. It is from these three sources, *Brain*, *Heart* and *Diagnosing*, that the following account is drawn.

The main problem for an organization in achieving viability is the extreme complexity and uncertainty exhibited by its environment. Thanks to Ashby (1964) cybernetics can provide some understanding of this difficulty and ways of dealing with it. Ashby takes the credit because he provides a measure of complexity—'variety', the number of possible states a system is capable of exhibiting—and because of his formulation of the famous Law of Requisite Variety; 'only variety can destroy variety' (1964, p. 207). So in order to become or remain viable, an organization has to achieve requisite variety with the complex environment with which it is faced. It must be able to respond appropriately to the various threats and opportunities presented by its environment. The exact level at which the balance of varieties should be achieved is determined by the purpose that the system is pursuing. Thus if an organization wishes to be successful at the purpose of selling bathroom suites, there will be a certain level of variety in the environment, relating to demand for particular styles and colours, which it will need to match with its own productive apparatus. Variety, then, is a subjective measure—a measure of relevant states given some defined purpose. And far from this making it inadequate for scientific work, as has been claimed (Rivett, 1977), it secures for the concept an extremely useful heuristic role in examining actual and potential variety balances. This said, the point remains that the potential variety of the environment always threatens to overwhelm that of the system. The same holds true for managers facing the massive potential variety of the operations they control. Complexity, therefore, has to be carefully managed. This is described by Beer as 'variety engineering' and he sets out a number of strategies (involving variety attenuation and variety amplification) that can be used in order to balance these variety equations in a satisfactory way (Beer, 1981, pp. 230–1). Beer's VSM can legitimately be seen as a sophisticated working out of the implications of Ashby's Law of Requisite variety in organizational terms.

With this in mind, it is now possible to elaborate on the structure of the VSM. According to Beer, all viable systems need to possess five functions, which he calls Systems 1–5. The System 1 of an organization consists of the various parts of it directly concerned with implementation. Each part of System 1 should be autonomous in its own right, so that it can absorb some of the massive environmental variety that would otherwise flood higher management levels. This means the parts themselves must be viable systems and must exhibit the five functions—the model is 'recursive'; the structure of the whole is replicated in each of the parts. System 1 has some special primacy in Beer's VSM because it consists of other viable systems and because it *produces* the viable system of which it is part (Beer, 1985, pp. 91 and 128). The management 'meta-system', Systems 2–5, emerges from the need to facilitate the operations of System 1, and to ensure the suitable adaptation of the whole organization. System 2, coordination, is necessary to ensure that the various elements making up System 1 act in harmony. System 3 is a control function ultimately responsible for the internal stability of the organization. It must ensure that System 1 implements policy effectively. System 4, or the intelligence function, has two main tasks. First, it switches information both ways between the 'thinking chamber' of the organization, System 5, and the lower-level Systems. Second, it must capture for the organization, all relevant information about its total environment. System 4 is the point in the organization where internal and external information can be brought together. As such, Beer proposes that it house the 'operations room' of the enterprise; a real 'environment of decision' in which all senior meetings are held. System 5 is responsible for policy. One of its most difficult tasks is balancing the sometimes antagonistic internal and external demands placed on the organization, as represented by the requirements of System 3 and System 4 respectively. System 5 must also represent the essential qualities of the whole system to any wider system of which it is part.

Much attention is also given in the model to the information channels linking Systems 1–5 and the organization and its environment. These channels, and the necessary 'transducers' translating information when it crosses system boundaries, must be designed according to the requirements of requisite variety. A special 'algedonic' (pain/pleasure) filter is employed to separate out particularly important signals which may require the intervention of senior management. Finally, there is particular concern about the nature of the information which flows around the various connections. This will often, given the importance of negative feedback for control, be information about how the different parts of the organization and the organization as a whole are doing in relation to their respective goals. Achievement in most organizations is measured in terms of money; the criterion of success being the extent to which immediate profits are maximized and costs minimized. However, this is not regarded as satisfactory by Beer. It ignores how well the organization is doing in terms of preparing for the future, perhaps by investing in research and development, and in terms of resources like employee morale. Instead, Beer advises adopting three levels of achievement (actuality, capability and potentiality) which can be combined to give three indices (productivity, latency and performance) expressed in ordinary numbers. These can be used as comprehensive measures of performance in relation to all types of resource throughout the organization.

The cases favourable to and against Beer's VSM can now be argued on the basis of both 'premises' and 'purposes'.

The case for the VSM

Premises

We are arguing here that the VSM can provide premises to underpin efficient and effective managerial action.

A useful starting point is to stress the generality of the model—the variety of contexts in which it has been found useful. This stems from its very nature. The recommendations endorsed in the model do not tightly prescribe a particular *structure*; they relate more to a systems essential *organization*, to use a distinction drawn by Varela (1984). They are concerned with what defines a system and enables it to maintain its identity, rather than with the variable relations that can obtain between the components integrating particular systems. The VSM lays down a minimum set of necessary relations that must obtain if a system is to continue long in existence. It does not try to provide a detailed blueprint for design. As a result it has been found to be applicable to small organizations (Espejo, 1979; Jackson and Alabi, 1986), large firms (Beer, 1979), training programmes (Britton and McCallion, this volume), industries (Baker, Elias and Griggs, 1977), local government (Beer, 1974) and national government (Beer, 1981, pt. 4). This is, of course, a tribute to its managerial applicability and significance. I can testify that it was the only management model capable of integrating into a book six diverse contributions to a seminar series about the management of transport systems (Keys & Jackson, 1985). From a detailed analysis of these various case studies it is possible to pick out those features of the VSM which serve it most advantageously when it is used to assist management practice. Five of these deserve attention here.

First, the model is capable of dealing with organizations the parts of which are both vertically and horizontally interdependent. The notion of recursion enables the VSM to cope with the vertical interdependence displayed in, say, a multinational company which itself consists of divisions, embracing companies, embracing departments, etc. In the VSM, as we saw, the parts of System 1 of an organization must be viable systems in their own right and must possess their own Systems 1–5. The organization being considered will, at a higher level of recursion, be simply an implementation subsystem of another viable system. The generality of the VSM and its applicability at different system levels allows elegant diagrammatic representations of management situations to be constructed, and acts as a great variety-reducer for managers and management scientists. Lower-level systems, which will inevitably appear as 'black boxes' at high levels of recursion, can become the focus of detailed interest in their own right with only a slight adjustment of attention. The Metapraxis consultancy company take full advantage of this in arrangements for the presentation of control information to senior managers (Preedy and Bittlestone, 1985). The notion of recursion is not unique to Beer's writings—Parsons' AGIL schema is applied at different system levels (see Hamilton, 1983)—but only in the VSM is it incorporated into a usable management tool.

Horizontally interdependent subsystems—the parts of System 1—are integrated and guided by the organizational meta-system, Systems 2–5. The hoary old problem of centralization versus decentralization is dealt with in the VSM by allowing to the subsystems as much autonomy as is consistent with systemic cohesiveness. The meta-system will intervene only to prevent one of the parts acting in such a way that it could threaten what the whole system is trying to achieve. The degree of systemic cohesiveness required is, according to Beer, a computable function of purpose. It follows that the nature of the systems purpose should be the guide to settling the balance between centralization and decentralization. There are some close parallels between Beer's account of this issue and the 'contingency theory' approach to differentiation and integration offered by Lawrence and Lorsch (1969).

Second, the model demands that attention be paid to the sources of command and control in the system. System 5 is responsible for policy in the VSM. In this it is aided by System 4 which collects relevant environmental information and brings it together with details of internal performance. System 5's policy function will often involve balancing internal and external demands, as represented in the organization by the desire of System 3 for stability and the bias of System 4 for adaptation. The System 3–4–5 inter-relationship, as described by Beer, shows interesting similarities with Thompson's (1967) well-known discussion of the administrative process. Britton and McCallion (this volume) note other parallels with Thompson's work.

The role of System 4 deserves special attention here. It is a development function which, in the light of threats and opportunities in the environment, can suggest changes to systemic purpose and consequent alterations of organizational structure. It cannot therefore be argued that the VSM simply takes existing organizational structures for granted and that this limits its ability to ensure organizational change. Cybernetic models based upon the negative feedback mechanism did tend to put the emphasis on stability at the expense of change. Beer's model, based on an equation of organisational and environmental varieties according to some purpose and with an institutionalized development function, is capable of ensuring structure elaboration. Only the system's *organization*, its source of identity and viability, must remain as prescribed by the VSM.

In relation to 'command and control', the autonomy of the parts of System 1 should again be noted. These subsystems are viable in their own right with their own relations with the outside world and their own localized managements. The restrictions on their autonomy imposed by Systems 2 and 3 are only such as to ensure overall systemic cohesiveness. In Beer's model, therefore, the source of control is spread throughout the architecture of the system. This allows the self-organizing tendencies present in all complex systems to be employed productively. Problems are corrected as close as possible to the point where they occur. Motivation should be increased at lower levels. Higher management should be freed to concentrate on meta-systemic functions. The importance of encouraging self-organization and freeing management for 'boundary management' activities has been well-documented in the literature of socio-technical systems theory (Rice, 1958). It is also one of the main planks of the 'St.

Gallen School' of management cybernetics. H. Ulrich (1984), Malik and Probst (1984), and Probst (1984), all offer reasons why it should promote greater efficiency.

Third, the model recognizes that information (in the service of planning) is the true cement holding organizations together. It offers a particularly suitable starting point for the design of information systems, as indeed has been convincingly argued by Espejo and Watt (1978) and Espejo (1979). Most designs for information systems are premised upon some taken-for-granted model of organization—usually the outdated, 'classical', hierarchical model. It takes a revolutionary mind to reverse this, to put information processing first and to make recommendations for organizational design on the basis of information requirements, as revealed by the Law of Requisite variety; yet this is what Beer succeeds in doing with the VSM. Galbraith (1973), with his model of the organization as an information-processing system, achieves a similar reversal, but his work is not as theoretically well-grounded as Beer's. Add to this Beer's formulation of the three indices of performance, and his insights on communication channel and transducer design, and it is clear that the potential of his work in the field of information-system design is only just beginning to be tapped.

Fourth, the organization is represented as being in close inter-relationship with its environment; both influencing it and being influenced by it. Waelchli (1985) praises the 'Ashby–Beer' model because it depicts organizations and environments as reciprocally adjusting to each other. The organization does not simply react to its environment but can proactively attempt to change the environment in ways which will benefit the organization. Morgan (1983b) sees dangers in this proactive aspect of cybernetics. 'Cybernetics as technique' may aid organizations achieve predetermined goals in the short term, but it can also lead them to damage the field of relationships on which they depend. This is ultimately self-defeating and could bring ruin on the organization as well as the environment. Morgan has little cause to worry about the role of the VSM. Full account is taken of 'cybernetics as epistemology' which, as Morgan says, reveals organizations and environments to be mutually dependent and as evolving through a process of mutual influence and adjustment. There is as much emphasis in the Beer model on 'viability', upon surviving within and developing a set of relationships, as upon goal-seeking.

Finally, the VSM can be used to make specific recommendations for improving the performance of organizations as systems. Beer's model can be employed to assist with the design of new organizational systems, which should be constructed so as to ensure that they adhere to the cybernetic principles elucidated in the VSM. The most ambitious attempt to employ the model in this way—Project Cybersyn, involving the regulation of the Chilean social economy—is described by Beer in *Brain of the Firm* (1981, Pt. 4). Other examples are given in *The Heart of Enterprise* (1979). The VSM can also be used in a 'diagnostic' mode to monitor the 'health' and 'vulnerability' of actually existing organizational systems. These will, naturally, already exhibit characteristics of viable systems. But, to use Beer's phrase, some of them 'creak' (Beer, 1983, p. 117). A system of concern can be compared with the VSM to check that its structures and processes support an underlying organization capable of ensuring survival and effectiveness. Discovery of any of the following features would be regarded by Beer as a mark of vulnerability and as a threat to the organization's continued existence:

(1) Mistakes in articulating the different levels of recursion so that the system is not organized to ensure viability at each of its 'hierarchical' levels of operation.
(2) The existence of organizational features which, according to the VSM, are additional and irrelevant to those required for viability.
(3) Systems 2, 3, 4 or 5 showing a predominant concern with their own interests rather than with the well-being of the organization as a whole. They are demonstrating 'pathological autopoiesis' (Beer, 1979, pp. 411–12).
(4) There are certain elements needed to perform functions shown as vital in the VSM which are either absent or not working properly. Systems 2 and 4, in particular, are often weak in organizations.
(5) System 5 is not representing the essential qualities of the whole system to the wider system of which it is part.
(6) The communication channels in the system and between the system and its environment do not correspond to the information flows shown to be necessary in the VSM and/or are not designed according to the requirements of the law of requisite variety.

These guidelines for diagnosis are included to show how a detailed check on the operational effectiveness of any organization can be made using the VSM. From such a check specific recommendations can be developed for improving performance. Further discussion on diagnosis, elaboration of the above points and examples can be found in a number of sources (Beer, 1984, 1985; Espejo, 1979; Jackson and Alabi, 1986; Keys and Jackson, 1985; Clemson, 1984).

We have now, I hope, constructed a forceful argument to the effect that Beer's model supplies premises of great value to managers in the design and operation of goal-seeking, adaptive systems. The VSM is often criticized (Rivett, 1977; Checkland, 1986) for offering a simplistic picture of the organization, based upon mechanical or organismic analogy. In fact, it provides a highly sophisticated organizational model. In the course of the analysis we have often had reason to mention the work of other organizational theorists—Parsons, Lawrence and Lorsch, Thompson, socio-technical thinkers, Galbraith. The knowledge encapsulated in the VSM fits well with the most advanced findings of modern organizational science. Moreover, it integrates these findings into an applicable management tool that can be used to recommend specific improvements in the functioning of organizations. Perhaps even more significant, the VSM is underpinned by the science of cybernetics. This ensures that its use generates enormous explanatory power compared with the usual analyses carried out in organization theory. Organization theorists, driven by positivism, cling to perceived relationships between surface phenomena as the source of their insights. Cybernetics allows an explanation of such perceived relationships to be extracted from consideration of processes at work at a deeper, structural level. For example, socio-technical thinkers find that delegating control to 'autonomous work groups' improves the efficiency and effectiveness of organizations by improving performance in the groups themselves and by freeing managers for 'boundary management'. The VSM can provide a 'scientific' explanation of this in terms of requisite variety.

Purposes

The argument in this section will be that the VSM favours and facilitates the emergence of shared purposes as well as permitting reflection on how purposes are actually derived.

An initial point that can be made is that the existence of a well-proven management tool like the VSM provides guidance on what purposes can be achieved. It does this in two ways: first, by enriching our conception of what it is possible to do; and second, by restraining the pursuit of useless options. The VSM, for example, opens up the possibility of combining economic planning and coordination with decentralized decision-making. Within a short time of its election in Chile, the Allende government took control of more than 300 firms, accounting for almost 60 per cent of the country's industrial production (Espejo, 1980). This obviously posed massive problems for efficient and effective control. At the same time, the government was committed to allowing workers in the factories full participation in management. The existence and use of Beer's model allowed considerable progress to be made in both these areas in an exceptionally short time, and in the context of a form of economic coordination which might, in another context, prove more acceptable than the market mechanism employed in capitalist countries and the centralized bureaucratic planning of East European communist states. One apparent option the VSM seems to foreclose is the dream of abolishing the division of labour. The VSM points to the need for selective attention to be given to the different functions that have to be fulfilled in a viable system. In an organization of any complexity this is going to require specialization. The VSM insists that designs for social systems must be scientifically based. They must not seek to flout cybernetic laws of viability which, as Beer has it, viable systems '. . . have set up . . . themselves . . .' (1983, p. 117).

Delimiting feasible purposes in this way can make an important contribution to bringing about shared perceptions and goals. So, too, in many circumstances, can the decentralization of control that is such an important feature of the structure of the VSM. Beer advocates decentralization of control because it promotes efficiency. It also happens to be consistent with human dignity and freedom. That it promotes efficiency follows from the implications of the Law of Requisite variety. The parts must be granted autonomy so that they can absorb some of the massive environmental variety that would otherwise overwhelm higher management levels. The parts of a system, therefore, are allowed to carry out their tasks with minimal interference. The only degree of constraint exercised is that necessary for overall systemic cohesion and viability, and this constraint facilitates the exercise of liberty rather than limits it. If less control were exercised the result would not be greater freedom for the parts, but anarchy. This would inevitably bring in its wake more severe and unpredictable constraints on liberty as the uncoordinated actions of the parts interfered unpredictably with one another. The constraints imposed on the parts of System 1 by the meta-system should be regarded as being like the laws enacted in a democratic society. We do not regard laws against assault and theft as infringements of our liberty because they increase our freedom to go about our normal business unhindered. The degree of autonomy granted to the parts by the VSM is the maximum possible if the system as a whole is to continue to exist—and the parts, of course, are assumed to benefit from the continuance of the system.

The VSM therefore advocates decentralized control and provides cybernetic guidelines for its effective operation. While this does not necessarily include 'worker' participation in the determination of the actual purposes to be pursued, it is an important step in its own right. There are many reasons for believing that industrial democracy needs to begin at the level of the shop-floor, with control over the task. Emery and Thorsrud (1969), in a famous study conducted for the Norwegian Employers Federation and the Norwegian Labour Organization, concluded that the attempt to extend industrial democracy should begin with the experience of work itself. Having representatives at the board level tends to make very little difference to the quality of the work environment as it is perceived by most workers, and to the emergence of shared purposes. The Norwegian Industrial Democracy Project which followed on from this report was based on socio-technical systems theory and concentrated on participation at shop-floor level. Moreover, while this form of participation does not necessarily include worker involvement in the determination of overall purposes, it does not preclude it. And experience of being in control on the shop-floor might encourage demands for more extensive power (Bosquet, 1972).

We have gone a long way with Beer in accepting that decentralization of control promotes efficiency, is consistent with humanitarian concerns and contributes to shared perceptions and goals. It is now necessary to make clear that the logic of this argument depends completely on the parts within the system being in agreement with the goals that the system is pursuing, presumably because they have had some part in their formulation. These are the only circumstances in which the argument holds up. Only then can it be guaranteed that the autonomy granted will be used in ways which promote efficiency rather than disruption, and only then is it justifiable to assume that the parts stand to gain from the continuity of the system; the constraints on autonomy becoming fully legitimate because systemic cohesion and viability must be maintained in order for it to seek purposes which all are agreed should be pursued. Inevitably then we are driven on, in this argument, to question the VSM about how overall purposes should be determined. Beer has got to answer that they should be determined by all those with a legitimate claim on the relevant system.

This is in fact the line he takes. The arrangement of the elements, Systems 1–5, in the VSM should not be regarded as hierarchical. These five major elements are all of equivalent importance. Indeed, if any has special primacy it is System 1, '. . . because it consists itself of viable systems' (Beer, 1984, p. 128). Systems 1–5 all support one another in the production and preservation of a viable system. System 3 '. . . is not constructed as a box to house people with better suits and bigger cars than anyone else. That they do have these things is simply the result of a general acquiescence in the hierarchical concept' (Beer, 1985, p. 92). Similarly, System 5 has an important role to play in the maintenance of the viable system—balancing internal and external demands and representing the system to any wider system of which it is part—but this is no more important than the roles played by the other elements. Just because System 5 is labelled 'policy' does not imply that it is solely responsible for deciding the purposes of the enterprise. Whose power then does 'the board' embody? The law says that of the shareholders. But, in Beer's view, the board '. . . also embodies the power of its workforce and its managers, of its customers, and of the society that sustains it. The board metabolizes the power of all such participants in the enterprise in order to survive' (Beer, 1985, p. 12). System 5 should represent what Beer (1984) calls '. . . the essential qualities of the whole system'. If then the 'stakeholders' in a system have agreed about the purposes to be pursued, and those purposes are embodied in System 5, the VSM offers a means of pursuing the purposes efficiently and effectively with only those constraints on individual autonomy necessary for successful operation. The model depends for its full and satisfactory operation on a democratic milieu—ideally perhaps on a president who, when System 5 is represented during an explanation of the workings of the VSM, can exclaim 'At last, el pueblo!' (Beer, 1981, p. 258).

Of course, in the real world, the purposes of systems very rarely reflect in a direct sense the wishes of those with legitimate claims on them. Rather, as Beer (1985, p. 99) acknowledges, purposes emerge as a compromise which reflects the power of the various groups involved. This means that we must take very seriously the possibility that the VSM could be used autocratically by a powerful group to control the behaviour of other individuals and in pursuit of some goal for which there is not general support. Beer is ready to do this, accepting that the risk of subversion does exist, '. . . since what amplifies regulatory finesse may do so for good or ill. In this cybernetic approaches mirror advances in all other branches of science' (Beer, 1983). However, we are not completely at the prey of such a possibility. Two preventative strategies are available.

First, in applying the VSM, the analyst is asked to identify the 'compromises of purpose' that have emerged and to consider whether he or she can settle for these (Beer, 1985, p. 99). He or she can determine how 'authoritarian' the enterprise is by considering whether the compromise on purpose is biased towards the purposes of the whole system rather than towards the purposes of System 1 (Beer, 1985, p. 100). He or she is also encouraged to consider whether System 5 does truly represent the interests of the system of which it is part or whether it claims to be something else (Beer, 1985, p. 131). Presumably, if the analyst dislikes the results of these investigations, he or she can cease to enhance the regulatory finesse of the relevant organization. Beer's cybernetic concepts allow us to establish exactly how a particular organization's purposes are being determined.

Second, certain features already present in the VSM, or which can easily be incorporated in use, act as 'immunological systems' which help minimize the risk of subversion (Beer, 1983). The VSM is essentially decentralizing in terms of the exercise of control. This is highly unlikely to appeal to an autocrat—as Beer (1983) remarks, the Pinochet regime in Chile did not use Cybersyn. Spreading the source of control in a system inevitably gives some power to those at lower levels. If the cybernetic message that information is control gets across—and an educational programme to this end was begun in Chile—then a powerful weapon for resistance to dictatorship is put in the hands of the populace. Spreading the source of control is not the same as spreading policy-making, but the two are linked.

As has been remarked, greater control at the workplace can lead to demands for a say in the actual setting of enterprise goals. It can act as a profound educational experience. This can be enhanced, as again happened in Chile (Beer, 1981, pt. 4), if attempts are made to help those at lower levels of recursion to understand policy issues debated at higher levels.

With these comments on how the VSM can be safeguarded against 'subversion', we can rest content that the strongest case possible has been made for the significance of the model in relation to the determination of managerial purposes.

The case against the VSM

Premises

The case against the VSM must now be put in place. Our concern initially is with whether the VSM provides premises to underpin efficient and effective managerial action.

We can begin to throw doubt on this if we consider whether the kind of intricate monitoring and control systems, implied in the VSM, can actually be made operable in organizations. The VSM does not rest on simple first-order negative-feedback systems—as we have seen there are possibilities for changing goal-state as 'threats and opportunities' occur in the environment; but much of the information flowing around the various connections in the VSM is inevitably meant to service such systems. It is information about how the different parts of the organization, and the organization as a whole, are doing in relation to their respective goals. It is not illegitimate, therefore, to consider as relevant some of the problems Sutherland (1975) associates with the application of first-order negative-feedback devices to control 'indeterminate' systems—such as organizations. As Sutherland argues, in 'indeterminate' systems there are potentially a huge number of variables to monitor, and choice of key variables (if indeed there are such 'key' variables) becomes difficult. Moreover, the permissible limits within which these variables can operate and still give satisfactory performance becomes extremely broad. The number of acceptable, alternative system states is very great. Then there are problems measuring how well a particular variable is actually performing: subjective judgement is involved. Finally, even when something is deemed to be going wrong, the likelihood of there being an appropriate programmed response available will be slight. Almost all such instances in 'indeterminate' systems will require the intervention of a decision-maker using judgement to the best of his or her ability.

Defenders of the VSM might, of course, argue that the kind of precise measurement and programmed response, which Sutherland regards as impossible to operate in indeterminate systems, is not a requirement imposed by the model. The VSM, after all, grants maximum discretion to 'low-level' parts in the matter of control and employs ordinary numbers on its measurement scales to make it easier to register changes in variables like employee morale and commitment to research and development. Nevertheless, much of the apparent power of the elaborate information and control systems, detailed in the VSM, dissolves when the difficulties of matching the models demands to real-world organizations are recognized.

A rather similar point is made by Checkland (1980), but specifically in relation to how instructive the Law of Requisite variety is when applied to organizations. The implications of this law, of course, form the very basis of the VSM. To Checkland (1980) the Law '. . . indicates, unexceptionally, that the control system of any entity which is going to survive in a changing environment must have a trick up its sleeve in response to every trick the environment can play on it . . .'. And, indeed, it is easy to see how it might appear unexceptional to the manager of a business organization facing an infinite number of possible 'tricks' and with an equally infinite number of possible responses at his or her disposal. Ashby is clear, according to Checkland, that the Law of Requisite Variety is a law of logic only. It must follow from this that its application and relevance to particular systems will require further investigation. Clearly, a physical machine that does not obey Ashby's Law is unlikely to survive, but, as Checkland asks, '. . . are organizations machines?'.

We are at the point in the argument now when we are beginning to expose the VSM to the assumptions of another paradigm—and it is as well to recognize the fact. For Checkland (1980), the VSM takes the organization to be like a machine set up to carry through some purposes. But, at best, this is only a partial representation of what an organization is. It is a representation, moreover, which misses the essential character of organizations—the fact that their component parts are human beings who can attribute meaning to their situations and can therefore see in organizations whatever purposes they wish, and make of organizations whatever they will. Because of this, it is as legitimate to regard an organization as a social grouping, an appreciative system or a power struggle as it is to see it as a machine (Checkland, 1980, 1986). Checkland (1980) prefers the 'phenomenological' perspective in which organizations '. . . are perhaps not machines at all but *processes* in which different perceptions of reality are continuously negotiated and renegotiated'. Dachler (1984), in the same vein, believes that the conscious and reflective nature of the elements of social systems separates them absolutely from all other system types. The interpretative property possessed by humans requires that social systems be understood as sense-making and self-organizing processes. Organismic approaches (and Dachler classes the VSM as 'organismic') profoundly restrict our ability to learn about social systems and prevent the generation of important insights into how they work. Ulrich (1981) draws a distinction between *purposiveness* and *purposefulness* to establish a similar argument. The VSM is purposive, being concerned with the effectiveness and efficiency of means or tools employed to achieve some end. Social system models should be purposeful; respecting the self-reflective individuals who participate in and are affected by social systems, and facilitating their awareness of the purposes being served.

Let us pursue these important points here only insofar as they concern the premises of managerial action. The implication is that a manager seeking to promote the efficiency and effectiveness of his or her enterprise by concentrating effort on its logical design as an adaptive goal-seeking system (as recommended by the VSM) is seriously misplacing his or her energies. Social organizations can exist with and perform well while employing a host of apparently illogical structures. Rather, the manager would be better off trying to bring about an accommodation between the various 'appreciative systems' operating in the enterprise, as Checkland's (1981) 'soft systems' methodology suggests; or with creating the conditions for the involvement in the goal-setting process of all those concerned and affected by a systems behaviour, as Ulrich's (1983) 'critical systems heuristics' advances. The gist of this argument, which indicates a completely different set of priorities for managers, is well-captured by Thomas Watson, writing about his experiences at IBM:

> 'Consider any great organization, one that has lasted over the years—I think you will find that it owes its resiliency not to its form of organization or administrative skills, but to the power of what we call *beliefs* and the appeal these beliefs have for its people. . . . In other words, the basic philosophy, spirit, and drive of an organization have far more to do with its relative achievements than do technological or economic resources, organization structure, innovation and timing. All these things weigh heavily in success. But they are, I think, transcended by how strongly the people in the organization believe in its basic precepts and how faithfully they carry them out.' (Quoted in Peters and Waterman, 1982, p. 280)

Interestingly enough, the importance of shared values and beliefs in securing the long-term viability of organizations is being increasingly recognized in cybernetics. Gomez and Probst (1985) have discussed the role of 'corporate culture', as maintained by the Curia, in ensuring the longevity and autonomy of the Catholic church. IBM's strong corporate culture is also cited by them as a reason for its success. Gomez and Probst go so far as to identify the 'beliefs' of a social entity with its essential 'organization' in Varela's terms. It is these beliefs which must be kept invariant over time if the entity is to maintain itself. Everything else about it—including its structure—can undergo alteration. Beer, of course, is not unaware of the need for a degree of shared purpose in an enterprise. System 4 is charged with providing a system with the self-awareness that should make this possible (Beer, 1985, p. 115), and System 5 must represent the purposes of the whole system if viability is to be maintained. However, specific mechanisms aimed at achieving shared values and beliefs appear, from the phenomenological point of view, to be absent, since Beer clearly cites the source of viability of a system (its 'organization') in its structural arrangements for handling complexity and concentrates on these, viewing corporate culture as a secondary, emergent property.

The VSM can therefore be seen as an insufficient 'control' model for organizations, because it leads managers to neglect their fundamental role as 'engineers' of an organization's corporate culture. Attempts to *implement* recommendations for change stemming from such an inappropriate device may entail even more direct dysfunctional consequences for an enterprise. In particular, it is claimed, disastrous consequences are likely to follow because of the 'conservative' effect modelling and designing an enterprise according to the VSM is likely to have. This will now be considered.

At the initial stage of a study using the VSM, the analyst is free to consider modelling the system from the point of view of a variety of different purposes. At this stage—and contrary to what some critics claim (for example, Checkland, 1986)—the VSM does not require existing organizational boundaries to be taken for granted. Creative thinking is possible. Indeed, Beer countenances users of the model against assuming '. . . that every division or department shown as depending from the boss is a viable system in its own right—and therefore an operational element in System One' (Beer, 1979, p. 204). The actual modelling, however, proceeds by choosing, or taking for granted, one set of purposes and suggesting a particular design to achieve these. Inevitably an enterprise so designed will become institutionalized in the pursuit of these particular purposes. Control devices will be in place to encourage conformity with sub-goals supporting these purposes and to discourage any form of 'deviancy'. Change in goal-state is still possible, as System 4 recognizes and System 5 responds to threats and opportunities in the environment. But the enterprise is essentially robbed of an exceptionally important source of constructive change—internal change stemming from individual deviancy, group conflict, etc. The VSM, as Ulrich (1981, p. 35) remarks, is unable to deal with the '. . . important capability of social systems to change their goal-state and structure in a stable environment. . .'. As this quotation suggests, particular structures are likely to become 'fixed' as well. Although the VSM allows that a variety of structures might prove capable of supporting a viable system, those which are represented in the modelling and are initially fabricated are likely to become binding.

De Zeeuw (1986) sees the 'control paradigm' (of which the VSM would be an example) as exhibiting two steps which combine to inhibit the constructive development of organizations. First a system is modelled, boundaries are defined and taken as binding, and, when the system is actually set up, these boundaries become maintained by actual use. The model '. . . will start to act like a contract, with a conservative effect' (de Zeeuw, 1986, p. 139). Second, the model should make possible better prediction and control and hence 'steering'. The result is that the purposes built into the model can continue to be effectively served by the external operation of the 'steering mechanisms'. De Zeeuw sees this as leading to an 'increasing dominance of history'. Both individual and collective competence are likely to suffer as alternative possibilities are neglected.

In terms of Checkland's (1981) soft-systems methodology, applying the VSM amounts to taking one out of the multitude of possible perceptions of a system (one 'root definition' in the jargon), modelling it and 'diagnosing' the system on the basis of the results obtained. The debate that could and should occur about the purposes to be served by the system and about what changes are 'feasible and desirable', given the various viewpoints operating, is effectively proscribed. The Britton and McCallion study (this volume) using the VSM, of a vocational training network in New Zealand, shows this feature. So does the Tripp, Pearson and Rainey (1986) study, using the model, of the American Air Force Logistics Command. The opportunity to discuss the purposes to be served by the systems of concern appears to be missed in these examples. Furthermore, as has been argued, application of the VSM can prevent future organizational learning. So much more is it important to know exactly what goals have been institutionalized in any designed system—as we shall see in the next section.

Purposes

Considerable criticism has been levelled at the VSM because of its perceived autocratic implications (Rivett, 1977; Checkland, 1980; Adams, 1973). It is believed that, when applied, the VSM inevitably serves the purposes of narrow elite groups. Much of this criticism is, in fact, seriously misplaced. Thus, Checkland (1980) charges the Beer model with offering the prospect of a 'negative utopia' in which the operational units have only that freedom which is compatible with systemic cohesion. This, apparently, ought to scare us, because the system is monolithically defined. But, of course, it need not be monolithically defined—its purposes *could* be determined by all those with a legitimate stake in the affairs of the system. Adams (1973) paints Beer's work in Chile as representing a fully elaborated and computerized model of a tyranny. But, of course, the top controller in Chile—the Allende government—had been democratically elected, and the VSM operated simply as a tool for putting into effect the wishes of the electorate. It will not therefore be argued that the VSM *inevitably* serves authoritarian purposes. There is nothing to prevent the application of the VSM to democratic organizations in which all participate fully in the process of goal-setting. The model might improve the efficiency and effectiveness of these organizations as well.

Of course, it is not the intention here to let the VSM 'off the hook' in respect of 'purposes' simply because it does not inevitably serve the interest of narrow elite groups. It was determined much earlier in this chapter that a management science model might reasonably be expected to do the exact opposite of this and invoke design specifications that favour and facilitate the emergence of shared purposes. And we might expect such a model to provide material to furnish a debate about the nature of the purposes being served. Criticism of the VSM on the question of purposes can certainly be reconstructed around these points. However, there are two possible lines of argument which, if they could be established, would seriously weaken any critical assault on the VSM's handling of 'purposes'. First, it might be argued that, far from serving autocratic purposes, the VSM is, in fact, genuinely supportive of the emergence of shared purposes. Second, it could be argued that, while the VSM does no more than serve all purposes equally—increasing efficiency and effectiveness whatever the nature of the organization involved—this is all that can realistically be expected of any management model. Beer (1983, 1985) employs both these arguments. We shall therefore proceed to build the case against the VSM by taking each of these arguments in turn and attempting to counter it.

The first argument comes down to saying that the VSM is best suited to serving democratically arrived at purposes. This is the case only if we are prepared to accept as valid Beer's demonstration that there is a happy correspondence between the demands of viability and the requirements of democracy. Beer (1985) argues that an authoritarian System 3 would prevent the necessary autonomy developing in the parts of System 1 to ensure the viability and effectiveness of an organization. Similarly, System 5 should, according to Beer (1984), represent the essential qualities of the whole system. If it does not do so, it will endanger the continued existence of the system. However, any examination of 'real-world' organizations must surely lead us to doubt that there is any such convenient correspondence between the demands of viability and the requirements of democracy. There are far too many authoritarian System 3s and unrepresentative System 5s around in apparently 'viable' systems for Beer's argument to be convincing. It is overoptimistic to believe that, in the long run, autocratic systems will be too inefficient and ineffective to survive. Robbed, then, of a comforting coincidence, we are forced to confront the very real possibility that the VSM can be turned to autocratic use.

It is the operation of power in organizations that ensures that the VSM's best intentions can get distorted. The existence of power relationships in social systems is well-recognized in cybernetics (Buckley, 1967; Busch and Busch, 1985). Beer himself acknowledges, in *Diagnosing* (1985, p. 91), the unfortunate effects the exercise of power can have in viable systems. But acknowledgement is scarcely enough in relation to such a pervasive aspect of organizational life. In an organization disfigured by the operation of power, many of the features of the VSM that Beer sees as promoting decentralization and autonomy, instead offer to the powerful means for maintaining control and consolidating their own positions. The notion of 'levels of recursion' and the arrangement of Systems 1–5, with development and policy located in Systems 4 and 5, take on an hierarchical significance. The 'autonomic management' system embedded in systems 1–3 appears as a magnificent structure for ensuring conformance to externally established plans and procedures. The 'algedonic' mechanism, which in a democratic milieu can be presented as registering 'cries for help' from below, appears as a device for alerting higher management levels that lower levels are not doing their bidding. Anybody who doubts that a model such as the VSM can be used in an autocratic manner should study how the UK National Coal Board (NCB) employs MINOS (Mine Operating System) to reinforce centralized control in the mining industry. Burns, Newby and Winterton (1985) describe MINOS as '. . . a highly centralized, hierarchically organized system of remote control and monitoring in mines'. A product of systems engineering, MINOS nevertheless has much in common with the VSM: the overall system is split into subsystems each with clearly defined sub-objectives; there are computerized information systems feeding different levels of recursion; and intricate monitoring and control devices are constructed. It is therefore interesting that the Management Information System (MIS) which is at the heart of MINOS is used solely to reduce the miners' job control and increase management control, and to service highly centralized, autocratic planning:

> 'It is the MIS that links the subsystems together into the overall system, and is therefore the key element in the NCB's strategy. However, the colliery level of the MIS is not the top of the hierarchy: it feeds its information into a national computer network that allows the NCB, and by implication the government, to compare performance over time, between collieries, between shifts, between faces, and between areas. Planning and control can thus be exercised rapidly and directly from the highest levels of management. The information collected is used by the CPU (Central Planning Unit) in its Strategic Model to construct scenarios based on the policy choices before management. So-called 'uneconomic

pits' can be identified according to the criteria in force at a particular time, and the list can be revised at a moment's notice should different criteria be applied.' (Burns, Newby and Winterton, 1985, p. 104)

It may be thought unreasonable to charge the VSM with guilt by association, but this example does demonstrate the misuse to which 'cybernetic' models can be put.

The MINOS example, of course, flouts the VSM's injunction that maximum autonomy should be granted to System 1 elements. If this is made the litmus test of 'proper' application, perhaps we can then guarantee the model against autocratic use. Unfortunately the evidence, even on this point, is not convincing. Nichols (1975), for example, sees no reason to believe that increased control on the shop-floor will lead to demands for more extensive power. The granting of such autonomy to workers can be interpreted not as a step on the road to 'industrial democracy', but rather as the imposition of a more sophisticated, but equally compelling, management control technique. Goals are still determined at higher levels; it is just that, in seeking to achieve these goals, indirect rather than direct control procedures are employed—subordinates are encouraged to control themselves. Used 'intelligently' by management this can contribute to worker subjugation by ensuring their 'subjective' as well as 'objective' subordination. Workers are encouraged to believe they possess freedom, but this is only the limited freedom to control themselves in the service of someone else's interests.

Beer might of course claim that, when power operates, and the effects discussed in the above are produced, then the VSM is not being used 'legitimately'. But then there will be very few situations (workers' co-operatives, democratic socialist states?) when its use could be regarded as legitimate. In any case, this argument is not fully developed by Beer. There are no *explicit* statements outlining the circumstances in which the VSM can be 'properly' employed.

This takes us on to the second line of argument that could deflate any critical assault on the VSM's handling of purposes. This argument accepts that there is a great risk of subversion and that, realistically, even if 'immunological systems' are incorporated, the model *can* be used for good or ill. But it goes on to suggest that, in this, '. . . cybernetic approaches mirror advances in all other branches of science' (Beer, 1983). In other words, it is a risk that simply cannot be avoided. The important question to ask here, perhaps, is whether scientific advances which are to be applied in the management context, to the design of social systems, *should* mirror advances in other branches of science. To this question, guided by the able arguments of Ulrich (1981, 1983), we can give a firm answer: no.

According to Ulrich, the fact that Beer's model lends itself to authoritarian usage stems from Beer's conception of his task, which can be categorized as 'tool design'. The VSM seeks to refer the formulation of purposes to some 'irrational' process of political decision and to leave for the system designer the task, simply, of providing expert, scientific tools which can be used in pursuit of these purposes. This is anti-democratic because the public is effectively excluded from debate about ultimate purposes ('informed' debate about such issues being apparently not possible), and from debate about 'means' (since questions about these are settled according to expertise). Against 'tool design' Ulrich sets 'social system design'. In social system design the separation between ends and means is denied—all ends can be seen as means to other ends and all means have normative implications of their own. The separation of function between political agencies and experts must therefore be replaced by critical interaction. There must be a debate conducted according to a model of rational discourse—'. . . a model that can guarantee an adequate translation of practical needs into technical questions, and of technical answers into practical decisions' (Ulrich, 1983, p. 78). This is compatible with democracy since all issues must now be subject to 'informed' debate. None is reserved for technical expertise. The public can therefore be fully involved in the process of conducting and monitoring the debate. Ulrich (1983) has suggested a procedure for guiding social system design which he calls 'critical systems heuristics'. This should enable the presuppositions (values and metaphysical beliefs) that enter into existing or potential system designs to be interrogated, and should enable all those involved or affected by such designs to take part in open debate about them. It is not my concern here to discuss critical systems heuristics, but three distinctions used by Ulrich (1981, 1982) to distinguish 'social system design' from 'tool design' do enable us to undestand further the weaknesses of the VSM with respect to 'purposes'. I therefore conclude this section with a discussion of these distinctions.

The first distinction is between 'intrinsic control' and 'intrinsic motivation'. It reinforces the point that the VSM provides no mechanisms for the democratic derivation of purposes. Ulrich argues that, while Beer's model is capable of generating a degree of 'intrinsic control' (spreading the sources of control throughout the architecture of the system), it cannot generate 'intrinsic motivation' (distributing the source determining the system's goal-state and purpose throughout the system). Only the top controller is in a position to change the system's basic orientation according to its own particular whims. Although the top controller in the Chile example—the Allende government—had been democratically elected, there is no requirement that this should be the case. For Ulrich, the model provides tools which are capable of being misused and, in all likelihood, will therefore be misused. This aspect of the VSM is a reflection of the 'organic' paradigm within which Beer's cybernetics operates. This shows an advance on the 'mechanistic' paradigm, which for so long dominated cybernetic thinking, because control, at least, is internally managed. It is not, however of sufficient sophistication for social systems because it implies that—as with organisms—goal-setting should be a privileged function of higher-order levels of the system.

The second and third distinctions establish further that the VSM is unable to facilitate debate about the nature of the goals pursued. The second is between 'purposiveness' and 'purposefulness'. The VSM is orientated to purposiveness, concerning itself with the effective and efficient design of means or tools, and failing to assist individuals involved in or affected by system designs to reflect upon the ends or purposes being served. One consequence of this was discussed in the last section. The third distinction is between the 'syntactic' and 'semantic–pragmatic' levels of communication. The syntactic level is solely concerned with whether a message is well-formed or not, in the sense of whether it can be 'read'. This matter can be dealt with by information-processing machines. The semantic and pragmatic levels are concerned respectively with the meaning and the significance of messages for the receiver—they inevitably involve people. The concept of variety, which underpins the VSM, operates only at the syntactic level. It is an information-theoretic measure of complexity, referring '. . . to the number of distinguishable states that a system or its output (the 'message' it sends out) can assume at the syntactic level' (Ulrich, 1981, p. 35). This is severely restricting, as can be seen as soon as we consider what criterion of 'good' management is being invoked. For Beer, apparently, good management can be no more than management that establishes requisite variety between itself and the operations managed, and between the organization as a whole and its environment. This goes against the reasonable assumption that good management must also concern itself with the nature of the purposes being served and the meaning and significance of these for participants in the enterprise. The VSM, therefore, fails to facilitate any discussion about the goals to be pursued.

The substitution of a syntactic criterion of good management for a semantic–pragmatic criterion is general throughout cybernetics and seriously constrains theoretical development and practical relevance. We can see this, for example, in the debates surrounding the 'self-organizing' capabilities of social systems. Because it potentially leads to increased information-processing capacity, there is strong pressure to see the promotion of autonomy at lower levels in systems as being of unquestionable benefit. This idea, of course, cannot possibly be sustained. The phenomenon of self-organization has been studied in organization theory since the Hawthorne experiments (Roethlisberger and Dickson, 1939), and is well-known to have deleterious as well as positive effects. This is grudgingly accepted in the management cybernetics literature. The VSM, as we know, employs a meta-system to ensure that the autonomy granted to the parts of System 1 does not threaten overall systemic cohesion. Haken (1984) discusses the need for some sort of command structure to limit the variety potential of system parts as necessary. Gomez (1982) sees managers as 'catalysts', channelling the inner dynamics of systems to bring them to a desired state. But once some limitation on the autonomy of the parts is accepted, the question immediately arises as to the basis on which this can be justified.

Geyer and van der Zouwen (1986) base an edited collection of papers on 'sociocybernetics' around this problem/paradox: 'How do you reconcile the fact that social systems tend to steer themselves with the fact that there is often a necessity to plan and steer them from the outside?' The answer is, of course, that you *cannot solve this paradox within the cybernetic paradigm*. You are driven outside the cybernetic paradigm to seek some semantic–pragmatic criterion of 'good' management. For only on the basis of the meaning and significance of purposes for concerned individuals can decisions be taken which impose limitations on their autonomy. Beer admits as much in relating the degree of systemic cohesion required to the nature of the purpose being pursued. Ben-Eli and Probst (1986) battle with the 'paradox' in cybernetic form, but eventually, and suddenly, are forced to switch to Ackoff's concept of development as a basis for resolving it. While cyberneticians eschew a semantic–pragmatic criterion of good management, they will continue to battle fruitlessly with paradoxes such as this and theoretical development will be constrained. Meanwhile, in the 'real world', no such problem exists. The issue of how much autonomy to grant to the parts is settled as a matter of 'managerial' convenience—in the interests of those who possess power.

The managerial significance of the VSM with regard to 'purposes' can therefore be portrayed as slight. When it comes to bringing about change in social systems we need what De Zeeuw (1985) calls 'multiple actor design involving values'. Other methodological approaches, such as Checkland's 'soft systems methodology' and Ulrich's 'critical systems heuristics', fulfil this need more adequately than the VSM.

Conclusion

We have been engaged in 'reflective conversation', detailing the case for the VSM and the case against it. The purpose of this was to assist reflection on the managerial significance of the VSM in terms of 'premises' and 'purposes'. The justification for spending time on reflective conversation is that it should produce insight into the nature, strengths and limitations of the approach examined. It is thus possible to gain a more realistic appreciation of the usefulness of that approach and, possibly, will enable refinements to be made. What then have we learned about the managerial significance of the VSM?

Its advocates would claim that it is of immeasurable value to managers trying to design and operate goal-seeking, adaptive systems. It provides a sophisticated organizational model which embodies great explanatory power and is readily applicable. Further, it favours and facilitates the emergence of shared purposes by insisting that viable organizations should possess representative policy-making bodies and that control should be decentralized. It even allows the analyst to see when power is operating to frustrate democratic decision-making. To its critics, however, the VSM is of dubious value even as a tool for increasing efficiency and effectiveness. The emphasis it places on organizational design may preclude proper attention being given to the generation of shared perceptions and values; to 'organizational culture'. Further, the imposition of a particular design may become fixed and prevent necessary adaptation. In practice, the VSM can easily be turned into an autocratic control device serving powerful interests. It lends itself to this as it provides no mechanisms either for the democratic determination of purposes or for facilitating debate about the nature of the purposes served.

There is, of course, a danger that the learning obtained from reflective conversation, rather than contributing to reflection, will simply reinforce existing prejudices. Jackson and Willmott (1987) have pointed out that the success of reflective conversation seems to presuppose that the conditions that make it necessary in the first place will somehow disappear. Those locked into particular paradigms will suddenly begin to listen to the advocates of alternative positions and become amenable to open discussion. Further, they will be able to set aside their strongly held theoretical and normative presuppositions, listen with equanimity and evaluate fairly the debate that ensues. Since these prerequisites are unlikely to be met, there can be no guarantee that reflective conversation will be productive. It is up to the reader to use the material presented here in a constructive way.

That said, one final attempt will be made to help ensure constructive debate. It *is* possible to point out certain areas where the concerns evinced by the VSM and by the critics of the model touch. Perhaps attempts at rapprochement should focus on these areas.

From the point of view of management cybernetics there is recognition of the importance of shared, or at least compatible, perceptions and values to organizational viability. As was noted, System 4 in the VSM is supposed to provide the organization with a model of itself which could help common appreciations to develop. System 5 is supposed to represent the essential qualities of the whole system. Espejo (1987), in an important statement of the theoretical underpinnings of management cybernetics, recognizes the need to take into account the aims, values and appreciations of the human beings that make up social systems. He points out, however, that the understanding and behaviour of individuals in organizations is constrained by the structure of the communication channels available to them. It is here, in the communication channels made available by a particular organizational structure, that systemic/structural constraint can be located. However, by organization design and structural adjustment, it should be possible to increase the autonomy of individual viewpoints. The role cybernetics reveals for management science is to bring about appropriate structural change as a means of making appreciative processes more effective. In this way the 'phenomenological' and 'structural' positions can be seen as complementary rather than contradictory.

From the point of view of the 'interpretive' critics of the VSM, there is some recognition of the existence of systemic-structural constraints. Checkland provides a 'formal system model', bearing some resemblance to the VSM, against which 'conceptual models' can be checked to see if they are defensible. This is not meant to be prescriptive, '. . . for it is absolutely not the intention of the methodology to diminish the freedom of actual human activity systems to be, if they wish, irrational or inefficient'. Nevertheless, it comprises management components which '. . . arguably have to be present if a set of activities is to comprise a system capable of purposeful activity' (Checkland, 1981, p. 173). Another 'soft systems' thinker, Ackoff, goes much further and presents (1983) a diagrammatic representation of a 'responsive decision system' which, if adopted, is supposed to increase the learning and adaptive capabilities of organizations. This has many features in common with the VSM. Even if it is clearly at a supportive level, therefore, some cybernetic input is accepted in soft-systems approaches.

The difference in emphasis is still pronounced: the VSM concentrating on systemic/structural constraints, the critics on processes of negotiation between different viewpoints and value positions. Perhaps, though, the basis for constructive dialogue can be discerned. I hope this chapter will assist in making such dialogue fruitful. The VSM has much to contribute to improving management practice. If it is to do so most beneficially, its strengths must be clearly understood and advertised. But there must also be recognition and discussion of its limitations.

References

Ackoff, R. L. (1983). 'Beyond prediction and preparation', *Journal of Management Studies*, **20**, 59–69.

Adams, J. (1973). 'Chile; everything under control', *Science for People*, **21**, 4–6.

Ashby, W. R. (1964). *An Introduction to Cybernetics*. Methuen: London.

Baker, W., Elias, R. and Griggs, D. (1977). 'Managerial involvement in the design of adaptive systems', in *Management Handbook for Public Administrators* (ed. Sutherland, J. W.). New York: Van Nostrand Reinhold, pp. 817–42.

Beer, S. (1966). *Decision and Control*. Chichester: John Wiley.

Beer, S. (1972). *Brain of the Firm*. Harmondsworth: Allen Lane.

Beer, S. (1974). *The Integration of Government Planning*, Study for the Government of Alberta.

Beer, S. (1975). *Platform for Change*. Chichester: John Wiley.

Beer, S. (1979). *The Heart of Enterprise*. Chichester: John Wiley.

Beer, S. (1981). *Brain of the Firm*, 2nd edn. Chichester: John Wiley.

Beer, S. (1983). 'A reply to Ulrich's "Critique of pure cybernetic reason: the Chilean Experience with cybernetics" ', *Journal of Applied Systems Analysis*, **10**, 115–19.

Beer, S. (1984). 'The viable system model: its provenance, development, methodology and pathology', *Journal of the Operational Research Society*, **35**, 7–25.

Beer, S. (1985). *Diagnosing the System for Organisations*. Chichester: John Wiley.

Ben-Eli, M. V. and Probst, G. J. B. (1986). 'The way you look determines what you see, or self-organisation in management and society', in *Cybernetics and Systems '86* (ed. Trappl, R.). Dordrecht: Reidel, pp. 277–84.

Bosquet, M. (1972). 'The prison factory', *New Left Review*, **73**, 23–34.

Buckley, W. (1967). *Sociology and Modern Systems Theory*. Englewood Clifs, N.J.: Prentice-Hall.

Burns, A., Newby, M. and Winterton, J. (1985). 'The restructuring of the British coal industry', *Cambridge Journal of Economics*, **9**, 93–110.

Burrell, G. and Morgan, G. (1979). *Sociological Paradigms and Organisational Analysis*. London: Heinemann.

Busch, J. A. and Busch, G. M. (1985). 'Sociocybernetics and social systems theory', in *Systems Inquiring* (Proceedings of the SGSR international conference, Los Angeles) (ed. Banathy, B.), pp. 544–53.

Checkland, P. B. (1980). 'Are organisations machines?', *Futures*, **12**, 421–4.

Checkland, P. B. (1981). *Systems Thinking, Systems Practice*. Chichester: John Wiley.

Checkland, P. B. (1986). 'Review of "Diagnosing the System" ', *European Journal of Operational Research*, **23**, 269–70.

Clemson, B. (1984). *Cybernetics: a New Management Tool*. Tunbridge Wells: Abacus Press.

Dachler, P. (1984). 'Some explanatory boundaries of organismic analogies for the understanding of social systems', in *Self-Organisation and Management of Social Systems* (eds. Ulrich, H. and Probst, G. J. B.). Berlin: Springer-Verlag, pp. 132–47.

Espejo, R. (1979). 'Information and management: the cybernetics of a small company', working paper 125, University of Aston Management Centre.

Espejo, R. (1980). 'Cybernetic praxis in government: the management of industry in Chile, 1970 to 1973', working paper 174, University of Aston Management Centre.

Espejo, R. (1987). 'From machines to people and organisations: a cybernetic insight on management', in *New Directions in Management Science* (eds. Jackson, M. C. and Keys, P.). Gower, Aldershot: Gower, pp. 55–85.

Espejo, R. and Watt, J. (1978). 'Management information systems: a system for design', working paper 98, University of Aston Management Centre.

Emery, F. E. and Thorsrud, E. (1969). *Form and Content in Industrial Democracy.* London: Tavistock.

Galbraith, J. R. (1973). *Designing Complex Organisations.* Reading, Mass.: Addison-Wesley.

Geyer, F. and van der Zouwen, J. (1986). *Sociocybernetic Paradoxes.* London: Sage.

Godel, K. (1962). *On Formally Undecideable Propositions.* New York: Basic Books.

Gomez, P. (1982). 'Systems-methodology in action: organic problem-solving in a publishing company', *Journal of Applied Systems Analysis,* **9,** 67–85.

Gomez, P. and Probst, G. J. B. (1985). 'Organisational closure in management, pt. I: Complementary view to contingency approaches', *Cybernetics and Systems,* **16,** 703–10.

Habermas, J. (1972). *Knowledge and Human Interests.* London: Heinemann.

Habermas, J. (1974). *Theory and Practice.* London: Heinemann.

Habermas, J. (1979). *Communication and the Evolution of Society.* London: Heinemann.

Haken, H. (1984). 'Can synergetics be of use to management theory:', in *Organisation and Management of Social Systems* (eds. Ulrich, H. and Probst, G. J. B.). Berlin: Springer-Verlag, pp. 33–41.

Hamilton, P. (1983). *Talcott Parsons.* Chichester: Ellis Horwood.

Jackson, M. C. and Alabi, B. O. (1986). 'Viable systems all!: a diagnosis for XY Entertainments', working paper 9, Dept. of Management Systems and Sciences, University of Hull.

Jackson, M. C. and Keys, P. (eds.) (1987). *New Directions in Management Science.* Aldershot: Gower.

Jackson, N. and Willmott, H. (1987). 'Beyond epistemology and reflective conversation—towards human relations', *Human Relations,* **40,** 361–80.

Kast, F. E. and Rosenzweig, J. E. (1985). *Organisation and Management: a Systems and Contingency Approach,* 4th edn. New York: McGraw-Hill.

Keys, P. and Jackson, M. C. (eds.) (1985). *Managing Transport Systems: a Cybernetic Perspective.* Aldershot: Gower.

Kuhn, T. (1970). *The Structure of Scientific Revolutions,* 2nd edn. Chicago: University of Chicago Press.

Lawrence, P. R. and Lorsch, J. W. (1969). *Developing Organisations: Diagnosis and Action.* Reading, Mass.: Addison-Wesley.

Malik, F. and Probst, G. J. B. (1984). 'Evolutionary management', in *Self-Organisation and Management of Social Systems* (eds. Ulrich, H. and Probst, G. J. B.). Berlin: Springer-Verlag, pp. 105–20.

Morgan, G. (ed.) (1983a). *Beyond Method.* Beverley Hills: Sage.

Morgan, G. (ed.) (1983b). Cybernetics and organisation theory: epistemology or technique', *Human Relations,* **35,** 345–60.

Nichols, T. (1975). 'The "Socialism of Management": some comments on the new "Human Relations" ', *Sociological Review,* **23,** 245–65.

Peters, T. J. and Waterman, R. H. (1982). *In Search of Excellence.* New York: Harper & Row.

Preedy, D. K. and Bittlestone, R. G. A. (1985). 'O.R. and the boardroom for the 1990s', *Journal of the Operational Research Society,* **36,** 787–94.

Probst, G. J. B. (1984). 'Cybernetic principles for the design, control and development of social systems and some afterthoughts', in *Self-Organisation and Management of Social Systems* (eds. Ulrich, H. and Probst, G. J. B.). Berlin: Springer-Verlag, pp. 127–31.

Rice, A. K. (1958). *Productivity and Social Organisation.* London: Tavistock.

Rivett, P. (1977). 'The case for cybernetics', *European Journal of Operational Research*, **1**, 33–7.

Roethlisberger, F. J. and Dickson, W. J. (1939). *Management and the Worker*. Cambridge, Mass.: Harvard University Press.

Sutherland, J. W. (1975). 'Systems theoretic limits on the cybernetic paradigm', *Behavioural Science*, **20**, 191–200.

Thompson, J. D. (1967). *Organisations in Action*. New York: McGraw-Hill.

Tripp, R. S., Pearson, J. M. and Rainey, L. B. (1986). 'Cybernetic approach to meaningful measures of merit', *Cybernetics and Systems*, **17**, 183–209.

Ulrich, H. (1984). 'Management—misunderstood societal function', in *Self-Organisation and Management of Social Systems* (eds. Ulrich, H. and Probst, G. J. B.). Berlin: Springer-Verlag, pp. 80–93.

Ulrich, W. (1981). 'A critique of pure cybernetic reason: the Chilean experience with cybernetics', *Journal of Applied Systems Analysis*, **8**, 33–59.

Ulrich, W. (1983). *Critical Heuristics of Social Planning*. Berne: Paul Haupt.

Varela, F. J. (1984). 'Two principles for self-organisation', in *Self-Organisation and Management of Social Systems* (eds. Ulrich, H. and Probst, G. J. B.). Berlin: Springer-Verlag, pp. 25–32.

Vickers, G. (1967). *Towards a Sociology of Management*. London: Chapman & Hall.

Waelchli, F. (1985). 'The proactive organisation and the Ashby–Beer model', in *Systems Inquiring* (Proceedings of the SGSR international conference, Los Angeles) (ed. Banathy, B.). pp. 863–72.

Zeeuw, G. de (1986). 'Social change and the design of enquiry', in *Sociocybernetic Paradoxes* (eds. Geyer, F. and van der Zouwen, J.), London: Sage, pp. 131–144.

The Viable System Model: Interpretations and Applications of Stafford Beer's VSM
Edited by R. Espejo and R. Harnden
© 1989 John Wiley & Sons Ltd

16

The VSM: an ongoing conversation . . .

Raul Espejo and Roger J. Harnden

Rather than attempting to summarize the other contributions to this volume, the final chapter offers a way forward. The editors offer their insight into what is entailed for an effective implementation of the model. This involves 'bedding' methodology within the context of a particular epistemology.

Introduction

We hope that the preceding chapters in this volume will have provided the reader with new insights into management cybernetics. Each of the contributions can be taken as standing on its own and as representing a particular interpretation and use of the model. Perhaps, therefore, only the reader who perseveres and makes the effort to integrate all the various viewpoints will succeed in gaining a rich insight into the unfolding conversation about management cybernetics.

This chapter does not provide a summary or synopsis of the contents of the book. Instead, it is our own personal interpretation of the overall 'state of the game'. It is our own conversation about the conversation, rather than a description of such a conversation. The views expressed are ours alone, though naturally they have been influenced by the ideas expressed in the other chapters.

Rather than attempting to grapple with or to undermine other approaches and social disciplines—those of the psychologist, sociologist, accountant, economist and so on—managerial cybernetics addresses issues concerning strengths and weaknesses in *organization structure*. But unless firmly bedded within some particular well-defined discourse, a phrase such as 'organization structure' may echo in a void or appear all things to all persons. This volume was intended to help bring forth such discourse, prepare such a context, so that the student or general reader might more readily perceive a concrete research strategy.

There have been, and are, critics who complain about the focus upon organization structure of management cybernetics, as if such a focus denied or excluded the insights and skills from other disciplines and perspectives. The chapter by Mike Jackson refers to several such critical stances. We do not want directly to counter or defend cybernetics against such attacks, but to lay the ground for an understanding of just what is entailed in attempting to utilize a 'cybernetic approach' in management science.

Debate concerning managerial cybernetics seems to be relatively uninspiring, and where it does occur it tends to be conducted in a hysterical rather than productive fashion. Most important of all, it is generally absent from the pages of publications in the field of organization theory. When cybernetic issues are addressed, there is often misinterpretation about what the perspective involves, or distortion of both its intent and the effects of its practice.

The various chapters in this volume open a positive and creative space for discussion and debate, but perhaps more is required. We see our task at this juncture as being to *make visible* the philosophy and methodological principles underlying our own work in management cybernetics. With this in mind, we have structured the chapter as follows:

(1) First we offer some background remarks about the way Stafford Beer's work on management cybernetics has tended to be interpreted.
(2) These remarks trigger the need for a discussion about the epistemological grounding of models.
(3) This is followed by an explication of our usage of terms such as 'organization' and 'structure'. We describe a perhaps novel relation between organization and structure, and unfold an explanation of what is meant by the term 'viability'. In our view, this discussion should lay the foundations for future work.
(4) We explore and highlight the significance of *methodology*, as distinct from questions about the *logical* coherence of the model. Implementation will be seen to entail using the model as a *generative mechanism* in a community of observers.
(5) The *human dimension* of the model is considered, by matching it against a statement of what we consider to be the significance of *management science*.
(6) The exploration of these epistemological and methodological issues deepens an understanding of the VSM. Further, it demonstrates that an

effective implementation of the model depends on making visible the interdependence of *autonomy* and *distributed regulation*.

Some background remarks

One of the most plaintive cries from Stafford Beer over the years has been his denial that the VSM is a metaphor—a denial that a 'viable system' functions *as if* it were the central nervous system. This point is made forcefully by Beer in his first contribution to this book. Because of the prevailing positivistic mood, there have tended to be two directions to the way such pleas have been interpreted, both leading to confusion.

The first, from within the positivistic camp, has tended to go: 'Well, if it isn't a metaphor then it must be a literal picture of how the perfect system would run', followed by either 'How wonderful!' or 'How awful!', depending on the ideology of the interpreter.

The second direction, from within a softer, more hermeneutic camp, once more seizes upon the notion that the distinction between 'model' and 'metaphor' resides in their respective claims for accuracy or literal representation of the 'real world', and unfolds as follows: 'Although this chap Beer is a decent enough fellow' (or not, as the case might be!) 'he is incapable of coping with the messy, fuzzy quality of the world we find ourselves in, and he wants to reduce its immense richness and complexity to a one-dimensional, set-theoretic, intrinsically totalitarian caricature of reality, by creating it in the image of his own model'.

The logical structure of Beer's model contains immediate counters to such interpretations; for example, his attempt to break from accusations that the model is one-dimensional by the inclusion of the mechanism of 'orthogonality' as a kind of bifurcation between autonomous levels. But these have not cut much ice in the literature, where they have tended to be ignored.

If naivity of the model is not what is being stressed, then often it is Beer's own political naivity which is held up to question. In other words—'All very nice and well for Beer, a clever and benevolent enough fellow, to have tried to formulate a set of principles to make the working of social organization more effective in terms of the well-being of their component actors (i.e. their 'eudemony'), but doesn't the guy realize that the world isn't like that! Look what happened in Chile!' (for an introduction to this story, see Beer, 1981).

Such suspicions, though voiced in the fashion of caricature, do carry weight, and do influence the way in which people approach the work.

This, in spite of the fact that there have been many other productive interventions carried out by Beer himself as well as others, and that if one looks at the events surrounding the last months of president Allende's government, it is impossible to unearth the spectre of Stafford Beer looming in the background as a destabilizing factor.

Is the work overly abstract, unrelated to reality, politically or philosophically naive? Relatively few people have developed a deep insight of it. Its paradigmatic relevance in the context of modern organizational theory has remained unrecognized. This volume has been intended to clarify matters somewhat, and this particular closing chapter is intended to comment on the relevant issues from a somewhat different angle, perhaps suggesting the way ahead.

Models and epistemology

Possibly it is easier today to bring to light issues concerning organization structure than it was when Stafford Beer commenced his search for the underlying invariances entailed in any viable system. This is largely due to a ground movement away from the goal of literally describing the phenomena of the world (i.e. positivism), towards an acceptance that such a description can never be completely neutral or freed from constraints imposed by conceptual, cognitive, cultural and other filters (e.g. Checkland, 1982; Feyerabend, 1978; Johnson-Laird, 1983; Kuhn, 1970; Rorty, 1980; Varela, 1979; Winograd and Flores, 1986). In the social sciences at least, we are forever burdened with the 'prejudice' of a historical location or situation (Gadamer, 1975), an 'episteme' which limits codes and linguistic practices available to address and explicate issues (Foucault, 1974).

People such as Humberto Maturana or Ernst von Glasersfeld go even further, suggesting that the same consensual constraints which prejudice or burden our insight and approaches to matters in the social sciences apply to the whole domain of human knowledge, including the pure sciences themselves (Maturana, 1978; von Glasersfeld, 1987). It is this last strand of thinking that has particularly informed our own interpretation of managerial cybernetics.

Strictly speaking, even when we address so-called reality, as soon as we start to make comparisons and so on, we distance ourselves from it and find ourselves actually comparing *mental models* (Johnson-Laird, 1983). It is never the case that someone neutrally points out objective phenomena and compares them. In an odd sort of way such mental models actually obscure that which they are held to refer to in the 'natural world' (Maturana, 1988). Indeed, the notion of 'natural world' is itself a model constituted by and in turn constituting a further set of models. The richness and density of a consensual domain (a culture, a language) does not entail, as we would like to believe, a deeper grasp of an objective Reality, but instead alters the *patterns of our orienting* in terms of reality, opening up new possibilities and foreclosing others.

It is often held that a model—whether conceptual or physical (an idea or a model airplane)—is a description of some phenomenon in the real world. Now, while one must of course accept that a formalized model must be inspired by phenomena outside its own terms of reference (else it would not be a model), this is *not* to say that it is a *description* of the phenomenon that it claims to name.

One way of looking at any model, including the VSM, is as a means of 'gathering' descriptions that might themselves concern non-intersecting phenomenal domains, under an umbrella of intersection. Quite simply, models enable diverse people with different mindscapes to have conversations about diverse matters. A model does this by providing a context or mood which directs discourses in particular paths, upon acceptance of such a model as a common *convention*.

A model is expected to provide a setting, a common frame—in other words, it is expected to *make visible a set of constraints*, within which certain problems can be enunciated in a particular way, and certain problems solved.

Let us be clear about this. A model is a *convention*—*a way of talking* about something in a manner that is understandable and useful in a community of observers. It is not a description of reality, but a tool in terms of which a group of observers in a society handle the reality they find themselves interacting with.

Thus, the annual flooding of the Nile Valley in ancient Egypt required a particular cosmological model in order to explain (make sense of) the annual flooding. This cosmological model did not portray or mirror a natural world: it did not set out to trace the source of the flow of water and the actual fluctuations of the river level. What it did was to provide the context for a rational framework, a pattern, in terms of which otherwise arbitrary events made sense to the observer. The same is the case with models concerning quantum mechanics, molecular biology, society and all else.

The point is that we can utilize models either knowingly or unknowingly. But—and this is crucial to the paradigm underlying and embracing these issues—we cannot function as sentient human beings except by the continual use of models of one sort or another. The individual who claims direct access to Reality is either a fraud or a mystic. Each of us has elements of both in our make-up, for fraud and mysticism are not absolute polar opposites, but are both situated along the dimension of consensual interaction in the human social domain. This is the message of Christ, Buddha,

Krishnamurti, Muhammad, Lao Tzu and many others. But whatsoever, an individual may never *communicate* what is accessed to another individual, except in terms of models. This is not a limitation, but is precisely the *motor for the generation of a consensual domain*. A consensual domain is none other than the play of a particular set of intersecting models.

Linguistic communication is forever bound within *secondary* distinctions, that in a peculiar fashion obscure rather than highlight the phenomenon which had been initially indicated by an individual's primary act of distinction. For the *effects* of primary acts of distinction are 'frozen' in what we call models. Whether or not these models coordinate our interactions with other people depends on agreements, and such agreements are secured by an act of faith rather than by rationality. One is *seduced* to accept a viewpoint in the first instance. What is communicated in terms of the boundaries of a consensual domain, 'travels' by means of models, whether linguistic, mathematical, material or aesthetic. These all gain their power as orienting tools in a linguistic domain, because of an initial *willingness* to share an area of commonality (i.e. seduction). Anything which cannot be modelled in some form or another in terms of this commonality remains mute or opaque within this particular consensual domain. We are not talking of formal modelling alone. A word that is recognizable within a particular context is a model. Just think about it for a moment.

For any observer, models are a necessary means of reducing the variety of the universe we find ourselves in. Indeed, an observer *is* an observer by being an active element in a web of relations that constitutes a consensual domain in respect of the models it allows to be generated by its terms.

In respect of our consensual interaction, in so far as it exists, we have brought forth a set of interlocking models that resonate in a domain of commonality. The participants *share* a particular domain of models. Such models are not shared because they represent fixed, static descriptions of an external Reality, but because they are the *source of commonality*. Implicitly it is accepted that these models are *generative* statements, even though the details of the statements may not yet be understood. What has happened is that a set of relations in a particular linguistic context is indicated (i.e. a linguistic region) which, provided you are coupled to the same linguistic domain, you will re-cognize or find familiar. They 'ring a bell' for you.

If you do not share this linguistic region with ourselves, we can shout until the cows come home and you will not know what we are talking about. What we do, then, presuming we have the inclination to continue the conversation, is to widen the boundaries, relax the conditions—perhaps we make a sketch, mould a shape in clay, indicate by a series of gestures the crucial features of the phenomenon we are intending should serve as this particular means for our consensual dance with each other, until 'something makes sense' to you in terms of the linguistic region we have now brought forth.

In other words, we change the texture or 'shape' of our model, until it takes some form which inspires from yourself the creative 'Ah, ha . . .' (Koestler, 1975), followed by some construct which we accept as mimicking what we intended by the original model (see, for example, Pask, 1975). Such a model is generative rather than descriptive, because it does not require an origin or source. Instead, it *generates its own origin or source backwards*, so to speak, in terms of our participation in what we all consider to be a coherent conversation marked by our mutual agreement (for useful discussion of this process, see Dell, 1986).

This source itself takes on its status or character as 'origin' within the confines and constraints of a specific consensual region—a particular set of linguistic and social behaviours, in terms of which it is accepted and recognized as a source phenomenon (i.e. it is mutually brought forth and defined as having a particular quality which is embraced as consensually valid by its persistence, marked by our continued conversation in terms of it).

Towards a consensual domain about organization structure

What do we mean when we talk of the organization structure of a social enterprise, or else refer to the social organization in terms of its organization structure? As we ourselves use the word organization, we are referring to the set of relations between people which characterize a social situation as a unity of a particular type in the eyes of the observer. This is regardless of the actual participants who can be any as long as this set of relations is maintained. 'Structure' refers to the actual relations and elements in a specific social situation, which *realize* a given organization from the perspective of an observer.

A simple example should make this distinction clear, and anticipate the unfolding argument of the chapter. Let us use one of the examples Beer himself uses in *Heart*—'hospital' (Beer, 1979). In our particular culture/ society, there are minimally a particular set of activities or social and linguistic relations which the observer will require to witness, before granting that something named 'Hospital' is indeed a hospital. This minimum set will not itself be fixed or rigid, but will indicate a strong degree of consensus or agreement about the function and purpose of a hospital—perhaps this will gather around the notion of healing and comfort. Different individuals will interpret the details of such a process of healing and care somewhat differently. Some may stress high technology, while others may stress nurse/patient relations; but there will be a tight overlap, at least there will be in this consensual region. That is indeed what makes it the consensual domain it is. The word hospital conjures up an area of consensus about which to converse on the subject of hospitals.

Whereas 'organisation' concerns the set of relations that for an observer make this social system recognizably a hospital, and states nothing about any single instance—colour of this building, the actual personnel in this location as opposed to that, staff/patient ratio, list of names of personnel—'*structure*' does refer to one instance of the class-type 'hospital'. It concerns this particular hospital, with these administrative blocks, the actual communication channels existing between them, named actors and their precise mode of interaction. No longer is there minimally an unkown set of actors linked by some recognizable set of relations. Now we focus upon a concrete set of elements, and the actual set of relations that we observe as existing between them.

The constitution of phenomena such as social enterprises, together with their unfolded structures and organization, does not reflect a one-to-one mapping with anything 'out there' in the objective realm. Nor indeed do any consensual terms describe other so-called 'natural' phenomena. The actual shape or form you personally distinguish 'exists' according to how you yourself, upon someone's prompting (or that someone upon your prompting), arrive at an *understanding* about the status of whatever it is that you are conversing.

At a given moment, in a particular instance, what we *interact* with are structural phenomena—in other words, the actual relations and actual elements which realize a perceived situation. However, what we *re-cognize* is the 'fit' of such a structure with an organization.

In respect of social enterprises, this point comes out very clearly in the course of reading the chapters in this volume. The reader witnesses and recognizes the actuality of a particular instance (*this* firm, *this* institution), while at the same time finding him or herself matching the instance with the class type (the set of firms, of institutions). The readers, as do the authors, flit between structure and organization. The point is, that no reader's individual interpretation of structure will match that of any other reader. With or without the aid of Beer's model, we would each of us unfold or describe the system under consideration (the 'system-in-focus') somewhat differently from the description we read.

However, had we chosen to adopt the conventions offered by the VSM, as is the case with the authors in this volume, the chances are that the unfolded descriptions would have followed convergent rather than divergent paths. This does not mean that a variety of phenomena is systematically reduced to one (i.e. that just one structure is observed, one monolithic viewpoint dominates); but that a *consensual domain emerges* in which a common range of questions can be asked about a similar range of relations in the system subject to our enquiry.

Particular *structures* find a coherent relationship with one another (individual viewpoints), by the provision of a re-cognizable *organizational context*. It is the establishment of such a *context* which promotes or generates coherent debate about shared areas of concern, whether that debate takes place in the public domain as discussion and argument, or in our own private domain as constructive reflection.

The taxonomy being used presently (i.e. 'organization' and 'structure') is significant, not in that it reverses or otherwise tinkers with the conventional usage of the terms, but in that it lodges the phenomena of structure and organization in a quite novel domain. Instead of them being distinct, observable phenomena 'out there' (as, for example, it is given that a skeleton is the structure for body processes); they are now being considered as terms in our *coupling*, as cognitive beings, with the situational domains we find ourselves coupled to.

In other words, they are *handles*, by the use of which we coherently find our way about real-world situations—they help us *orient* ourselves in relation to others, by providing the means to encourage us to generate our own unique universe, which is braided with the universes of all other beings we find ourselves structurally coupled with, in this consensual region. Perhaps the key handle to our functioning in our human cognitive space is the mode of switching from structural to organizational perception and conception—*from the particular to the general and back again*. In a social domain this is what membership entails.

When we turn back to Beer's model, we discover that the structure of an organization (its 'organization structure') is the particular set of relations observed as existing between components or elements, by which the situation we are indicating (the system-in-focus) realizes the organization of which it is *perceived* as being a class member. This is a matter of *identity*. In practice, an organization cannot but be realized in terms of an actual set of relations and elements, in other words, an actual structure. This structure is how the particular social system is brought forth as a particular member of a class of organization, and can be witnessed. The system-in-focus can thence be indicated as being such an organization. Thus, an observer might look at a particular unidentified social form and declare 'It's a school!'. Conversely, on inspecting a particular social form identified as a school, s/he might exclaim 'No way!'.

It is class identity as expressed within some consensual domain which confers re-cognizability and supplies the terms of reference for our interaction in respect of social enterprises. It is not survival *per se* of some objective enterprise. The terms of its perceived organization is what enables us to identify a particular social phenomenon as belonging to this or that type, as fulfilling this or that function. The function fulfilled by a particular instance, realized through an actual structure, is always of far greater variety than the relations suggested by its organization. This particular village school does not just educate children—it imbues a special flavour, confers an added quality to this community, acts sometimes as the bridge club, communal meeting place and so on.

But, provided that we can re-cognize its organization, or provided that we confirm or give witness to the fact that, whatever individual quirks and variations, it continues to perform the same understood function for our community, then we will continue to call and treat it as a school. In other words, its organization will be perceived as viable. In this respect, viability turns out to involve the continued positive grounding of a stated identity in a field of understanding and expectations as expressed within a consensual domain. It does not concern survival 'out there'.

This is what Beer intends, when he comments that a hospital is 'what it does', rather than what it is said to do. Although the comment was originally offered in somewhat positivistic language (for the hospital itself does not *do* anything), Beer's meaning was that, if in this particular consensual domain, the structure is no longer recognized as functioning in terms of its previously attributed organization, then, whatever an observer chooses to *call* it, it will have ceased to be a member of the class it previously was considered an instance of. Another example (i.e. another structure) might offer pleasant music, debates, parties and so on, yet still be felt to realize those particular relations necessarily entailed in class membership of a particular organization, those conferring the identity 'hospital'. It may still be observed to heal, tend, care for the sick and wounded.

Such observers, as witnesses, might decide to change the rules of the game—might choose to get rid of the class-type 'hospital' and replace it with the class-type 'place of care and understanding'. Such changes are not at all trivial, not at all isolated within language itself, but couple perceptions to the shared reality that such perceptions bring forth. This level of change, or shift in perspective, is able to take place coherently because it is located within some higher-order logical dimension (reference signal; Powers, 1973)—in this case, the logical dimension of control (regulation) in human activities.

Thus the VSM is not specifically concerned with the viability of some particular social organization or some particular organizational form. Instead, it is concerned with the *possibilities for viability of social systems*, in respect of a community of observers. This viability depends to a greater or lesser extent upon the *play of control* as it is distributed throughout the social domain. The VSM explicates a particular insight into how such a *system* of control might be conceived, and the 'fit' of actual social practices with such an insight. Finally, the VSM is concerned with the means by which such a system might emerge with the full participation of the organization's members rather than be imposed 'top down'.

Given a re-cognizable organization in a particular consensual domain (i.e. a class-type), a specific instance (structure) will, to a lesser or greater extent, be witnessed as realizing the goal or intention bedded within the consensually expressed identity of the class-type. Perceptions can and do change (e.g. Feyerabend, 1975), and class-type can be reallocated although this may often involve a degree of trauma (e.g. Kuhn, 1970). However, there is no intrinsic virtue in the survival of a particular class-type (organization). Class-type is continually being matched or fitted to the consensual domain in which it gains positivity at any instant.

This is the importance of Beer's concept of economy in real-time. It relaxes the need to collapse organizational processes into fixed institutional forms. By building up flexibility in the *relations* between the situational participants, it avoids the imperatives of rigid, inflexible, time-lag dominated structures. Of course, the model does not do this 'in itself'—there is no open door to paradise! But the VSM provides a handle for the generation of a richer discourse upon, and a deeper understanding of, the emergence and evolution of social forms in the course of the interaction of individual human beings.

The relevance of methodology

The evolution of management cybernetics in recent years, as made apparent in this volume, evidences a shift in tone away from preoccupation with developing a methodology that would capture insight into a neutral, objective domain, towards a methodology that explicitly acknowledges the *subject-dependency* of any domain under human study. This shift is away from understanding models as descriptive or prescriptive, towards interpreting them as providing a *generative mechanism within a community of observers*.

This is not at all to say that the shift is away from objectivity and towards subjectivity. To repeat, it is literally and specifically away from objectivity and towards *subject-dependency within a consensual domain*: (within any one particular cultural and social setting). No-one can escape the fundamental grounding of all discourse, scientific or otherwise, within a prevailing 'paradigm', however much we may strive for 'revolution' (Kuhn, 1970).

As we throw away the stick meant to support us in our journey toward certainty, we do not fall into helpless confusion tossed about a sea of relativity, with no landmarks or coastlines to guide us; but we start to be able to conceptualize a new domain of landmarks—more accurately, perhaps, we start to rethink just what it is that we actually mean by such a term as 'landmark'.

The reason why this historical shift might make it easier to explicate the notion of organization structure is that it changes the emphasis of the approach. There is a move away from a demand for, and apparent need of, literal representations in order to successfully apply algorithms (see examples given in Chapter 3), towards the notion of mental models and heuristic strategies coherently working off one another in a consensual domain. Evidence of such a shift can be seen in Russell Ackoff's idea of 'participative planning', and in Peter Checkland's 'soft systems method-ology' (Ackoff, 1974; Checkland, 1981).

Such approaches do not at all abandon the claim to *coherently* open up the social domain to our observation as participant actors, but deny that such coherence needs to be chained to any one unidimensional perspective, any one Truth. In other words, it has gradually become acceptable to state the obvious—that it is just not true that there is 'one best way' of looking at social phenomena, or that we should strive towards 'the best way' (see, for example, Morgan, 1986).

No model 'describes' a real-life world, nor does it prescribe such a world (see the discussion in Chapter 14). The VSM is no exception to this rule. Rather, it is generative of collaborative linguistic interactions in a consen-sual domain. It brings forth a coherent means of orienting with one another in terms of a world that is essentially opaque to us. In a particular consensual domain, it formalizes a set of relations (a logic), in terms of which we may or may not choose to find our way about in respect of such a world.

Such a consensual region of language provides a medium for coherent linguistic interaction about a particular set of human interests—namely, the complex of issues concerning the relations between human individuals in the course of their constitution of social forms with distinct and perhaps conflicting claims. The Viable System Model is a template for methodology.

As a model for doing things—changing situations, it is hoped, for the better—the VSM is an aid to management in its quest for change—change as a promise, and change as a threat. *Management cybernetics* suggests how to minimize the chaos that inevitably threatens in the course of move-ment and flux, and to orchestrate a coherent interaction between all the actors involved in change so that coherent evolution can take place. The VSM is concerned with the complexity that emerges in the course of the interaction of human beings in collaborative ventures—in social enter-prise.

One of the significant moves of this volume has been to shift the emphasis away from the social enterprise as a thing 'out there', in some manner like a tree or a telephone box, to the enterprise as a dynamic phenomenon that is continuously constituted and reconstituted in the trajectory of its existence—in the manifestation of 'viability'. In other words, the quality 'viability', as explained above, is not some absolute value attributed and validated by the expert, who then does his best to put this value into place and maintain it, whatever the cost. Viability is observer-defined, crucially defined by the actual individuals engaged in 'effective' participation in the realization of an organization.

There has been a good deal of confusion on this point. The model does not present the practitioner with a description of the perfect firm, the ideal social system, a Utopian society. It offers no blueprint for the solution of humankind's ills. It does not point to short-cuts along the route to Utopia.

In the *logical* sense one might say that the VSM is descriptive of a particular set of functional invariances necessary for some organization to be viable once the observer has ascribed a purpose to it. However, in the *methodological* sense, precisely because systems do not have purposes of their own but are ascribed them by an observer, it might be considered as being prescriptive. But it is important to grasp just what it is prescriptive of.

It is prescriptive of the need to have the participation of multiple viewpoints in order for the complexity of a social enterprise to be at least partly expressed. These viewpoints need to exchange *different appreciations* of complexity in order to pin down useful references to support their conversations and coordinate their activities in the organization. The VSM is generative of such collaborative interaction and participation. This is the insight into *distributed* regulation and planning (see below; also Espejo and Garcia, 1984).

Formally speaking, the VSM is a set of logically coherent abstract ideas that may be interpreted and used very differently by different people. That the model itself is logically coherent does not imply any moral imperative. It is the ways in which it is interpreted and used in practice that grants it an ethical connotation, or makes it relevant or not to social activities. So, perhaps it is not so much that a more formal development of the model itself is necessary, but that there is a need for a sharper focus upon methodology.

Up to now, discussion has tended to concentrate on how it is commonly imagined the model has been implemented by Beer and others, instead of subjecting to scrutiny specific methodologies used in given instances. As a result, publications about the VSM have tended to confuse logical with methodological issues, and this has distracted attention from focusing on actual strengths and weaknesses of the *model-in-use*.

To take one example. Some people treat a single Viable System Model of an organization as if it were a description or representation of how things are or should be outside in the 'real world'. Indeed, there are people who interpret any model in such a manner. Based on such an approach, they try to force the 'real world' into the straitjacket of the single model. As made apparent by several of the chapters of this book, this approach is not only naive, but generates a mismatch with reality. If this were the only way to apply the VSM, we would not need to defend the model against its detractors—the model would be useless and would not attract detractors!

An *application* of the VSM requires the interpretation of the *expression of various distinct viewpoints* within the system-in-focus, and this is where it is different from most other models. What is required by the analyst, for any effective study and implementation of coherent change, is to be able to bring forth different VSMs about the same situation—in other words, to *generate a multisystemic expression* of the social enterprise.

The efficacy of the model in practice depends absolutely on this diversity of perspective. Such diversity, gathered in a particular accessible and visible manner, provides a rich reference that facilitates the orienting of participants in an interactive situation. Instead of seeking a 'neat conclusion' or a 'tidy' implementation, the primary function of the VSM is as a source of inspiration, for the analyst and the participants alike. It is not the logical coherence of the model *per se* that represents such an inspiration. The inherent logical consistency merely provides grounding necessary for the methodology to *generate a forum* that enables protagonists in a conversation to enhance their degrees of freedom for action in a given organizational context.

Our views on these matters do not represent a school of thought, so much as suggest patterns that have been generated for ourselves in the course of our study of these problems, both from the theoretical and the methodological perspectives. Specifically, it is our intention to help provide a handle for the practitioner who aims not merely to contemplate the world passing by, but intends to leap into the messy flow with an enhanced repertoire of tools that meld the theoretical domain to the practical one.

The human dimension of management science

Much of the debate concerning the VSM has centred on the ideological or political implications of the use of models which are perceived as being either descriptive or prescriptive of reality. Much contemporary critical philosophy would insist that claims to be descriptive are inherently prescriptive. Such debate as that brought to light by the Frankfurt School, and by thinkers such as Michael Foucault, Anthony Giddens, Jurgen Habermas or Herbert Marcuse, is an important and necessary part of a politically conscious society given the freedom for reflection, and highlights the danger of confusing the model with reality.

However, it seems to ourselves that such overtly sociological, philosophical or political analyses have rarely come to grips with the issues tackled by the seminal thinkers of the systems and cybernetic tradition. It was not a radical philosopher who stressed Korzybski's statement that 'the map is not the territory', but a systems thinker (Bateson, 1973). Unfortunately, the most serious attempts to discuss Beer's own work fluctuate between being either polemics or else esoteric mazes (Checkland 1980; Ulrich, 1983; Zeleny, 1986). This is especially true of the most weighty of these works.

Change is problematical, but the problematical human dimensions, such as ethics and responsibility, are not to be feared and avoided, nor surreptiously slipped on to the table when no one is looking. Rather, they are to be explicitly confronted and explored. However, management science, unlike moral philosophy, is about acting and doing and the repercussions of these in the social domain. What this means is that the management scientist *makes a space* for such issues as ethics, rather than diving into a detailed consideration of them. The challenge, then, is not to have all the answers to such questions, so much as to anticipate the need for their expression. Thus the remit is to explicitly make visible the domain within which such questions will be able to take place.

So the sort of question asked by ourselves as management scientists is how to reconcile the problems of individual autonomy in the face of the apparently inescapable demands of an 'organizational' world? This is something we are faced with in every minute of our waking lives, and it is something that no responsible social actor can comfortably put to one side in order to 'make life easy'.

In particular, our own insight is that management is not about 'making life easy', but rather is about clarifying the problematical issues of *organizational* existence in a way that such issues may be dissolved as problems as they are made visible, and leave uncluttered for reflection whatever problematical areas remain. Problems as such will never of course disappear once and for all. *Readiness* to cope with uncertainty is all we can ask for. And one of the ways that we conceive of allowing for the development of such readiness is by making visible the different effects of various social structural forms. We make no ethical claims beyond this modest proposition.

A management *science* is not about keeping things the same or maintaining the status quo, however desirable and attractive this might at first sight appear in the 'uncommitted' world of abstraction. In the social domain, every individual viewpoint has, in one form or another, vested interests in particular social, political, intellectual structures. Management science is about using the generative power of science and any related methodologies, in order to coherently orchestrate the change that is happening anyway.

To clarify this, we return to a consistent theme of this chapter. Scientific explanations are not simply reflections of the 'way things are and ought to be', but are *generative* of hitherto unthought ways of experience and novel insights about the order of things. They are generative in a community of observers, in which they give rise to willingness to *witness* novel relations. They are not uncovering or discovering absolute facts 'out there' in some fixed, neutral, objective domain.

The results of such explanations and accompanying methodologies are never absolute, untouchable pillars of stone, holding aloft some sacred cow, as pillars in ancient Egypt were taken to support heaven. Though coherent in terms of the argument or methodology, such results are *ad hoc*. They are conclusions which upon their own emergence themselves become further problems to be explained by new generations of scientists, new groups of interests, with different prejudices, different sets of mindscapes.

A *science* of management does not seek ways to hang on to power, to stop the world or shape it according to a particular ideology—all too often the sentiments felt and expressed by ambitious managers. Management science sets out to invest the domain of 'management' with a visible and coherent set of *methodologies*. Such methodologies do not involve claims for truth or objectivity. They entail coherent and consistent ways of doing things in the context of the distinct flows of discourse which constitute social forms and processes. As such, these 'management' methodologies are not the prerogative of a class 'manager', but are the prerogative of any individual with interest in social interaction—in other words, every single one of us!

History itself will never cease to be a play of power, a threat of conflict. Discourse itself will never finally be emancipated from its own 'tyranny', the claims of its own 'organizational closure'. Social codes can never evade the problematic of 'power/knowledge' (Foucault, 1980; Barnes, 1988). But possibly there are ways and means of *making visible* the flows of such deep issues, without making claims to infringe on the expertise of other disciplines. This is no abdication of responsibility, but an explicit recognition of the limits to the claims being made for one particular approach. Only by acknowledging such limits is it possible to mark out the territory within which a particular discourse gains the power to generate coherent insight. The explicit and visible expression of such consensual territory in terms of a community of observers is perhaps the furthest one can go along any 'right' road.

Autonomy as distributed regulation

In respect of the VSM, the issues explored by this chapter might beckon to the interested reader, but not in their familiar guise of political theory, philosophy or sociology. The issues permeate the model itself, as indeed the work for the Allende government in Chile and the consequent controversy amply attest (Ulrich, 1983). But they permeate in a unique fashion, a point that is perhaps most clearly exemplified in the debate about control and autonomy.

A central problem for the social sciences, one rarely faced head-on, is the existence of two dimensions that are often conflated and treated as one. For want of better terms we shall call this a matter of individual freedom. But what are actually involved are the twin dimensions of the 'acategorical', existential freedom of the unique human individual on the one hand, and on the other hand the relative freedom or autonomy, of social forms and structures, including the social actor.

This tension between the needs of the individual and the needs of the various groups which constitute the social domain, is not one that will ever be neatly dissolved. Indeed, the tension is a generative condition for human existence. What can perhaps be done, is to make visible or explicit such a tension, in order to enable the emergence of the variety of structures that might realize any particular organization. Thus may we encourage maximum flexibility and choice. This is in contrast to approaches that seek to secure organizational hegemony by demanding that a system be realized through only one structural form (something which is held to be the accepted organizational norm).

This distinction between the interpretation of a situation based in a range of structural forms and the interpretation of a situation in terms of one specific structure, is not integral to the model, as the fascinating chapter on natural history by Richard Foss makes clear (Chapter 6). In order to forestall criticism that might be levelled against the model itself, it should be said that it is only the structural autonomy in the concerned domain granted by the class of natural systems that Foss discusses, which allows them their fantastic evolutionary success. The Viable System Model helps Foss to nudge these mechanisms into light.

However, in general terms, and most significantly in the management context, the issue and problematic of autonomy is paramount. It is here that Beer's model has attracted most attention, both by those applauding what they see as its intrinsic democratic quality, and by those who see the model as an example par excellence of the letting loose of technocracy, and at worst, tyranny.

One of the most simplistic and unkind reactions to Beer's work in the past, has been to suggest the inevitability of such a hegemony of interests, whether in an actual political situation, or in respect to the evolution of the ideas themselves (e.g. Ulrich, 1983; Zeleny, 1986). Perhaps one reason for such interpretations is that many critics believe that the only way to *use* the VSM is by *inflicting* one particular structure upon the situation at hand (generally that model most compatible with the views of the people in power), and getting everyone within the enterprise to conform to this uniformity.

However, the methodological views expressed above together with the design of related tools (see for instance Beer's last contribution to this book; or Espejo and Garcia, 1984) explicitly reinforce the need to *distribute* information and regulation throughout the organization.

The argument is simple. Requisite variety for control cannot exist in a totalitarian system. Dissemination of regulation cannot be top-down; it does not enable more efficient control from above. At least it does not do so if viability is sought in a changing environment. The only way regulation can be disseminated is by its own closure as such closure is effected by *voluntary* participation in the realization of a shared vision or an explicitly stated purpose. It must be a form that allows for the development of peoples' potentials and provides a trigger for their own motivation.

Of course, it is possible for a practitioner to utilize the VSM to inflict one viewpoint alone over all others in a situation-in-focus. However, such a practitioner will be attempting to contradict the 'prescriptive' imperative of the model—the requirement for a *multisystem* with the consensual capacity to generate viable social systems. The result itself is unlikely to be a viable system, since such organization will lack the flexibility necessary to amplify its own variety *vis-á-vis* a changing environment.

Hence, using the VSM does not assume the agreement of all participants with *one purpose*; participants at all structural levels ascribe *purposes of their own* in terms of the enterprise as they themselves see it (i.e. 'viewpoints'; used as in Chapter 13). They are not referring to one objective social enterprise 'out there', but are *bringing forth a consensual domain* in which to develop coherent conversations in terms of the organization structure which their own contributions constitute.

If the VSM is used to facilitate the coherent participation of 'all' members of the enterprise (i.e. *disseminated regulation*), then the process encourages the effective contribution of all members in constituting on-going, dynamic structures that realize the viability of the particular social system in real time. From the point of view of 'communication and control', this is the only way of distributing regulation and distributing power. Once it is understood that the VSM is a generative mechanism, rather than some descriptive or prescriptive tool, it is a contradiction in terms to accuse it of being inherently totalitarian.

Summing up

The relevance and significance of the VSM does not concern constraining the world to fit the model. On the contrary, the model is a *pointer for understanding and action.* Hence our efforts in this closing chapter, to suggest a coherent epistemological perspective in order to appreciate and understand the Viable System Model. On the one hand, we have offered our own interpretation of the model, explicating insights into organization, structure and viability. On the other hand, from the methodological point of view, we have stressed its generative, rather than descriptive or prescriptive nature. In a sense the model has been presented as an instrument to focus debates about the social nature of human activity.

In such a context, this instrument can generate deep insights. The explicit mechanisms suggested by the model itself allow one to develop discourse concerning the closure of functional relations and their recursive nature. This permits one to use the VSM to unfold the full diversity of social structures, in a uniquely powerful manner. In this paradigm, the analyst does not find him or herself trapped within notions dictated by arbitrary organizational units or institutions, which often have little to do with dynamic situational relationships. At the same time, the observer is not forced to deal with a proliferating complexity far beyond the means of effective control in the course of his/her appreciation of such relationships.

The language offered by the VSM is a language that reflects the interactions of people in human activity systems and that develops appreciation about the costs entailed by the forms (i.e. structures) of these interactions (i.e. limits). We have stressed that the VSM is not a tool for one management level alone (e.g. the corporate level), but for all levels; it is tool to support conversations about the management of complexity *spread throughout the enterprise.* In using the model for diagnostic and design purposes, this view has important methodological implications. By definition, no one viewpoint is ever granted the privilege of seeing the full complexity of human activity. Such activity is a function of the conscious participation of all the individuals involved in it.

At this point we leave the conversation to the reader. . . .

References

Ackoff, R. (1974). *Redesigning the Future*. London: Wiley–Interscience.

Barnes, B. (1988]). *The Nature of Power*. Cambridge: Polity Press.

Bateson, G. (1973). *Steps to an Ecology of Mind*. London: Paladin.

Beer, S. (1979). *The Heart of Enterprise*. Chichester: John Wiley.

Beer, S. (1981). *Brain of the Firm*, 2nd edn. Chichester: John Wiley.

Checkland, P. (1980). *Are Organisations Machines?*, *Futures*, **12**, 421–4.

Checkland, P. (1981). *Systems Thinking, Systems Practice*. Chichester: John Wiley.

Dell, P. F. (1986). *In Defense of Lineal Causality*. *FAM Process*, **25**, 513–21.

Espejo, R. and Garcia, O. (1984). 'A tool for distributed planning', in *Proceedings— Orwellian Symposium and International Conference on Systems Research, Information and Cybernetics*. Baden Baden.

Feyerabend, P. (1978). *Against Method: Outline of an Anarchist Theory of Knowledge.* London: Verso.

Foucault, M. (1974). *The Order of Things: an Archaeology of the Human Sciences.* London: Tavistock.

Foucault, M. (1980). *Michel Foucault: Power/Knowledge—Selected Interviews and Other Writings 1972–1977.* Brighton: Harvester Press.

Gadamer, H. (1975). *Truth and Method.* London: Sheed & Ward.

Johnson-Laird, P. (1983). *Mental Models—Towards a Cognitive Science of Language, Inference and Consciousness.* Cambridge: Cambridge University Press.

Koestler, A. (1975). *The Act of Creation.* London: Picador, Pan Books.

Kuhn, T. (1970). *The Structure of Scientific Revolutions,* 2nd (enlarged) edn. Chicago: University of Chicago Press.

Maturana, H. R. (1978). 'Biology of langauge: the epistemology of reality', in *Psychology and Biology of Language and Thought—Essays in Honor of Eric Lenneberg* (eds. Miller, G. and Lenneberg, E.). Notre Dame, India: Academic Publishers.

Maturana, H. R. (1988). 'Ontology of observing: the biological foundations of self-consciousness and the physical domain of existence', in *Texts in Cybernetic Theory,* Workbook for the Fall Conference of the American Society for Cybernetics, Felton, CA, 18–23 October.

Maturana, H. R. and Varela, F. J. (1980). *Autopoiesis and Cognition: the Realization of the Living.* Dordrecht: Reidel.

Maturana, H. R. and Varela, F. J. (1987). *The Tree of Knowledge.* Boston: Shambhala.

Morgan, G. (1986). *Images of Organization.* London: Sage.

Pask, G. (1975). *Conversation, Cognition and Learning.* Amsterdam: Elsevier.

Powers, W. (1973). *Behavior: the Control of Perception,* Chicago: Aldine.

Rorty, R. (1980). *Philosophy and the Mirror of Nature.* Oxford; Basil Blackwell.

Ulrich, W. (1983). *Critical Heuristics of Social Planning.* Berne: Paul Haupt.

Varela, F. J. (1979). *Principles of Biological Autonomy.* Oxford: North Holland.

Von Glasersfeld, E. (1987). *The Construction of Knowledge—Contributions to Conceptual Semantics.* Salinas, CA: Intersystems.

Winograd, T. and Flores, F. (1986). *Understanding Computers and Cognition: a New Foundation for Design.* Norwood, N.J.: Ablex.

Zeleny, M. (1986). 'The law of requisite variety: is it applicable to human systems?', *Human Systems Management,* **6**, 269–71.

Index